T0093387

# INTERDISCIPLINARY RESEARCH ON CLIMATE AND ENERGY DECISION MAKING

This book explores the role and importance of interdisciplinary research in addressing key issues in climate and energy decision making.

For over 30 years, an interdisciplinary team of faculty and students anchored at Carnegie Mellon University, joined by investigators and students from a number of other collaborating institutions across North America, Europe, and Australia, have worked together to better understand the global changes that are being caused by both human activities and natural causes. This book tells the story of their successful interdisciplinary work. With each chapter written in the first person, the authors have three key objectives: (1) to document and provide an accessible account of how they have framed and addressed a range of the key problems that are posed by the human dimensions of global change; (2) to illustrate how investigators and graduate students have worked together productively across different disciplines and locations on common problems; and (3) to encourage funders and scholars across the world to undertake similar large-scale interdisciplinary research activities to meet the world's largest challenges.

Exploring topics such as energy efficiency, public health, and climate adaptation, and with a final chapter dedicated to lessons learned, this innovative volume will be of great interest to students and scholars of climate change, energy transitions and environmental studies more broadly.

**M. Granger Morgan** is the Hamerschlag University Professor of Engineering in the Department of Engineering and Public Policy (EPP) at Carnegie Mellon University. He also holds appointments in the Department of Electrical and Computer Engineering and in the Heinz College of Information Systems and Public Policy. He holds a PhD in Applied Physics from the University of California at San Diego. He was the founding Department Head in EPP, a job he held for 38 years. He is a member of the U.S. National Academy of Sciences and of the American Academy of Arts and Sciences.

# RESEARCH AND TEACHING IN ENVIRONMENTAL STUDIES

This series brings together international educators and researchers working from a variety of perspectives to explore and present best practice for research and teaching in environmental studies.

Given the urgency of environmental problems, our approach to the research and teaching of environmental studies is crucial. Reflecting on examples of success and failure within the field, this collection showcases authors from a diverse range of environmental disciplines including climate change, environmental communication and sustainable development. Lessons learned from interdisciplinary and transdisciplinary research are presented, as well as teaching and classroom methodology for specific countries and disciplines.

**Institutionalizing Interdisciplinarity and Transdisciplinarity**
Collaboration across Cultures and Communities
*Edited by Bianca Vienni Baptista and Julie Thompson Klein*

**Interdisciplinary Research on Climate and Energy Decision Making**
30 Years of Research on Global Change
*Edited by M. Granger Morgan*

**Transformative Sustainability Education**
Reimagining our Future
*Elizabeth A Lange*

For more information about this series, please visit: www.routledge.com/Research-and-Teaching-in-Environmental-Studies/book-series/RTES

# INTERDISCIPLINARY RESEARCH ON CLIMATE AND ENERGY DECISION MAKING

## 30 Years of Research on Global Change

*Edited by M. Granger Morgan*

*with Ahmed Abdulla, Jay Apt, Inês Azevedo, Ann Bostrom, Wändi Bruine de Bruin, Elizabeth Casman, Hadi Dowlatabadi, Mike Griffin, Tim McDaniels, Joshuah Stolaroff, Brinda Thomas, Parth Vaishnav, and a cast of several dozen*

Routledge
Taylor & Francis Group

LONDON AND NEW YORK

from Routledge

Cover image: Jay Apt

First published 2023
by Routledge
4 Park Square, Milton Park, Abingdon, Oxon OX14 4RN

and by Routledge
605 Third Avenue, New York, NY 10158

*Routledge is an imprint of the Taylor & Francis Group, an informa business*

© 2023 selection and editorial matter, M. Granger Morgan; individual chapters, the contributors

The right of M. Granger Morgan to be identified as the author of the editorial material, and of the authors for their individual chapters, has been asserted in accordance with sections 77 and 78 of the Copyright, Designs and Patents Act 1988.

All rights reserved. No part of this book may be reprinted or reproduced or utilised in any form or by any electronic, mechanical, or other means, now known or hereafter invented, including photocopying and recording, or in any information storage or retrieval system, without permission in writing from the publishers.

*Trademark notice*: Product or corporate names may be trademarks or registered trademarks, and are used only for identification and explanation without intent to infringe.

*British Library Cataloguing-in-Publication Data*
A catalogue record for this book is available from the British Library

ISBN: 978-1-032-36098-0 (hbk)
ISBN: 978-1-032-36100-0 (pbk)
ISBN: 978-1-003-33022-6 (ebk)

DOI: 10.4324/9781003330226

Typeset in Bembo
by Newgen Publishing UK

In memory of Lester B. Lave

# CONTENTS

# PREFACE

Research at most universities gets done in single academic departments and is focused on problems that fall comfortably within the boundaries of a single discipline. Indeed, it is often performed in groups named after single professors like "the Jones Group" or "the Smith Group."

Lots of universities talk about how important it is to do interdisciplinary research but most of them find that actually doing it is extraordinarily difficult, both because junior faculty have difficulty getting promoted and tenured in traditional departments, if they spend time collaborating with colleagues outside of their discipline, and because senior faculty, having gotten set in their ways, are uncomfortable breaking out of their well-established patterns to take on new things.

For 30 years, an interdisciplinary team of faculty and students anchored at Carnegie Mellon University, joined by investigators and students in a number of other collaborating institutions across North America, Europe, and Australia (see map in Appendix 3), have worked together to better understand the global changes that are being caused by both human activities and natural causes. This book tells their story.

Each chapter is written in the first person by the people who were involved in the work. For readers who would like to see more technical results, references and abstracts for a number of the papers the authors talk about are listed and briefly described at the end of each chapter. A list of many of the publications that have come from our work is provided in Appendix 2.

Our hope in writing this book is three-fold: (1) to provide a semi-technical account for interested readers of how we have framed and addressed a range of the key problems that are posed by the human dimensions of global change; (2) to suggest some strategies that can be effective to coordinate and manage networks of scholars working together across different disciplines and locations on a common problem; and (3) to encourage funders and scholars across the world to undertake similar large-scale interdisciplinary research activities to address the world's largest challenges.

While our publisher thought it was too vague, our original title for this book was *Uncertainty Is No Excuse*. When politicians and other decision makers don't want to make a tough choice because it runs against their short-term political or economic interests, they often argue that more research is needed to resolve uncertainties before they can act. Sometimes that is actually true. However, for the decision to control the power plant emissions that cause acid rain, which is discussed in Chapter 1, and for the decision to limit the emission of carbon dioxide and the other greenhouse gases that cause climate change, which is the focus of the balance of this book, modern method of decision research combined with a careful characterization of the known science can provide clear insight about how to proceed.

In such situations, uncertainty is no excuse for not taking action.

# ACKNOWLEDGMENTS

The work discussed in this book has been supported by a variety of sources. The most important has been core support to three interdisciplinary Centers provided by the US National Science Foundation (NSF) that our group won by responding to competitive calls for proposals. The grants and cooperative agreements that resulted are:

- The Center for the Integrated Study of the Human Dimensions of Global Change (HDGC): SES-9022738; SES-9209783; BCS-9218045
- The Climate Decision Making Center (CDMC): SES-0345798; DMS-9523602; SBR-9521914
- The Center for Climate and Energy Decision Making (CEDM): SES-0949710 and SES-1463492

We are deeply grateful for all that this NSF support has made possible.

In our applications to NSF, we have always argued that we would use the funding from NSF as core support and would seek additional sources of funding that would allow us to expand our efforts. In that regard we thank the following:

- Brookhaven National Laboratory
- Lawrence Berkeley National Laboratory
- The Alfred P. Sloan Foundation
- The Doris Duke Charitable Foundation
- The Electric Power Research Institute
- The Gordon and Betty Moore Foundation
- The John D. and Catherine T. MacArthur Foundation
- The U.S. Department of Energy

- The U.S. Environmental Protection Agency
- The William and Flora Hewlett Foundation

We also thank the years of support and encouragement provided by our several participating institutions,[1] especially Carnegie Mellon University and the University of British Columbia. Throughout the life of our three Centers, we have been fortunate to receive outstanding secretarial and administrative support from Patti Steranchak, who frequently anticipates and addressed our needs before we even figured out we have them. Others who have supported the Centers' work have included Barbara Bugosh, Tracy Farbacher, Terri Jones, Rachel Reolfi, Deborah Scappatura, Meryl Sustarsic and Lucas Valone. Jenni Miller provided excellent editorial assistance in finalizing this manuscript. Primary author Granger Morgan is deeply grateful to his wife Betty, who has been his best friend, biggest critic, and the love of his life for more than 60 years.

Granger Morgan is listed as the author of Chapter 5. However, because that Chapter reports on extensive work by Ann Bostrom and Wändi Bruine de Bruin, we list them as co-authors of this book. Not listed as co-authors, but also colleagues who made major contributions to the work of our Centers, are Baruch Fichhoff and Ed Rubin.

## Note

1 For a list of all the Universities and other organizations that have participated in our Centers, see Appendix 3.

# 1

# TOOLS TO ANALYZE UNCERTAINTY

*Granger Morgan*

On a September afternoon in 1978, Max Henrion, a young man with a British accent, walked into my office at Carnegie Mellon University. He was a PhD student in our School of Urban and Public Affairs (SUPA) and had been trying to find a faculty member who was interested in supervising a thesis that involved building a tool to make it easy to characterize and deal with uncertainty in quantitative policy analysis. None of the regular SUPA faculty were interested, so he'd come to me because, while my main appointment was in engineering, I had a SUPA affiliation.

I had just returned from a summer spent at Brookhaven National Laboratory out on New York's Long Island where, in the days before ubiquitous Internet and email, I had been able to hide for much of the summer in order to work on a computer model to assess the health impacts of the sulfur air pollution that is produced by coal-fired power plants. While I worked on the modeling, my wife Betty and our two young kids spent lots of time on the beach, where I often joined them on the weekends. When they weren't at the beach, the kids tooled around the Brookhaven lab site on bicycles or romped in giant piles of fire suppression foam that the laboratory's fire department produced for the entertainment of the lab's children.

While today it's pretty common, in the early 1970s, nobody had done much quantitative analysis of the health and environmental impacts of energy systems. During a course I had taught, a few years earlier at the University of California in San Diego, I had asked students to try to do some simple calculations to estimate the health and environmental impacts of making electricity by burning coal. Two students, Barbara Barkovich and Alan Meier, did an especially nice job, so I had asked them to continue working with me. The three of us ended up publishing a paper together that was titled, "The social costs of producing electric power from coal: A first-order calculation." The phrase "first-order" is just a fancy way of saying "approximate." Since we didn't know the precise values of lots of things,

DOI: 10.4324/9781003330226-1

in most cases we reported ranges of values (low to high). Doing that work had gotten me interested in trying to improve the descriptions of uncertainties about the health and environmental costs of energy and in assessing other technical systems.

The biggest costs from making electricity from coal are the health impacts that result from air pollution. We knew from the work of economists Lester Lave and Eugene Suskin (as well as others) that sulfur air pollution made people sick and sometimes caused them to die earlier than they would have without the pollution. The technical word for this is "premature mortality."

But how the sulfur in coal got converted into pollution in the atmosphere, how it was carried along by the winds and exposed people on the ground, and how much that exposure actually hurt people was all pretty uncertain. With my colleague Sam Morris at the Brookhaven Laboratory, I had spent several months that summer constructing a model of these processes in a computer language called Fortran. Unlike many other models that people were building at the time, the model I had built didn't assume that we knew precisely all the different values that were needed – things like how quickly sulfur dioxide gas gets turned into tiny particles called aerosols, how rapidly the gas and those aerosols fall out or are washed out of the atmosphere, how people got exposed by breathing that pollution, and how doing that hurts their health. In my model, I described uncertainty about each of the steps using something called probability distributions. These are a mathematical way of displaying the odds of what the value of important numbers in the model might actually be. Sam and I and our graduate students had interviewed a number of leading atmospheric scientists and health experts to get their best estimate of the range of values (basically their betting odds) on the value of key numbers we needed for our model. We used those probability distributions in our model.

The model I had written consisted of many hundreds of lines of Fortran computer code that got put into a computer using large decks of punch cards. Writing and debugging that program had been a lot of work, but I was on the verge of having results which my students and I could turn into a paper to publish in a scientific journal.

Max had arrived at SUPA after a somewhat circuitous academic career that had involved studying physics and statistics at Emmanuel College at Cambridge University and then completing an MS degree in design research at the Royal College of Art in London. Despite his ventures into design, Max's early training in the natural sciences had stuck with him and, like me, he had become increasingly concerned that people performing analysis on problems in public policy were largely ignoring uncertainty.

Max was a computer whiz. What he proceeded to show me that afternoon was a prototype of a computer system that would allow someone to complete in a day or two much of the work on which I had just spent two laborious months of programming. Max called his new system "Demos," and while the early version was slow, it clearly held enormous potential. He explained the problems he'd been having in trying to find somebody on the SUPA faculty to supervise his PhD and asked if I would be willing to work with him on Demos. I could immediately see that this was a match made in heaven, and I quickly agreed to be the advisor for his PhD.

Over the next several years, Max and I worked together both to elaborate and refine Demos and to develop and demonstrate a wide range of strategies for describing and analyzing uncertainty. Often, research on a subject such as the health effects of air pollution only provides part of what a decision-maker needs in order to make an informed decision. In many situations, the experts who have done the research have a better understanding of the uncertain science than is apparent from the research papers they publish. Building on the work of a number of previous researchers, Max and I, along with several co-workers, developed and demonstrated various ways to interview those experts and get them to supply additional information about their understanding and about the uncertainties in what they knew.

In the early 1980s, the air pollution modeling community was having big arguments about how the sulfur pollution in the exhaust plumes of large power plants is carried along by the wind, converted to fine particles, and lost from the atmosphere. While it was clearly important to improve understanding of these issues, they were also making the argument that without such improvements, it would be difficult to develop good estimates of the magnitude of the health effects that this air pollution was causing.

My Brookhaven Lab colleague Sam Morris and I had gotten interested in better understanding how much the uncertainty in the atmospheric science contributed to the uncertainty about health damages as compared with the uncertainty contributed by how well the processes that cause health damage were understood. To better explore this issue, we conducted very detailed face-to-face interviews with nine scientists who were expert in atmospheric science and seven scientists who were expert in health effects. In these interviews, we asked general questions about the science, but we also asked the experts to give us numbers to describe their uncertainty.

For example, for a specific well-defined quantity, such as the rate at which a chemical reaction proceeds, we'd start by getting experts to tell us the largest value they could imagine the quantity could take on. If they said 10, we'd ask, "Can you tell us any story about how it could end up being 11? If they could tell us a plausible story about how it could be 11, we'd suggest that they should increase their "upper bound." After working for a while on how large the value could turn out to be, we'd then do the same thing to get them to estimate how small the value could turn out to be. After getting the expert to tell us the range in which the number could fall, we'd begin to ask questions like, "What is the chance that the value will turn out to be greater than 5?" or "What is the chance that the value will turn out to be less than 3?" By asking a series of such questions, we could build up a curve, called a "probability distribution," that told us just what chance each expert thought there was that the quantity could take on each of the possible values across the range between the smallest and largest. We could put these probability distributions into our computer model and run them through the model we had built to get an answer, again in the form of a distribution.

This process of getting experts to put numbers on their uncertainty is called "expert elicitation."[1] It is obviously not a substitute for doing serious research to get more precise answers, but research often takes a lot of time, so in the meantime, expert elicitation is a strategy that under some circumstances can help get better values for use in doing analysis on important decision problems. As later chapters explain, over the years that followed we used similar methods to quantify the uncertainty that was involved in a range of other important problems, including many of the uncertainties about climate change.

In 1990, Max and I summarized much of what we have learned about describing and analyzing uncertainty in a book called *Uncertainty: A Guide to Dealing with Uncertainty in Quantitative Risk and Policy Analysis.*

At about the time that Ronald Reagan became president, many people in the United States were becoming concerned about the possibility that sulfur and nitrogen air pollution could be causing "acid rain" and that this could be damaging the ecosystems of pristine lakes in places like the Adirondacks and northern Minnesota. The Reagan administration didn't want to impose any regulations, so they argued that there was a lot of uncertainty and that nothing should be done until the problem was better understood. They started a big research program called the National Acid Precipitation Assessment Program (NAPAP). In addition to doing basic science about air pollution, how it might make lakes more acidic, and how this could affect sensitive species like young lake trout, the NAPAP program also set out to build a set of models designed to figure out whether acid rain was making lakes more acidic, and if so, how serious that might be.

The Department of Energy operates a number of large research laboratories across the country, and they approached the problem of acid rain by asking each laboratory to build a different, very detailed part of a giant model, arguing that later this whole family of models would be plugged together and this large collection of models would be able to answer the question, "Do we need to regulate acid rain?" If this strategy sounds to you a little bit like a stalling tactic, you're probably not wrong.

Our colleague Ed Rubin, together with Max and a number of others, suspected that a much simpler approach, that used all the tools we had developed to deal with uncertainty, would be better able to answer the question about whether there was a need to regulate acid rain. Using the Demos computer system, they built a model they called ADAM, short for "Acid Deposition Assessment Model." In building ADAM, Ed, Max, and their colleagues followed the advice (often attributed to Albert Einstein) to make their model "as simple as possible but no simpler." Unlike the large fancy models that were being built by the national labs, Ed, Max, and their colleagues only put as much detail into ADAM on all the different parts of the acid rain problem as was necessary to answer the basic question. Because it covered all the important pieces of the problem of deciding about acid rain, ADAM is what's now often called an "integrated assessment" model.

Once the ADAM model had been finished and used to explore the issues, it became pretty clear that acid rain was posing serious environmental risks and that the United States should undertake regulatory actions to limit the emissions of air pollution that were damaging our pristine lakes. It may not surprise you that government funding for the ADAM project ended soon after their analysis reached that conclusion. The NAPAP program continued on for a few more years before it finally petered out as growing scientific evidence and political will finally resulted in the nation taking serious steps to reduce acid air pollution. After it had wrapped up, Ed Rubin wrote a paper on NAPAP in which he argued: "While the NAPAP effort made significant scientific contributions to the study of acid deposition, key gaps are found in the assessment of benefits and costs most relevant to policy decisions. Lessons learned from NAPAP may be helpful in avoiding similar problems in assessing emerging environmental issues such as global climate change." Unfortunately, I believe that many folks in government labs (and elsewhere) still have not learned these lessons.

Our early experience in using integrated assessment to examine an important environmental problem that involved a lot of uncertainty persuaded a number of us that the time had come to undertake a similar study for a much larger global problem: the looming issue of climate change. The next two chapters tell their story of how we set out to do that.

## Technical Details for Chapter 1

Here is the research paper and abstract that describe the Demos software system that Max Henrion developed for his PhD thesis. This system was designed to make it easy to describe and analyze uncertainty in many different kinds of problems:

Henrion, M., & Morgan, M. G. (1985). A computer aid for risk and other policy analysis. *Risk Analysis*, 5(3), 195 208.

**Abstract:** The use of appropriately designed computer aids for policy could improve the standards of risk analysis and other quantitative policy analysis in several important ways. They could make it easier to treat uncertainties more thoroughly and systematically than is now typical. To do this, they should provide a broad variety of techniques for representing uncertain quantities as ranges of alternative values or as probability distributions, for propagating uncertainties through a model, for analyzing and comparing the impacts of different sources of uncertainty, and for displaying results in various numerical and graphic formats. A nonprocedural modeling language allowing interactive editing of input values and model structure could encourage exploration and progressive refinement of models, and comparison of alternative formulations. The integration of model documentation and explanatory text within the computer representation could encourage maintenance of consistency between different versions of the mathematical structure and their descriptions. It could also allow interactive scrutiny of the model assumptions and sensitivities by outside reviewers. We describe a particular system, Demos, designed to provide these facilities and test their usefulness. The use of Demos is illustrated by an analysis of the risks and optimal control level for a hypothetical air pollutant, with uncertainty about the population exposure, health effects, and control costs. This example demonstrates progressive refinement of a model, and various kinds of parametric and probabilistic uncertainty analysis. Demos is now being used by a growing number of risk analysts, students, and policy researchers in a wide variety of tasks.

Demos is now distributed commercially as Analytica® by a company that Max founded called Lumina Systems (see www.lumina.com). On that website, you can find very nice worked-out examples and details on how the system works.

Here is the abstract of the paper I wrote with two students at University of California at San Diego on the social costs of electric power produced by coal plants:

Morgan, M. G., Barkovich, B. R., & Meier, A. K. (1973). The social costs of producing electric power from coal: A first-order calculation. *Proceedings of the IEEE, 61*(10), 1431–1442.

**Abstract:** A methodology is discussed for quantitatively computing the social cost, or external diseconomies, which result from the production of electric power in conventional coal-fired steam electric plans. With the available data, and our present level of understanding, it is possible to obtain preliminary numbers which place the social cost for the technology of the mid and late 1960s at ≥11.5±2 mills/kWh, somewhat more than the price of bulk power at the plant bus bar. In applying controls to limit the social costs, control costs are incurred. If the optimum level of control is taken as that level at which the sum of the social costs and the control costs is minimum, then we estimate the total social and control costs with optimum control as ≥4.5±1.5 mills/kWh and the costs of controlling to that level as ≥3±1 mills/kWh. These numbers will probably be reduced, and the optimum levels for controlling increased, as new technologies are developed. The paper is limited to a straightforward development of social costs. No attempt is made to develop policy implications or to draw broad conclusions on the basis of the cost which are derived.

This is the abstract of a research paper that reports on the first work described in this chapter on the health impacts of air pollution from coal-fired power plants:

Morgan, M. G., Rish, W. R., Morris, S. C., & Meier, A. K. (1978). Sulfur control in coal fired power plants: A probabilistic approach to policy analysis. *Journal of the Air Pollution Control Association, 28*(10), 993–997.

**Abstract:** The optimum level of sulfur pollution control for a coal fired power plant is the point where the sum of societal costs, due to pollution, and control costs is minimized. This basic microeconomic concept has been of limited practical value due to considerable uncertainty in estimating both costs. A probabilistic approach is used to characterize these uncertainties quantitatively for a hypothetical 1000 $Mw_e$ plant located near Pittsburgh, Pennsylvania. Only mortality effects within a distance of 80 km of the plant have been included. The results allow explicit consideration of attitude toward risk and appropriate level of investment to prevent deaths. Limitations of the findings are discussed. Implications are described for policy based on alternative sets of values and assumptions.

Here is another paper that used the results of a series of expert elicitations we ran with both air pollution experts and health effects experts:

Morgan, M. G., Morris, S. C., Henrion, M., Amaral, D. A., & Rish, W. R. (1984). Technical uncertainty in quantitative policy analysis—a sulfur air pollution example. *Risk Analysis, 4*(3), 201–216.

**Abstract:** Expert judgments expressed as subjective probability distributions provide an appropriate means of incorporating technical uncertainty in some quantitative policy studies. Judgments and distributions obtained from several experts allow one to explore the extent to which the conclusions reached in such a study depend on which expert one talks to. For the case of sulfur air pollution from coal-fired power plants, estimates of sulfur mass balance as a function of plume flight time are shown to vary little across the range of opinions of leading atmospheric scientists while estimates of possible health impacts are shown to vary widely across the range of opinions of leading scientists in air pollution health effects.

You can find a less technical description of much of the same work in:

Morgan, M. G., Morris, S. C., Henrion, M., & Amaral, D. A. (1985 Aug). Uncertainty in environmental risk assessment: A case study involving sulfur transport and health effects. *Environmental Science & Technology, 19,* 662–667.

**Paper Summary:** This paper explores two questions: 1) in estimating the average annual mass balance of sulfur in the plume of a large power plant, how much difference does it make which atmospheric science experts one consults about oxidation rates? 2) in estimating possible human mortality arising from exposure to this air pollution, how much difference does it make which health effects experts one consults? When estimating the health damages caused by a large power plant we found that it made very little difference which of the judgments by different air pollution experts we used, but we could obtain an enormous range of possible health damages depending upon which judgments by different health experts we used. The atmospheric scientists share similar disciplinary skills and a large, common literature. When asked about details of other experts' work, they gave specific comments that made it clear that they were intimately familiar with studies outside their immediate areas. In contrast, the health effects experts had a broader range of disciplinary backgrounds. Their literature is more diverse, incomplete, and ambiguous, and many experts were unfamiliar with details of studies outside their particular areas and there was less agreement about what evidence was important or relevant.

Here is the paper on the integrated assessment Ed and his colleagues built on acid rain:

Rubin, E. S., Small, M. J., Bloyd, C. N., & Henrion, M. (1992). Integrated assessment of acid-deposition effects on lake acidification. *Journal of Environmental Engineering, 118*(1), 120–134.

**Abstract:** An integrated assessment model is used to estimate $SO_2$ emission effects on regional lake acidification and fish viability in two regions of North America (Adirondack Park, New York, and the Boundary Waters region of northern Minnesota). An uncertainty analysis is employed to estimate the likely range of possible impacts. Based on emission projections for the United States and Canada, lake acidification in these two regions appears likely to improve slowly over the next two decades. An acid-rain control program will accelerate the recovery of acidic lakes at Adirondack Park, with a projected decrease over the no-control case of approximately 2–11% in the number of lakes below pH 5.5 and a 0.4–6% increase in the number of lakes potentially able to support brook or lake trout by the year 2010. For Boundary Waters, the expected improvements are negligible since deposition levels are relatively low. Our analysis does demonstrate a potential for larger or smaller improvements in these two regions, with lower probabilities of occurrence. Uncertainties in regional lake chemistry and aquatic biology dominate the overall uncertainty in acidification effects estimated for these two regions, within the limitations of the analysis.

Here is the paper by Ed Rubin evaluating the NAPAP program:

Rubin, E. S. (1991). Benefit-cost implications of acid rain controls: An evaluation of the NAPAP integrated assessment. *Journal of the Air & Waste Management Association*, *41*(7), 914–921.

**Abstract:** Concluding ten years of study, the U.S. National Acid Precipitation Assessment Program (NAPAP) recently issued its integrated assessment report designed to provide guidance to policy makers on the sources and effects of acid deposition, and the costs and benefits of alternative control measures. This paper focuses on an evaluation of the benefit-cost implications of acid rain controls as revealed by two of the five major questions addressed in the NAPAP assessment framework. While the NAPAP effort made significant scientific contributions to the study of acid deposition, key gaps are found in the assessment of benefits and costs most relevant to policy decisions. Lessons learned from NAPAP may be helpful in avoiding similar problems in assessing emerging environmental issues such as global climate change.

Finally, here are the details on the book that Max and I wrote about uncertainty:

Morgan, M. G., & Henrion, M. with Small, M. (1990). *Uncertainty: A guide to dealing with uncertainty in quantitative risk and policy analysis*. Cambridge University Press, New York, 332pp.

**Some notable writings by authors not affiliated with our Centers:**

Hordijk, L., & Kroeze, C. (1997). Integrated assessment models for acid rain. *European Journal of Operational Research*, *102*(3), 405–417.

Howard, R. A., & Matheson, J. E. (1984). The Principles and Applications of Decision Analysis, Strategic Decisions Group, 955pp. in two volumes.

Keeney, R. L. (1982). Decision analysis: An overview. *Operations Research*, *30*(5), 803–838.

Menz, F. C., & Seip, H. M. (2004). Acid rain in Europe and the United States: An update. *Environmental Science & Policy*, 7(4), 253–265.

Raiffa, H. & Schlaifer, R. (1961). *Applied Statistical Decision Theory*. Harvard University Press, 356pp.

National Research Council. (2010). Hidden Cost of Energy: Unpriced consequences of energy production and use. National Academies Press, 506pp.

vonWinterfeldt, D., & Edwards, W. (1986). *Decision Analysis and Behavioral Research*. Cambridge University Press, 604pp.

Watson, S. R. & Buede, D. M. (1988). *Decision Synthesis: The principles and practices of decision analysis*, Cambridge University Press, 299pp.

# Note

1 For details on this process of asking experts to give us probability distributions, see Chapter 4 as well as Morgan, M. G. (2014). Use (and abuse) of expert elicitation in support of decision making for public policy. *Proceedings of the National Academy of Sciences*, *111*(20), 7176–7184.

# 2

# LET'S DO THE SAME FOR CLIMATE CHANGE

*Granger Morgan*

With the success of the ADAM model behind us, we began to think that we should build a similar integrated assessment model to address the much more complicated issues of global climate change and its impacts.

In the winter of 1990–91, Ed Rubin, Lester Lave, and I published an article in *Issues in Science and Technology* titled, "Keeping Climate Change Relevant." *Issues* is a magazine published by the National Academies of Sciences, Engineering, and Medicine that is devoted to the discussion of critical issues in the area of science, technology, and public policy.

We began our article by recapping the history of the National Acid Precipitation Assessment Program (NAPAP), arguing that "the 10-year, half-billion-dollar inter-agency program to guide U.S. policy on acid rain control proved largely irrelevant to the effort to forge the new Clean Air Act… Although NAPAP won praise for its scientific accomplishments, the program failed in its primary mission—to provide policy-relevant information in a timely manner. Now, Government attempts to deal with the more difficult and far-reaching environmental issues associated with global warming appeared to be headed down the same ill-fated path."

After briefly reviewing the national and international situation, we asked, "How can a multibillion-dollar research program involving some of the best scientific minds be so predictably irrelevant? The lessons learned from NAPAP and the current plans for R&D in global climate change offer some answers."

What was missing, we argued, was "effective management tools for focusing research and improving its ability to address policy as well as fundamental science issues. By redirecting a tiny proportion of the program's billion-dollar budget to these ends Congress could vastly increase the timeliness and relevance of its findings." The best way to do that, we argued, was by undertaking "a series of comprehensive integrated assessments." We argued that multiple assessments were needed because the issues were too contentious and too important to be entrusted

DOI: 10.4324/9781003330226-2

to a single group. In the balance of that first article, we elaborated how we thought this should be done. Some months later, we published a letter outlining a shorter version of this argument in the leading scientific journal *Nature*.

If we were going to undertake a major program doing integrated assessment of climate change and its impacts, it was clear we would need serious funding for technical staff and graduate students. We were fortunate to be able to secure initial support from the U.S. National Science Foundation, the U.S. Department of Energy, and the environmental program at EPRI.[1]

I also knew that if we were going to mount a serious effort in integrated assessment, we'd need someone who could coordinate it and help provide intellectual leadership. So, with initial support in hand, I made a phone call to Hadi Dowlatabadi.

After growing up in Tehran and attending school in Scotland, Hadi had completed his PhD in physics, working on energy modeling at Cambridge University in England. Ed Rubin had gotten to know Hadi when he was doing a sabbatical year at Cambridge and invited him to come do a post-doc. From 1984 to 1986, Hadi was a post-doc in EPP at Carnegie Mellon. When he finished that, he took a position as a Fellow at Resources for the Future (RFF) in Washington, DC. RFF is the world's leading group of environmental and resource economists. They have always tried to have a few people on their staff who have deep technical expertise, and that was the role Hadi played. Then, in 1989, Hadi was awarded a Warren Weaver Fellowship by the Rockefeller Foundation and moved to New York to help them design their new program for global environmental leadership (LEAD).[2]

I had gotten to know Hadi during his two years at Carnegie Mellon. My most vivid recollection is of this irreverent guy who sat on the low console in the back of the conference room where I taught my EPP graduate core course, arguing and disagreeing (often just for the fun of it) with more or less everything I said. It was great! The course is designed to help our students develop their own critical assessment of the ideas and tools of policy analysis. It works best when we can have vigorous class discussions and air a wide variety of thoughtful views. Hadi had thought a lot about many of the analytical issues we discussed in that class, and his participation really perked things up and helped all of us to think harder and more critically about the topics we were discussing.[3] I developed a deep appreciation of his knowledge and intellect, and we became close friends.

So when I learned we were going to get funding to do integrated assessment of the climate problem, I went to work on persuading Hadi that he should come back to Carnegie Mellon to be Executive Director of the new center we were creating. Hadi had gotten the LEAD program into pretty good shape and was tired of the hassle and high cost of New York City, so after a bit of persuading, he agreed to join us.

Even before Hadi arrived, we had begun to scope out the problem by drawing and then refining a number of "influence diagrams." You may know the American spiritual song "Dem Bones" that goes, "The toe bone's connected to the foot bone, the foot bone's connected to the heel bone, the heel bone's connected to the ankle

bone…[all the way up to]… the neck bone's connected to the head bone…."
Influence diagrams are sort of similar. They are a simple graphical way of showing
what's connected to what and what influences what in a big system like the climate and its impacts. Some ecologists are fond of saying "everything is connected
to everything," which may be true, but the *strength* of those connections varies a
lot from one to the next. The trick in building useful models is to identify the
connections that are strongest and have the most influence on how a system works.
These are the ones to focus on. That was the insight that had made ADAM, our acid
rain integrated assessment model, so successful.

Once Hadi arrived, he and Lester Lave went to work to first do a very simple
"back of the envelope" analysis. Later, using the Demos system that Max Henrion
had developed, Hadi took the lead in building a series of increasingly more
sophisticated computer models that we called the ICAM models, where ICAM
stands for "Integrated Climate Assessment Model." So in retrospect, we called
Hadi's and Lester's first analysis ICAM-0. Since they did not know just how big the
impacts from climate change will be on U.S. GDP, they picked a range of plausible
numbers that might apply in the year 2040. Similarly, since they did not know just
how much it would cost to reduce emissions of $CO_2$, they also picked a range of
plausible numbers for that. The technical name for such a strategy is "parametric
analysis."

In the paper that Lester and Hadi published in the journal *Environmental Science
and Technology* in 1993, they concluded:

> At least for the next decade, the prior judgments that decision makers bring
> to the problem, the decision rules they employ (e.g., expected value versus
> minimizing maximum loss), could be far more important in controlling…
> policy conclusions…than the results of scientific discoveries over that period.

We believed that the central focus of any serious assessment of climate change and
its impacts had to be on describing and analyzing the most important uncertainties. So, while the rest of the faculty and doctoral students worked on studying and
refining a lot of the pieces we were going to need in our assessment, Hadi set out
to use Max's Demos system to build the ICAM modeling framework. In the next
chapter, Hadi outlines how that effort unfolded.

In 1995, the U.S. National Science Foundation (NSF) announced a competitive opportunity to create research centers to address issues related to the Human
Dimensions of Global Change. In the proposal we submitted, we argued:

> Climate change is not a new problem. Anyone who has been a regular
> attendee of the biannual meetings of the American Geophysical Union has
> been hearing about climate change in specialized technical sessions for more
> than a quarter of a century. What *is* new is the high national and international
> visibility and the powerful political overtones that global climate issues have
> now assumed.

Today climate change is clearly an important political issue. Given the current uncertainties it is impossible to know whether climate change will turn out to be a serious environmental problem. It has the potential to be very serious.

In addition, from our perspective, climate change is an important problem for another reason. It is prototypical of a class of problems in science and technology in which:

- There are a significant number of different elements whose interactions should not be ignored in planning research, performing policy analysis, or formulating policy responses.
- Uncertainty involving these elements and their interactions is large and unlikely to be fully resolved before potential impacts would be felt and policy choices will have to be made.
- Scientific and technical issues are of central importance and cannot be treated as a black box in the analysis of the social and policy dimensions of the problem.

Research in the Department of Engineering and Public Policy[4] has been substantially concerned with problems of this type for many years…

in addition to viewing climate change as an important problem in its own right, we view climate change as an excellent research testbed within which to extend and improve the theory (the repertoire of research strategies and techniques) for dealing with the growing set of policy problems which involve high uncertainty and in which scientific and technical issues are of central importance.

A central part of the work we proposed involved building and using an integrated assessment. We wrote:

The process of global climate change and its possible ecological, economic and social impacts involves interactions between physical, ecological, economic and social systems, each of which involves numerous uncertainties. We will undertake a systematic analysis of these interactions and the uncertainties, using the full range of advanced analytical techniques now available, and applying several new techniques. Emphasis will be placed on a synoptic view which attempts to integrate the many disparate pieces of science relevant to the problem of global climate change and weighs and compares the contributions of uncertainty from each. We will begin by systematically identifying a set of key policy questions. Then we will develop a hierarchically structured set of influence diagrams to describe the basic dependencies which are likely to be important in controlling the climate change outcomes relevant to this set of questions. We will perform an integrated assessment that.

- Consists of a coordinated set of analysis and models.
- Is iterative.
- Adopts analytical strategies and tools which are most relevant to the specific problems at hand, making full use of powerful modern computer capabilities when that is appropriate.
- Places emphasis on the analysis of uncertainties and on time dynamics, sometimes using Bayesian data windows and other filtering techniques in order to check results against available empirical evidence.

We will use our integrated assessment to develop recommendations for priorities in applied research based on an analysis of how a resolution of various uncertainties in our understanding of physical, ecological, economic and social systems and their interactions is likely to affect resolution of uncertainties in our identified set of significant policy questions and outcomes. This priority setting exercise will be directed at applied research. Given the potential for "surprise" it is important to maintain a portion of the research budget for basic research which is guided by conventional scientific paradigms. Finally, we will use the integrated assessment to perform policy analysis to explore key questions and significant outcomes.

After responding to many reviewer comments and successfully surviving a site visit by a group of outside experts who NSF had invited to come and critically assess what we were proposing to do, we got the good news: we'd won five years of major support to create a center. We called it the "Center for Integrated Study of the Human Dimensions of Global Change." We were off and running on what has turned into a quarter-century adventure in assessing the human dimensions of global change (HDGC).

## Technical Details for Chapter 2

We wrote three opinion pieces arguing that integrated assessment could help the world to do a better job of addressing the problems of climate change. Here is where they were published:

Rubin, E. S., Lave, L. B., & Morgan, M. G. (1991). Keeping climate research relevant. *Issues in Science & Technology, 8*(2), 47–55.

Lave, L. B., Dowlatabadi, H., McRae, G. J., Morgan, M. G., & Rubin, E. S. (1992). Uncertainties of climate change. *Nature, 355,* 197.

Dowlatabadi, H., & Morgan, M. G. (1993). A model framework for integrated studies of the climate problem. *Energy Policy, 21*(3), 209–221.

Here is the abstract of the NSF proposal that resulted in the creation of our first distributed research center for studying the Human Dimensions of Global Change (HDGC):

A set of integrated studies of problems in managing research and human responses to global climate change are proposed. While the elements of the work interact, for expository and administrative convenience the project is broken into three broad activities: integrated assessment with a particular focus on uncertainty (approximately 40% of the total effort), a set of cross-national policy assessments (approximately 30% of the total effort), and studies on public understanding and communication (approximately 30% of the total effort). While the climate change problem is important in the its own right, it is also prototypical of a class of problems and science, technology, and public policy in which: (1) there are a significant number of different elements who's interactions cannot safely be ignored in planning research, performing policy analysis, or formulating policy responses; (2) uncertainty involving these elements and their interactions is large and unlikely to be fully resolved before potential impacts would be felt and policy choices will have to be made; and (3) scientific and technical issues are of central importance and cannot be treated as a black box in the analysis of the social and policy dimensions of the problem.

Here is the paper describing the analysis that we call ICAM-0:

> Lave, L. B. and Dowlatabadi, H. (1993). Climate change: The effects of personal beliefs and scientific uncertainty. *Environmental Science & Technology*, 27(10), 1962–1972.

**Some notable writings by authors not affiliated with our Centers:**

> Manne, A., Mendelsohn, R., & Richels, R. (1995). MERGE: A model for evaluating regional and global effects of GHG reduction policies. *Energy Policy*, 23(1), 17–34.
> Schneider, S. H., & Dickinson, R. E. (1974). Climate modeling. *Reviews of Geophysics*, 12(3), 447–493.
> Schneider, S. H. (1989). The greenhouse effect: Science and policy. *Science*, 243(4892), 771–781.
> Wigley, T. M., Richels, R., & Edmonds, J. A. (1996). Economic and environmental choices in the stabilization of atmospheric $CO_2$ concentrations. *Nature*, 379(6562), 240–243.

Also see the several papers on integrated assessment that are listed at the end of Chapter 3.

## Notes

1 EPRI is the Electric Power Research Institute. It was started after the 1965 Northeast blackout and for many years received substantial funding from most of the investor-owned electric utilities across the U.S. More recently, as a result of the restructuring of the electric power industry, its scope has narrowed and its resources have been reduced. However, when it began to support our work on climate assessment, it still had substantial resources and a very broad mandate.
2 For details on the LEAD program, see www.lead.org
3 This course is still offered, and I have written a book that covers most of the issues we address:

> Morgan, M. G. (2017). Theory and Practice in Policy Analysis: Including applications in science and technology. Cambridge University Press, 590pp.

4 The Department of Engineering and Public Policy (or EPP) at Carnegie Mellon University has been the primary home for the work discussed throughout this book. See www.cmu.edu/epp/

# 3

# THE ICAM MODEL OF CLIMATE CHANGE[1]

*Granger Morgan and Hadi Dowlatabadi*

As explained in the previous chapter, our colleagues had been successful in building the ADAM integrated assessment model of acid rain that provided clear policy advice. In January of 1992, the two of us joined with Lester Lave and several other colleagues to argue, in a letter we published in the scientific journal *Nature*, that while continuing research on basic climate science and its impacts was important, what was really needed was integrated assessment that could provide guidance for shaping future research priorities and developing climate policy. We argued:

> If scientific research is to inform political decisions, administrators need to exercise tough control. And integrated assessment is needed to identify the priority research and coordinate individual projects... integrated assessments can spot the critical gaps in the current research agenda, discover research that isn't on target or is wasteful, and detect the mismatches between the inputs that each group is expecting and outputs that will be produced by other researchers.

We set out to build a computer-based integrated assessment model that we called ICAM (Integrated Climate Assessment Model). We decided that the key was not getting every little detail into such a model (like the National Lab folks had done in the case of acid rain) but instead to focus on those things that made the biggest difference and do that in a way that made it easy to describe and explore the implications of all the various uncertainties.

Once again, we turned to Demos, which is now called Analytica® so that is what we'll call it in this chapter. Two of the really nice features of Analytica® are that it uses influence diagrams to lay out and create the structure of the model, and it makes it easy to include uncertainty about all the important quantities in the model and carry them through the analysis so that the answers come out as probability

DOI: 10.4324/9781003330226-3

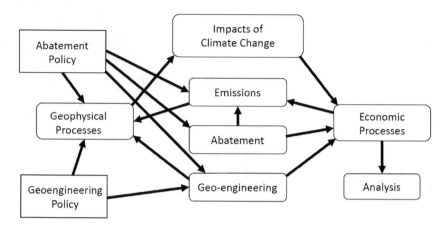

**FIGURE 3.1** Top-level influence diagram that shows the structure of ICAM-1.

Dowlatabadi, H., & Morgan, M. G. (1993). A model framework for integrated studies of the climate problem. *Energy Policy*, *21*(3), 209–221.

distributions – that is, "betting odds" on what the answer would turn out to be if we could ever resolve all the uncertainties.

The basic structure of our first model, ICAM-1, is shown in Figure 3.1. Another nice thing about Analytica® is that it is hierarchical: rather than making the diagram of the model all on one big sheet the size of a tabletop, it is easy to take sub-parts of the model and put them into separate bins or "nodes," sort of like being able to click on a link when you read text in Google and have it take you to more detailed explanations. The hierarchical nature of the model is illustrated in Figure 3.2.

The ICAM-1 model divided the world into two regions (the developed "north" and the less developed "south"). It was a simulation model that stepped through the future in 25-year time steps out to the year 2100. Hadi took the lead in building the model, and the rest of us worked on coming up with a lot of the numbers and other things he needed to put into the model. For example, as described in the next chapter, David Keith and Granger worked on getting climate experts to make judgments about how much warming would occur as the amount of $CO_2$ in the atmosphere increased. Hadi then put those results into ICAM so we could explore how the answers changed depending on which expert's judgments we used.

In the words that some people attribute to Albert Einstein, our strategy from the beginning was to "keep the model as simple as possible, but no simpler" and to place a heavy emphasis on describing and analyzing all the uncertainties. When we first presented ICAM-1 to some colleagues at MIT who were also building a much more complicated integrated assessment model that contained a full (if simplified) climate model, they were pretty dismissive. Like the National Lab folks working on acid rain described in Chapter 2, they wanted to get lots of science detail into their model. We were amused a few years later when they had figured out that they really

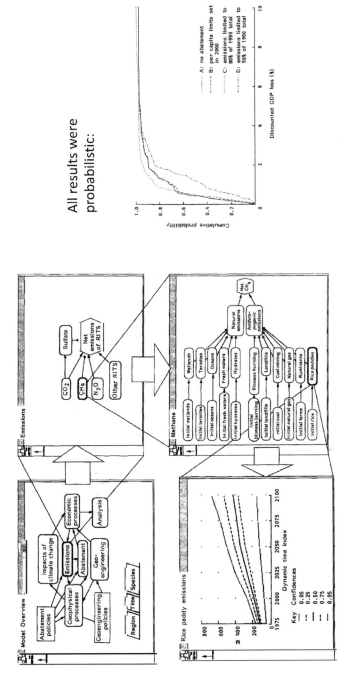

**FIGURE 3.2** Illustration of the hierarchical structure of ICAM–1. By clicking on a node such as the one labeled "Emissions," the user can move down a level in the model hierarchy to see the various sources of modeled emissions of greenhouse gases. Then, by clicking on "CH4" (methane), the user moves down another level in the model to see all the various global sources of methane. Clicking on "Rice paddies" and asking for a plot of a probabilistic time series results in the plot at the lower left.

Dowlatabadi, H., & Morgan, M. G. (1993). A model framework for integrated studies of the climate problem. *Energy Policy, 21*(3), 209–221.

needed to address uncertainty and then struggled to figure out how to do that in a model that, unlike ICAM, had not been designed from the ground up to include and analyze uncertainty.

In a 1993 paper that describes our work with ICAM-1, we wrote:

> The idea of an assessment framework is that the structure and assumptions of the sub-elements, and indeed even the relationship between those sub-elements, is not fixed. Rather it is the central focus of on-going analysis. Various alternative sub-models may be substituted, and their implications explored. These may be as simple as a few variables or they may involve reduced form or response surface descriptions of very complex models.
>
> Because uncertainty is so important, it should be easy to represent uncertain quantities as probability distributions, propagate probabilistic values through the models and perform various kinds of deterministic and stochastic sensitivity and uncertainty analysis…
>
> [we found] that the choice of decision rule plays a key role in the selection of mitigation policies, that given a decision rule, uncertainty in key variables can make it difficult or impossible to differentiate between the outcome of alternative policies and that the model parameters that contribute the most uncertainty to outcomes depend on the choice of policy, the discount rate and the geographical region being considered.

We believed that it was very important not to display answers as single numbers (best estimates). However, as we began to present results to non-technical audiences, we found that some people, who had a predisposed preference for what the answer should be, tended to cherry-pick our results. When Hadi did some briefings to Capitol Hill staff, some of those who thought climate change was a big problem latched on to the numbers at the high end of the range of numbers coming out of ICAM. Those who thought climate change was not a big deal latched on to the low end of the range.

After completing ICAM-1, under Hadi's leadership we went on to build ICAM-2, which divided the world into seven regions and treated 2000 variables as uncertain. We developed probability distributions from the literature and from the various interviews we were conducting with experts.

ICAM-2 was not just a single model. It contained lots of "logical switches" that allowed us to examine the implications of adopting alternative plausible model structures. These included the option of making alternative assumptions about things such as the response of the energy system as the planet warmed; the rate of discovery of new resources; the diffusion of new technologies; ecological responses; different models of how populations of different regions would grow; and the treatment of air pollution including aerosols.

Later, in ICAM-3, we added switches that allowed the user to choose alternative assumptions about things like the treatment of time preference (how people value something today or in the future); the global warming produced by different gases and very small particles (aerosols); the lifetime of things like big structures (called capital stock); whether or not "tax revolts" occur, etc. A view of the top-level influence diagrams for ICAM-2 and 3 is shown in Figure 3.3.

While we were doing all this, other groups were building integrated assessment models that looked for an "optimal climate policy." For example, at Yale, economist Bill Nordhaus and his colleagues built a series of simple but elegant economic models called DICE and RICE. These models were (and are) widely used in public policy circles. They make a single set of assumptions about how to model the problem and then look for an optimum climate policy. We don't think that this strategy makes sense for two reasons. First, there is not one "global decision maker." What may be optimal for us in the U.S. or for people in Europe may not be optimal for Africa, Australia, China, India, or South America. Each region faces very different situations and has many different cultures with different values and preferences. Second, when we started changing the assumptions about the form of the model in ICAM, we found that, within a very broad range, we could obtain almost any answer we wanted. For these reasons, we never chose to run any of the ICAM models to get a "best" overall policy, but we could and did run them to see how different policies played to the advantage or disadvantage of different regions around the world.

In building ICAM-3, Hadi created little national decision-makers (autonomous adaptive agents) that he put into each region of the world. He gave each one of these agents a set of decision rules (that could be changed with different settings of switches). The agent in each region observed the evolving model world and reacted according to those rules – for example, if at the start of a simulation we assumed that all regions had signed on to a global emissions control agreement based on carbon taxes, when taxes got too high in a simulation in which we'd switched on the option to allow tax revolts, some agents would decide that the taxes had gotten too high and take their region out of the global agreement.

We also explored the consequences of applying a number of different decision rules such as *minimize ecological impacts*, *minimize economic impacts*, and various combinations of those and others. In ICAM-2 and 3, we found that:

- We could get an *enormous* variety of answers depending on the range of plausible assumptions we made about the structure of the model and which regional decision-maker we considered.
- Almost no policy was optimal for all regions.
- Rarely was any single result better than all the others at all levels of confidence (i.e., none was stochastically dominant).

In a 1996 paper that discussed what we'd learned in four progressively more detailed rounds of integrated assessment, we argued:

> Integrated assessment is neither an end in itself, nor a one-shot proposition. The most useful results from doing integrated assessment will typically not be

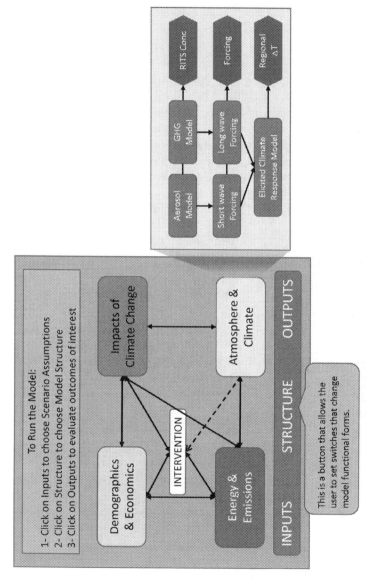

**FIGURE 3.3** Illustration of the top-level influence diagram for ICAM-2 and 3, together with a display one level down for the set of models of atmospheric composition and climate. In addition to treating roughly 2000 variables as uncertain, this model is populated with multiple switches that can be used to turn on and off a variety of alternative model functional forms.

Drawn by Hadi Dowlatabadi.

"answers" to specific policy questions. Rather they will be insights about the nature and structure of the climate problem, about what matters, and about what we still need to learn.

The Energy Modeling Forum (EMF) is a group run by John Weyant at Stanford that, for years, has conducted comparisons of various energy and environmental models (Hadi participated in several comparisons when the EMF was looking at integrated assessment models). In 2012, drawing on insights from over 35 years of EMF work, John reached a similar conclusion:

> …the avalanche of detailed quantitative results produced by the models has often tended to blind model users to the insights they can provide…[Harvard's Bill Hogan captures this view when he says] "it is not the individual results of a model that are so important; it is the improved user appreciation of the policy problem that is the greatest contribution of modeling."…the purpose of energy-environmental policy modeling is to develop insights…not precise numerical forecasts.

In describing our work on ICAM in a paper we wrote with Liz Casman in 1999, we noted that:

> … uncertainty about the appropriate functional form of different sub-models is sufficiently large, and the difficulty of constructing all plausible alternatives sufficiently great, that it is often best to report results paramet-rically across a set of combinations of different model structural assumptions, in much the same way that one reports the results of parametric sensitivity studies of coefficient uncertainty. For example, in an application of ICAM-2 designed to explore the probability that a specific carbon tax policy… would yield net positive benefits, we found that the probability ranged from 0.15 to 0.95 for the world as a whole, depending upon the structural assumptions made…

MIT economist Robert Pindyck is similarly pessimistic about the value of using integrated assessment models to search for optimal policies or to perform mean-ingful benefit-cost analysis in support of estimates of the social cost of carbon (see Chapter 18). He writes[2]: "…IAMs are of little or no value in evaluating alter-native climate change policies and estimating SCC. On the contrary, an IAM-based analysis suggests a level of knowledge and precision that is non-existent, and allows the modeler to obtain almost any desired result…" Unfortunately, in dismissing the utility of integrated assessment, we believe that Pindyck ignores the fact that integrated assessment can also be used to achieve better understanding of the complexity of a problem like climate change and guide future analysis and research. The process of building and using integrated assessment models has helped the climate community to better identify and understand many key complexities. Unfortunately, too many of those who have built or used such models have gone on to treat their outputs as making reliable predictions about the real world.[3]

## Technical Details for Chapter 3

Here is our first paper describing ICAM-1:

Dowlatabadi, H., & Morgan, M. G. (1993). A model framework for integrated studies of the climate problem. *Energy Policy*, *21*(3), 209–221.

**Abstract:** Establishing research priorities and developing and evaluating alternative policy options in the domain of climate change are activities that require a broad look across all the elements of the climate problem. Such broad integrated assessment should involve different analytical approaches in different parts of the problem. However, appropriate computer model frameworks, into which a variety of specific alternative results and sub-models can be inserted, can provide convenient vehicles for putting the pieces together. Such a framework is described. A simple version is used to illustrate how such frameworks can be used to perform uncertainty analysis, and how, together with expert judgment, they might be used in setting research priorities.

After completing ICAM-1 and ICAM-2, we summarized what we'd learned in this paper:

Morgan, M. G., & Dowlatabadi, H. (1996). Learning from integrated assessment of climate change. *Climatic Change*, *34*(3), 337–368.

**Abstract:** The objective of integrated assessment of climate change is to put available knowledge together in order to evaluate what has been learned, policy implications, and research needs. This paper summarizes insights gained from five years of integrated assessment activity at Carnegie Mellon. After an introduction, in Section 2 we ask: who are the climate decision makers? We conclude that they are a diffuse and often divergent group spread all over the world whose decisions are primarily driven by local non-climate considerations. Insights are illustrated with results from the ICAM-2 model. In Section 3 we ask: what is the climate problem? In addition to the conventional answer, we note that in a democracy the problem is whatever voters and their elected representatives think it is. Results from studies of public understanding are reported. Several other specific issues that define the problem, including the treatment of aerosols and alternative indices for comparing greenhouse gases, are discussed. In Section 4 we discuss studies of climate impacts, focusing on coastal zones, the terrestrial biosphere and human health. Particular attention is placed on the roles of adaptation, value change, and technological innovation. In Section 5 selected policy issues are discussed. We conclude by noting that equity has received too little attention in past work. We argue that many conventional tools for policy analysis are not adequate to deal with climate problems. Values that change, and mixed levels of uncertainty, pose particularly important challenges for the future.

This is a review Hadi wrote of a number of different integrated assessment models:

Dowlatabadi, H. (1995). Integrated assessment models of climate change: An incomplete overview. *Energy Policy*, *23*(4–5), 289–296.

**Abstract:** Integrated assessment is a trendy phrase that has recently entered the vocabulary of folks in Washington, DC and elsewhere. The novelty of the term in policy analysis and policy making circles belies the longevity of this approach in the sciences and past attempts at their application to policy issues. This paper is an attempt at providing an overview of integrated assessment with a special focus on policy motivated integrated assessments of climate change. The first section provides an introduction to integrated assessments in general, followed by a discussion of the bounds to the climate change issue. The next section is devoted to a taxonomy of the policy motivated models. Then the integrated assessment effort at Carnegie Mellon is described briefly. A perspective on the challenges ahead in successful representation of natural and social dynamics in integrated assessments of global climate change is presented in the final section.

Here is the paper on uncertainty in policy models that we wrote with Liz Casman:

Casman, E. A., Morgan, M. G., & Dowlatabadi, H. (1999). Mixed levels of uncertainty in complex policy models. *Risk Analysis, 19*(1), 33–42.

**Abstract:** The characterization and treatment of uncertainty poses special challenges when modeling indeterminate or complex coupled systems such as those involved in the interactions between human activity, climate and the ecosystem. Uncertainty about model structure may become as, or more important than, uncertainty about parameter values. When uncertainty grows so large that prediction or optimization no longer makes sense, it may still be possible to use the model as a "behavioral test bed" to examine the relative robustness of alternative observational and behavioral strategies. When models must be run into portions of their phase space that are not well understood, different sub-models may become unreliable at different rates. A common example involves running a time stepped model far into the future. Several strategies can be used to deal with such situations. The probability of model failure can be reported as a function of time. Possible alternative "surprises" can be assigned probabilities, modeled separately, and combined. Finally, through the use of subjective judgments, one may be able to combine, and over time shift between models, moving from more detailed to progressively simpler order-of-magnitude models, and perhaps ultimately, on to simple bounding analysis.

**Some notable writings by authors not affiliated with our Centers:**

Hope, C., Anderson, J., & Wenman, P. (1993). Policy analysis of the greenhouse effect: An application of the PAGE model. *Energy Policy, 21*(3), 327–338.

*Integrated Assessment Journal.* While it is no longer publishing, between 2000 and 2009 the journal published 9 volumes and 25 issues. Archives are at: https://journals.lib.sfu.ca/index.php/iaj/issue/archive

Nordhaus, W. D. (1993). Optimal greenhouse-gas reductions and tax policy in the "DICE" model. *The American Economic Review, 83*(2), 313–317.

Parson, E. A., & Fisher-Vanden, A. K. (1997). Integrated assessment models of global climate change. *Annual Review of Energy and the Environment, 22*(1), 589–628.

Rotmans, J. (1998). Methods for IA: The challenges and opportunities ahead. *Environmental Modeling & Assessment, 3*(3), 155–179.

Pindyck, R. S. (2013). Climate change policy: What do the models tell us?. *Journal of Economic Literature, 51*(3), 860–872.

Gillingham, K., Nordhaus, W. D., Anthoff, D., Blanford, G., Bosetti, V., Christensen, P., … & Sztorc, P. (2015). *Modeling Uncertainty in Climate Change: A multi-model comparison* (No. w21637). National Bureau of Economic Research.

Paltsev, S., & Sokolov, A. (2021). Scenarios with MIT Integrated Global Systems Model: Significant Global Warming Regardless of Different Approaches. In *World Scientific Encyclopedia of Climate Change: Case studies of climate risk, action, and opportunity Volume 2* (235–241).

## Notes

1 Some fragments of text in this chapter are reworked from chapter 11, The Use of Models in Policy Analysis, in Morgan, M. G. (2017). *Theory and Practice in Policy Analysis: Including applications in science and technology*. Cambridge University Press, 590pp.

2 Pindyck, R. S. (2013). Climate change policy: What do the models tell us?. *Journal of Economic Literature, 51*(3), 860–872.

3 Steve Rose and his colleagues at the Electric Power Research Institute have performed a particularly interesting analysis of the three different IAMs (DICE, FUND, and PAGE) used by the U.S. in estimating the social cost of carbon. They take each separate part of each model (estimate of future emissions; emissions to climate change; climate change to impacts and damages), drive them with identical inputs, and find that each yields very different outputs. See: Rose, S., Turner, D., Blanford, G., Bistline, J., de la Chesnaye, F., and Wilson, T. (2014). *Understanding the Social Cost of Carbon: A technical assessment – Executive Summary*, EPRI, 18pp.

# 4

# GETTING EXPERTS TO GIVE US THEIR "BETTING ODDS"[1]

*Granger Morgan*

There are many people who don't want to see society do serious things to address the problem of climate change because doing so would work against their short-term economic interests. These "climate deniers" often go to great lengths to emphasize the uncertainty surrounding climate science and the impacts that a changing climate will produce.[2]

There *is* uncertainty about some of the details of climate change and its impacts. But today there is really *no* uncertainty about three basic facts:

1.  Human activities such as adding more and more carbon dioxide and other greenhouse gas to the atmosphere and changing land use (e.g., cutting down forests and changing the surface of the earth) are warming the planet.
2.  Because the climate system is driven by heat, that warming is also changing many other aspects of the climate.
3.  If we keep on adding carbon dioxide to the atmosphere and making other big changes to the earth, the results will be irreversible for many centuries.[3]

These basic facts are firmly established in the scientific community. However, many of the *details* about climate change and its impacts remain uncertain. When we set out to do analysis of the climate problem and build models like the ICAM model described in Chapter 3, we believed that it was important to include a careful description of those uncertainties and to carry the uncertainty through the models so that the answers we got were not just single numbers but ranges of numbers that basically gave the "betting odds" on what the actual answers would be if we could learn the exact value of all the uncertain numbers we needed to feed into our models or use in our other analyses.

Beginning in the late 1950s, analysts began to develop a set of analytical tools to help people make decisions in situations in which there is a lot of uncertainty.

DOI: 10.4324/9781003330226-4

These tools go by the name of decision analysis.[4] One of the things that analysts using decision analysis needed to do was to ask experts to give their odds on what the actual value would be for numbers needed in the analysis. The process of asking for these odds is called "expert elicitation."

For example, if the study was for a company that was thinking it might want to introduce a new product, the analyst might ask questions about how much a key ingredient needed to make that product was likely to cost five years in the future. They also might ask how much of the product people were likely to buy.

With my colleague Sam Morris from Brookhaven National Labs and several graduate students, we had started using such expert elicitation in studies we were doing on the health impacts caused by coal-fired power plants.[5] When we began our first NSF-supported Center (HDGC), it seemed like an obvious idea to use a similar approach to get climate experts to give us their best estimates (their betting odds) about the likely value of numbers we needed in our integrated assessment model, ICAM.

Whenever you set out to ask experts about something like climate change, it is easy to come up with a very long list of questions you'd like to have answered. So the first thing you need to do is boil that long list down to the stuff you *really* need to know. This process can take many weeks. We typically try out the questions we want to ask and the other things we want the experts to do on a couple of more junior people – people like post-doctoral fellows who are working in the same field. In that way, we can figure out what is clear, what is confusing, and what strategies work best. In most cases, we start the interview with a general discussion of the state of the field, what is well known, where the big uncertainties are, and so on. We always make sure that one member of our interview team has recently done a careful review of the literature so that they can push back a bit on the answers we get and ask follow-up questions. We often also prepare summary materials to help the experts keep all the evidence in their mind.

We call the final list of questions, and other tasks we ask the expert to perform, the "protocol for the expert elicitation." We have sometimes joked that the ideal protocol is one that stops just short of the point at which the expert is so fed up with us that they are ready to punch us out. While some experts we have worked with started out being pretty dubious about the whole process, virtually all of them have told us at the end of the session (which frequently runs half a day or more) that they found the whole thing very interesting and that it made them think about the issues in some ways they had not previously considered.

In most cases when we have done an expert elicitation, we have traveled to meet the experts in their offices. One reason we do this is that we don't want the experts to give us off-the-cuff answers to our questions. We want them to give us their careful, well-considered expert judgments. If they are going to do this, they need to be able to look things up in their files, check the scientific literature, and perhaps even run one of their computer models. Of course, meeting the expert in their office also makes it more convenient for them to participate.

Table 4.1 lists all the expert elicitations we have conducted. Telling you about the details of everything we learned in these studies would get pretty tedious in a book like this. If you want to learn details about any of the specific studies listed in the table, there are references to the papers that describe the results from these studies at the end of this chapter.

While the elicitation protocols we have developed have involved a lot of different tasks, all of them have included asking experts to make judgments about the true value of some uncertain quantities we care about. Typically, we get their answers in the form of a curve called a "cumulative probability distribution." A simple way to display such results is with a diagram called a "box plot." Figure 4.1 shows an example. One advantage of box plots is that several of them can be placed side-by-side, making it easy to compare judgments made by several different experts.

Climate change is expected to have a big impact on the world's forests. In an expert elicitation we did in 1999–2000, we asked forest experts about possible changes in northern and tropical forests 500 years after a climate change caused by doubling the amount of $CO_2$ in the atmosphere. One of the quantities we asked about was changes in the amount of "biomass" above ground (i.e., the weight of all the trunks, branches, leaves, and needles) and below ground (i.e., the weight of all the roots).

Figure 4.2 shows what experts told us for northern forests. Notice that in several cases, experts told us they think there is some chance that there could be either

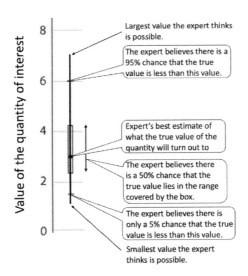

**FIGURE 4.1** Example of a "box plot," which is a simple way to display an expert's uncertain judgment about what the true value of some quantity we care about will turn out to be. The dot shows the expert's best estimate. The box shows the range of values in which the expert says there is a 50% chance the true value will lie.

Drawn by Granger Morgan.

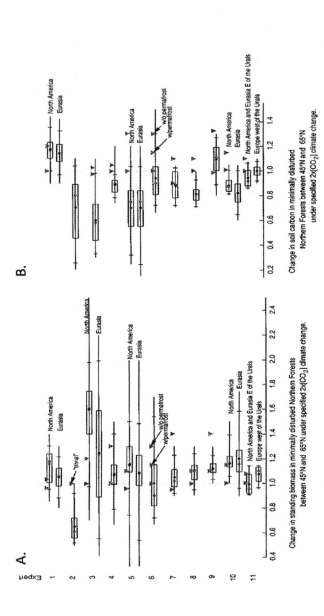

**FIGURE 4.2** Box plots summarizing judgments made by forest experts in 1999–2000 of (A, left) standing biomass and (B, right) soil carbon in minimally disturbed northern forests at least 500 years after doubling the atmospheric concentration of $CO_2$.[7] A value of 1 means no change, 2 means doubling, and 0.5 means a reduction by half. Notice that at the time we conducted the elicitation, experts were not even agreed about the sign of the change (i.e., increase or decrease). Except where it says Eurasia, all these results are for North America. Carbon dioxide in the air can have a fertilizing effect for many plants. The shaded triangles indicate the estimated range of response that would occur if the atmospheric concentration of $CO_2$ were doubled but there was no accompanying climate change. Figure from Morgan, Pitelka, & Shevliakova, 2001.

Morgan, M. G., Pitelka, L. F., & Shevliakova, E. (2001). Elicitation of expert judgments of climate change impacts on forest ecosystems. *Climatic Change, 49*(3), 279–307.

**TABLE 4.1** The set of expert elicitations we have conducted as part of the work of our NSF-supported climate Centers.

| *When we did it* | *Topics we asked experts about* | *Reference at the end of this chapter to the paper we published describing the results* |
|---|---|---|
| 1980–1 | Interviews with 9 air pollution experts and with 7 health experts to better understand and model the health impacts of the sulfur air pollution that comes from power plants that burn coal. | Morris, Henrion, Amaral, and Rish (1984); Morgan, Morris, Henrion, and Amaral (1985) |
| 1993–4 | Interviews with 16 leading U.S. climate scientists to ask about how much warming may happen and other uncertainties in climate science. | Morgan and Keith (1994) |
| 1999–2000 | Interviews with 11 leading forest experts (and 5 biodiversity experts) to ask about the impacts that climate change may have on tropical and northern forests. | Morgan, Pitelka, and Shevliakova (2001) |
| 2005–6 | Interviews with 12 leading oceanographers and climate scientists to ask about how climate change may influence the circulation of water and heat in the Atlantic Ocean. | Zickfeld, Levermann, Morgan, Kuhlbrodt, Rahmstorf, and Keith (2007) |
| 2005–6 | Survey of 24 leading atmospheric and climate scientists to explore how the direct and indirect ways in which high-altitude small particles in the atmosphere warm or cool the planet. | Morgan, Adams, and Keith (2006) |
| 2006–7 | Interviews with 18 experts about conventional and advanced technology for solar cells to explore how cost and performance may change over time. | Curtright, Morgan, and Keith (2008) |
| 2008–9 | Interviews with 14 leading U.S. climate scientists (4 who were the same as in the earlier study) to ask about how warming will change over time and other uncertainties in climate science. | Zickfeld, Morgan, Frame, and Keith (2010) |
| 2011–12 | Interviews with 16 nuclear engineers about how the cost and future performance of small modular nuclear reactors (SMRs) are likely to compare with the cost of existing large reactors. | Abdulla, Azevedo, and Morgan (2013) |
| 2017 | Interviews with 39 experts about how the cost and future performance of a kind of membrane for use in fuel cells. The fuel cells would be used to power automobiles. | Whiston, Azevedo, Litster, Whitefoot, Samaras, and Whitacre (2019) |

an increase *or* a decrease; that is, they were not even sure about the sign (positive or negative) of the change. Our experts expected more modest changes in tropical forests, but many were even less sure about the sign of those changes.

Beyond the specific results from these studies (a number of which were used in the ICAM model described in Chapter 3), we can draw several more general lessons from our work on expert elicitation. First, the physical scientists and engineers we have interviewed have been comfortable about giving us probabilities in the form of numbers. Many think this way naturally when they design the studies they conduct. They are perfectly happy to say things like, "I think there is an 80% likelihood that this oxidation rate will be more than 3% per hour," or "I think there is only a 30% chance that the cost will turn out to be less than $75." In contrast, many health experts adopt a much more qualitative approach to their work and get quite uncomfortable about assigning probabilities to possible future outcomes. Some haven't wanted to give us odds on things like a range of values for the slope of a health damage function (how much health damage goes up with larger exposure). Instead, they have wanted to just give a fixed number that they think is the right answer. This probably has a lot to do with the kinds of training the people in these different fields receive. This observation is not unique to our experience. For example, a report written for the White House and the Congress in 1997[6] by a group of medical professionals and lawyers argues that while scientists and economists should include quantitative estimates of uncertainty in their analysis, health experts should just give single numbers that are their best estimates. This of course makes no sense and is gradually changing, but we were really surprised when we first began to encounter the issue.

Another thing we have found is that the amount of uncertainty that experts express when we interview them individually is often significantly larger than when they work together to produce a consensus report. The "consensus panel" results produced by the experts who have participated in the reviews produced by the Intergovernmental Panel on Climate Change (IPCC) tend to display less uncertainty than the results we have gotten separately from each individual. For example, when we did the expert elicitation that produced the results in Figure 4.3, the IPCC consensus study of that time said the largest value that the climate sensitivity could have was 4.5°C. But the small numbers across the bottom of that figure show that the experts we talked with thought there could be between a 10% (expert 13) and a 35% (expert 12) chance that the value could be larger than the IPCC's maximum value of 4.5°C. We have seen the same sort of difference between experts' individual judgments about how much warming or cooling is caused by small particles in the upper atmosphere.

One strength of reporting expert judgments using numbers is that the results can make the range of experts' opinions across a field very apparent. A good example of this is provided by results we got in our interview with oceanographers about the future of circulation in the Atlantic Ocean. There is a big circulation pattern in the Atlantic that brings warm water north to the regions around Iceland and Greenland, where it then sinks deep down and goes south again. That warm water

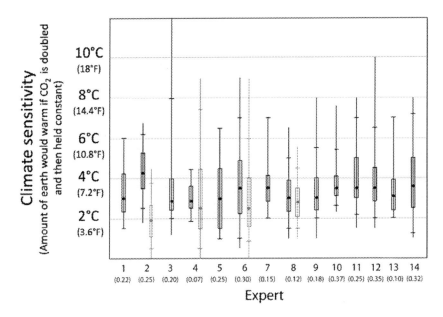

**FIGURE 4.3** Comparison of the values for "climate sensitivity" (the average amount that the earth would warm if the concentration of $CO_2$ in the atmosphere were doubled and then held at that value for hundreds of years) that experts gave us in 2005–6. The larger numbers across the bottom refer to the different experts. Four of them (experts 2, 4, 6, and 8) also participated in the expert elicitation we did 15 years earlier. Their earlier estimates are shown with the light gray box plots. The little numbers in parentheses under the expert numbers are the probability each expert gave to the possibility that the value of climate sensitivity is more than 4.5°C, the largest value the IPCC consensus study at that time had said it could have. These experts said there was between a 10% (expert 13) and a 35% (expert 12) chance that the value could be larger than the IPCC's maximum value. Figure from Zickfeld, Morgan, Frame, & Keith, 2010.

Zickfeld, K., Morgan, M. G., Frame, D. J., & Keith, D. W. (2010). Expert judgments about transient climate response to alternative future trajectories of radiative forcing. *Proceedings of the National Academy of Sciences, 107*(28), 12451–12456.

is very important for local climates. For example, southern England is at the same latitude as central Labrador. Labrador is pretty cold – icebergs and Arctic vegetation – while there are palm trees and other mild-weather plants growing in gardens in southern England. This difference in climate is caused by ocean circulation that goes by the technical name of the "Atlantic Meridional Overturning Circulation" (AMOC) and is popularly called the "Ocean Conveyer Belt." The AMOC moves heat from the tropics to more northern regions.

Paleontologists who study the climate of the very distant past have found that sometimes the AMOC has shut down. When that has happened, places like southern England got a whole lot colder. One of the questions we asked the oceanographers was what chance they thought there was that such a shutdown of the AMOC might begin to happen by 2100 if various amounts of warming had occurred by that date. Figure 4.4 shows the judgments made by the group of leading oceanographers and climate scientists we interviewed in 2005. The experts fall into two groups – a group that believes that if significant warming occurs, there is a pretty good chance that the AMOC will begin to shut down (upper set of lines) and a group that thinks there is not much chance the AMOC will begin to shut down (bottom set

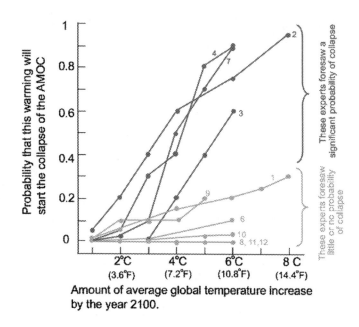

**Amount of average global temperature increase by the year 2100.**

**FIGURE 4.4** Estimates by different oceanographers made in 2005 of whether the large circulatory pattern in the Atlantic Ocean (the AMOC) might begin to shut down if the planet has warmed by different amounts by the end of the century. The experts fall into two groups. One group did not think there is much chance the AMOC will begin to shut down (bottom set of lines); the other group thought that if significant warming occurs, there is a pretty good chance the AMOC will begin to shut down (upper set of lines). The numbers that run from 0 at the bottom up to 1 at the top, along the left of the figure, are probabilities. For example, 0.4 means there is a 40% chance. The little numbers at the end of the lines indicate different experts. So for example, expert 2 is saying that if the planet warms by 3°C by 2100, there is a 40% chance that the AMOC will have begun to shut down; expert 3 is saying the warming would have to be 5°C for there to be a 40% chance of a shutdown. Figure modified from Zickfeld, Levermann, Morgan, Kuhlbrodt, Rahmstorf, & Keith, 2007.

Zickfeld, K., Levermann, A., Morgan, M. G., Kuhlbrodt, T., Rahmstorf, S., & Keith, D.W. (2007). Expert judgements on the response of the Atlantic meridional overturning circulation to climate change. *Climatic Change, 82*(3), 235–265.

of lines) with the kind of global warming that might have occurred by the end of the century.

Doing a good job on expert elicitation takes quite a bit of time and care. Good elicitations are neither something that can be done quickly nor should they be used as a substitute for doing the research to get the actual answers. However, when that research will take a long time they may be able to provide decision makers with some short-term insights. A few years ago, the journal *Proceedings of the National Academy of Sciences* asked me to write a general paper on the use (and abuse) of expert elicitations. Readers interested in learning more details about expert elicitation can download this paper for free at: www.pnas.org/content/pnas/early/2014/05/08/1319946111.full.pdf

The abstract of that paper reads:

The elicitation of scientific and technical judgments from experts, in the form of subjective probability distributions, can be a valuable addition to other forms of evidence in support of public policy decision making. This paper explores when it is sensible to perform such elicitation and how that can best be done. A number of key issues are discussed, including topics on which there are, and are not, experts who have knowledge that provides a basis for making informed predictive judgments; the inadequacy of only using qualitative uncertainty language; the role of cognitive heuristics and of over-confidence; the choice of experts; the development, refinement, and iterative testing of elicitation protocols that are designed to help experts to consider systematically all relevant knowledge when they make their judgments; the treatment of uncertainty about model functional form; diversity of expert opinion; and when it does or does not make sense to combine judgments from different experts. Although it may be tempting to view expert elicitation as a low-cost, low-effort alternative to conducting serious research and analysis, it is neither. Rather, expert elicitation should build on and use the best available research and analysis and be undertaken only when, given those, the state of knowledge will remain insufficient to support timely informed assessment and decision making.

## Technical Details for Chapter 4

Here are the references for the papers listed in the right-hand column of Table 4.1. Abstracts are included for the papers that are about climate science. The elicitation we did on how climate change may affect forests in the tropics and the far north is described in Chapter 6.

These first two papers describe results from the first expert elicitations we did on the health effects of air pollution:

Morgan, M. G., Morris, S. C., Henrion, M., Amaral, D. A., & Rish, W. R. (1984). Technical uncertainty in quantitative policy analysis—a sulfur air pollution example. *Risk Analysis*, *4*(3), 201–216.

Morgan, M. G., Henrion, M., Morris, S. C., & Amaral, D. A. (1985). Uncertainty in risk assessment. *Environmental Science & Technology, 19*(8), 662–667.

The next five papers report the results from expert elicitations we did on climate science:

Morgan, M. G., & Keith, D. W. (1995). Subjective judgments by climate experts. *Environmental Science & Technology, 29*(10), 468A–476A.

**Abstract:** Structured interviews using "expert elicitation" methods drawn from decision analysis were conducted with 16 leading U.S. climate scientists. We obtained quantitative, probabilistic judgments about a number of key climate variables and about the nature of the climate system. We also obtained judgments about the relative contributions of various factors to the uncertainty in climate sensitivity. We found strong support for the importance of convection/water vapor feedback and of cloud optical properties. A variety of questions were posed to elicit judgments about future research needs and the possible sources and magnitude of future surprises. The results reveal a rich diversity of expert opinion and, aside from climate sensitivity, a greater degree of disagreement than is often conveyed in scientific consensus documents. Research can make valuable contributions, but we interpret our results to mean that overall uncertainty about the geophysics of climate change is not likely to be reduced dramatically in the next few decades.

Morgan, M. G., Pitelka, L. F., & Shevliakova, E. (2001). Elicitation of expert judgments of climate change impacts on forest ecosystems. *Climatic Change, 49*(3), 279–307.

The abstract for this paper can be found at the end of Chapter 6.

Morgan, M. G., Adams, P. J., & Keith, D. W. (2006). Elicitation of expert judgments of aerosol forcing. *Climatic Change, 75*(1), 195–214.

**Abstract:** A group of twenty-four leading atmospheric and climate scientists provided subjective probability distributions that represent their current judgment about the value of planetary average direct and indirect radiative forcing from anthropogenic aerosols at the top of the atmosphere. Separate estimates were obtained for the direct aerosol effect, the semi-direct aerosol effect, cloud brightness (first aerosol indirect effect), and cloud lifetime/distribution (second aerosol indirect effect). Estimates were also obtained for total planetary average forcing at the top of the atmosphere and for surface forcing. Consensus was strongest among the experts in their assessments of the direct aerosol effect and the cloud brightness indirect effect. Forcing from the semi-direct effect was thought to be small (absolute values of all but one of the experts' best estimates were 0.5 $W/m^2$). There was not agreement about the sign of the best estimate of the semi-direct effect, and the uncertainty ranges some experts gave for this effect did not overlap those given by others. All best estimates of total aerosol forcing were negative, with values ranging between $-0.25$ $W/m^2$ and $-2.1$ $W/m^2$. The range of uncertainty that a number of experts associated with their estimates, especially those for total aerosol forcing and for surface forcing, was often much larger than that suggested in 2001 by the IPCC Working Group 1 summary figure (IPCC, 2001).

Zickfeld, K., Levermann, A., Morgan, M. G., Kuhlbrodt, T., Rahmstorf, S., & Keith, D. W. (2007). Expert judgements on the response of the Atlantic meridional overturning circulation to climate change. *Climatic Change, 82*(3), 235–265.

**Abstract:** We present results from detailed interviews with 12 leading climate scientists about the possible effects of global climate change on the Atlantic Meridional Overturning

Circulation (AMOC). The elicitation sought to examine the range of opinions within the climatic research community about the physical processes that determine the current strength of the AMOC, its future evolution in a changing climate and the consequences of potential AMOC changes. Experts assign different relative importance to physical processes which determine the present-day strength of the AMOC as well as to forcing factors which determine its future evolution under climate change. Many processes and factors deemed important are assessed as poorly known and insufficiently represented in state-of-the-art climate models. All experts anticipate a weakening of the AMOC under scenarios of increase of greenhouse gas concentrations. Two experts expect a permanent collapse of the AMOC as the most likely response under a $4\mathring{A}\sim CO_2$ scenario. Assuming a global mean temperature increase in the year 2100 of 4 K, eight experts assess the probability of triggering an AMOC collapse as significantly different from zero, three of them as larger than 40%. Elicited consequences of AMOC reduction include strong changes in temperature, precipitation distribution and sea level in the North Atlantic area.

Zickfeld, K., Morgan, M. G., Frame, D. J., & Keith, D. W. (2010). Expert judgments about transient climate response to alternative future trajectories of radiative forcing. *Proceedings of the National Academy of Sciences*, *107*(28), 12451–12456.

**Abstract:** There is uncertainty about the response of the climate system to future trajectories of radiative forcing. To quantify this uncertainty, we conducted face-to-face interviews with 14 leading climate scientists, using formal methods of expert elicitation. We structured the interviews around three scenarios of radiative forcing stabilizing at different levels. All experts ranked "cloud radiative feedbacks" as contributing most to their uncertainty about future global mean temperature change, irrespective of the specified level of radiative forcing. The experts disagreed about the relative contribution of other physical processes to their uncertainty about future temperature change. For a forcing trajectory that stabilized at 7 Wm$^{-2}$ in 2200, 13 of the 14 experts judged the probability that the climate system would undergo, or be irrevocably committed to, a "basic state change" as $\geq 0.5$. The width and median values of the probability distributions elicited from the different experts for future global mean temperature change under the specified forcing trajectories vary considerably. Even for a moderate increase in forcing by the year 2050, the medians of the elicited distributions of temperature change relative to 2000 range from 0.8–1.8 °C, and some of the interquartile ranges do not overlap. Ten of the 14 experts estimated that the probability that equilibrium climate sensitivity exceeds 4.5 °C is >0.17, our interpretation of the upper limit of the "likely" range given by the Intergovernmental Panel on Climate Change. Finally, most experts anticipated that over the next 20 years research will be able to achieve only modest reductions in their degree of uncertainty.

Finally, these three papers report results from studies of technologies that could be valuable in decarbonizing the energy system:

Curtright, A. E., Morgan, M. G., & Keith, D. W. (2008). Expert assessments of future photovoltaic technologies. *Environmental Science & Technology*, *42*(24).

Abdulla, A., Azevedo, I. L., & Morgan, M. G. (2013). Expert assessments of the cost of light water small modular reactors. *Proceedings of the National Academy of Sciences*, *110*(24), 9686–9691.

Whiston, M. M., Azevedo, I. L., Litster, S., Whitefoot, K. S., Samaras, C., & Whitacre, J. F. (2019). Expert assessments of the cost and expected future performance of proton exchange membrane fuel cells for vehicles. *Proceedings of the National Academy of Sciences*, *116*(11), 4899–4904.

**Some notable papers by authors not affiliated with our Centers:**

Anadon, L. D., Bosetti, V., Bunn, M., & Lee, A. (2012). Expert judgments about RD&D and the future of nuclear energy. *Environmental Science & Technology*, *41*(21), 11497–11504.

Baker, E., Bosetti, V., Anadon, L. D., Henrion, M., & Reis, L. A. (2015). Future costs of key low-carbon energy technologies: Harmonization and aggregation of energy technology expert elicitation data. *Energy Policy*, *80*, 219–232.

Budnitz, R. J, Apostolakis, G., Boore, D. M., Cluff, L. S., Coppersmith, K. J., Cornell, C. A., & Morris, P. A. (1998). Use of technical expert panels: Applications to probabilistic seismic hazard analysis. *Risk Analysis*, *18*(4), 463–469.

Clemen, R. T., & Winkler, R. A. (1999). Combining probability distributions from experts in risk analysis. *Risk Analysis*, *19*, 187–203.

Cooke, R. M. (1991). *Experts in Uncertainty: Opinion and subjective probability in science.* Oxford University Press, 336pp.

Cooke, R. M., & Goossens, L. L. H. J. (2008). TU Delft expert judgment data base. *Reliability Engineering and System Safety*, *93*, 657–674.

Cooke, R. M., Wilson, A. M., Tuomisto, J. T., Morales, O., Tainio, M., & Evans, J. S. (2007). A probabilistic characterization of the relationship between fine particulate matter and mortality: Elicitation of European experts. *Environmental Science & Technology*, *41*, 6598–6605.

US Environmental Protection Agency. (2011). Expert Elicitation Task Force White Paper. *Prepared for the Science and Technology Policy Council.* Available at: www.epa.gov/stpc/pdfs/ee-white-paper-final.pdf

Evans, J. S., Graham, J. D., Gray, D. M., & Sielken Jr., R. L. (1994). A distributional approach to characterizing low-dose cancer risk. *Risk Analysis*, *14*(1), 25–34.

Kahneman, D., Slovic, P., & Tversky, T. (eds.). (1982). *Judgment Under Uncertainty: Heuristics and biases.* Cambridge University Press, 555pp.

Mandel, D., & Barnes, A. (2014). Accuracy of forecasts in strategic intelligence. *Proceedings of the National Academy of Sciences*, *111*(30), 10984–10989.

Murphy, A. H., & Winkler, R. L. (1977). Can weather forecasters formulate reliable probability forecasts of precipitation and temperature?. *National Weather Digest*, *2*, 2–9(a).

O'Hagan, A., Buck, C. E., Daneshkhah, A., Eiser, J. R., Garthwaite, P. H., Jenkinson, D. J., Oakley, J. E., & Rakow, T. (2006). *Uncertain Judgments: Eliciting experts' probabilities.* John Wiley and Sons, 321pp.

Roman, H. A., Walker, K. D., Walsh, T. L., Conner, L., Richmond, H. M., Hubbell, B. J., & Kinney, P. L. (2008). Expert judgment assessment of the mortality impact of changes in ambient fine particulate matter in the U.S. *Environmental Science & Technology*, *42*, 2268–2274.

Spetzler, C. S., & Staël von Holstein, C-A. S. (1975). Probability encoding in decision analysis. *Management Science*, *22*(3), 340–358.

Tversky, A., & Kahneman, D. (1974). Judgments under uncertainty: Heuristics and biases. *Science*, *185*(4157), 1124–1131.

Wallsten, T. S., Budescu, D. V., Rapoport, A., Zwick, R., & Forsyth, B. (1986). Measuring the vague meanings of probability terms. *Journal of Experimental Psychology: General*, *155*(4), 348–365.

## Notes

1 Some fragments of text in this chapter are reworked from the following paper: Morgan, M. G., Fischhoff, B., Bostrom A., Lave, L., & Atman, C. J. (1992). Communicating risk to the public. *Environmental Science & Technology*, *26*(11), 2048–2056.
2 For an excellent discussion of all the efforts to which "climate deniers" have gone to keep the public confused about climate change, see Oreskes, N., & Conway, E. M. (2011). Merchants of doubt: How a handful of scientists obscured the truth on issues from tobacco smoke to global warming. Bloomsbury Publishing USA, 355pp.
3 Once carbon dioxide enters the atmosphere, much of it stays there for centuries, where it continues to warm the planet. See the discussion in Chapter 5.
4 For details on decision analysis and the people who first developed these tools, see chapter 4 in Morgan, M. G. (2017). *Theory and Practice in Policy Analysis*. Cambridge University Press, 590pp.
5 For an example of this work, see Morgan, M. G., Henrion, M., Morris, S. C., & Amaral, D. A. (1985). Uncertainty in risk assessment. *Environmental Science & Technology*, *19*(8), 662–667.
6 See: Presidential/Congressional Commission on Risk Assessment and Risk Management (1997). Risk Assessment and Risk Management in Regulatory Decision Making, 1997.
7 The paper specified the resulting climate change in detail.

# 5

# WHAT THE PUBLIC KNOWS ABOUT CLIMATE CHANGE

*Granger Morgan*

What the general public and opinion leaders know and think about climate change has a big effect on how society responds. Over the course of the 25 years that our NSF-supported Centers have run, public awareness and understanding of climate change and its impacts has changed quite a lot. I have often argued that to be able to engage in informed public discourse about climate issues, people need to know three simple facts:

1.  Burning coal, oil, and natural gas produces carbon dioxide that enters the atmosphere.
2.  Carbon dioxide in the atmosphere warms the earth, and that warming changes the climate.
3.  Once carbon dioxide gets into the atmosphere, much of it remains there for many hundreds of years.

Studies that Ann Bostrom and I did with several colleagues and graduate students between 1994 and 2010 show pretty clearly that, despite some continuing confusion and misunderstanding about some of the details, there has been real progress on public understanding of facts 1 and 2. More recent work that PhD student Rachel Dryden had done with me, Ann, and Wändi Bruine de Bruin shows that there is still widespread and profound misunderstanding about fact 3.

Before telling the story of these and other studies, I should go back and describe some earlier work on developing methods for risk communication that Ann did for her PhD with Baruch Fischhoff, me, and several other colleagues and PhD students.

In the early 1980s, the phrase "risk communication" became very popular. People from state and federal agencies, from private companies, and from nonprofit advocacy organizations all began giving speeches about how important it was to have good risk communication. But while lots of people were talking about it, and

DOI: 10.4324/9781003330226-5

a variety of people had ideas about how it should be done, nobody had conducted the necessary empirical studies to determine how good risk communications could actually be developed.

Finally, in 1987 the National Science Foundation issued a request for proposals to address these issues, and Baruch Fischhoff, Lester Lave, and I were fortunate enough to write the winning proposal. This support along with support from the Electric Power Research Institute launched us on a multi-year program of studies designed to understand the various aspects of risk communication and to create an empirically based approach to developing risk communication messages. While the project produced lots of very interesting applied social science (as well as five PhDs and over 30 refereed publications), at its heart it was an engineering enterprise. We set out to learn how to develop risk communications that people could understand and would find useful for the decisions they faced. Theory and experiments were important, but the ultimate proof lies in producing real risk communications, testing them, and demonstrating that they did a better job than other communications developed through more traditional ad hoc methods.

Developing a risk communication had traditionally been a two-step process:

1.  Find a specialist who knows a lot about the risk and ask them what they think people should be told.
2.  Give that information to a "communications expert" to have them turn it into a message.

If you were being really careful, you got the health or safety specialist to review the message before it was finalized, and you ran some tests after the message had been completed to see how well it was understood by the people it was intended to inform.

On the surface, this may seem like a pretty reasonable approach; however, two key things are missing: (1) the traditional method doesn't determine systematic-ally what people already know about the risk – peoples' knowledge is important because they interpret anything you will tell them in light of what they already believe; (2) the traditional method didn't determine systematically the specific information that people need to make the decisions they face.

This led us to develop a strategy we call "the mental model approach" to risk communication. People know a lot, and they interpret any new information using their existing knowledge and beliefs. For example, researchers have found that even science students who get good grades often add new knowledge on to their funda-mentally incorrect, naïve "mental models" about ideas like inertia for quite a long time before they finally recognize some inconsistencies and adopt a technically correct model. This means that if we want to develop a good risk communication, the first thing we need to do is find out what people already know and how they think about the risk.

Finding out what someone already knows about a risk (that is, their mental model) is easier said than done. We could administer a questionnaire, but people are smart. As soon as we start putting information in our questions, they are going to start using that information to make inferences and draw conclusions. Soon we're not going to know if the answers we are getting are telling us about the mental model the person already had before we started quizzing them, or about the new mental model that the person has been building because of all the information we've been supplying to them in our questions.

After quite a bit of work, we ended up developing a five-step process for creating a risk communication.

> *Step 1: Create an expert model:* For starters, we need to be able to describe the available scientific knowledge. To do that, we need to review current knowledge and determine the risks. Then we need to summarize it, focusing on what can be done about the risk. We represent this knowledge in the form of something called an "influence diagram" and get that reviewed by technical experts who have differing perspectives to be sure it is correct and balanced, revising it as needed.

> *Step 2: Conduct mental models interviews:* We conduct open-ended interviews with people to learn about their beliefs about the hazard, expressed in their own terms. Later stages of the interview should be shaped by the expert influence diagram so that we have covered all the potentially relevant topics. In the interview, we should allow for the expression of both correct and incorrect beliefs and ask follow-up questions to make sure that the respondents' intent is clear. After we have recorded the interviews, we need to develop an appropriate way to record all the topics and ideas people have raised (a coding scheme) and then code and summarize the results. We should analyze responses in terms of how well peoples' mental models correspond to the expert model captured by the influence diagram that was created in Step 1.

> *Step 3: Conduct a closed-form survey:* Some of the things people say will be just the unique ideas of a single individual. We need some way to weed these out and find the ideas and concepts that are widely shared. To do this, we create a questionnaire whose items capture the beliefs expressed in the open-ended interviews and the expert model. We administer it to larger groups of people drawn from the intended audience in order to estimate how prevalent the various beliefs are in the population.

> *Step 4: Prepare a draft communication:* Next, we use the results from the structured interviews and from the survey, as compared with the expert diagram, to identify a small number of things that are most important to communicate, including gaps in knowledge that are most important to fill and erroneous beliefs that are most important to correct. We draft a communication in appropriate lay language using things like organizers, supporting figures, and similar aids.

> *Step 5: Evaluate the communication:* Our experience makes it clear that nobody developing a risk communication ever gets things right the first time. So we need to test and refine the communication with individuals selected from

the population for whom the message has been developed, using one-on-one read-aloud interviews, focus groups, closed-form questionnaires, and perhaps problem-solving tasks. We repeat this process until the communication is understood as intended.

The tricky part of this method is Step 2. Suppose that I want to learn about the mental model that someone has for something, like radon in a home (the risk we used in much of our early work on risk communication). I could simply ask: "Tell me about radon in homes" and see what they say. In 1988, we tried running a few interviews like this and decided that the approach had enough promise to make it worthwhile to test the idea more broadly. We sent several engineering graduate students out to interview their non-technical friends and neighbors. When we transcribed the tapes, the results were hilarious. A few moments into most of the interviews, the non-technical person had figured out that they were being interviewed by someone who knew a lot about radon, and they had begun busily extracting information from them!

After posing a question such as "Tell me about radon in homes," most people can only talk for a few sentences before they run out of steam. However, those few sentences often contain five or ten different concepts. If the interviewer has been trained to keep track of all the things that were mentioned, they can then go on to ask follow-up questions on each one. For example, the interviewer might say, "You mentioned that radon can come in to a house through the cellar, tell me more about that." By systematically following up on all the concepts that have been introduced, a good interviewer can often sustain a conversation about the risk for 10 to 20 minutes, without introducing any new ideas of their own. Only in a later stage of the interview will the interviewer go on to ask questions about other key ideas that the subject did not bring up on their own.

Ann Bostrom played a central role in helping us to develop and refine this approach. She went on to conduct a series of mental model interviews with people[1] about climate change. Here is an example of the transcript of the opening few moments of one of those interviews:

ANN: I'd like you to tell me all about the issue of climate change.
PERSON ANN IS INTERVIEWING: Climate change, you mean global warming?
ANN: Climate change.
PERSON ANN IS INTERVIEWING: OK. Let's see. What do I know. The earth is getting warmer because there are a lot of holes in the atmosphere and this is global warming in the greenhouse effect. Umm... I really don't know very much about it, but it does seem to be true. The temperatures do seem to be kind of warm in the winters. They do seem to be warmer than in the past... and hmmm... that's all I know about global warming.

Notice the first thing the person being interviewed did after Ann asked her opening question was to ask for more information – which Ann did not provide. After that, the person being interviewed listed off a bunch of ideas, each of which Ann followed up on with questions like, "You said there are holes in the atmosphere, tell me more about that."

Here are two more examples of opening moments of interviews that Ann ran with two other people:

ANN: I'd like you to tell me all about the issue of climate change.

PERSON ANN IS INTERVIEWING: I'm pretty interested in it… the ice caps are melting—the hole in the ozone layer. They think pollution from cars and aerosol cans are the cause of all that. I think the space shuttle might have something to do with it too, because they always send that up through the earth, to get out in outer space. So, I think that would have something to do with it, too.

ANN: I'd like you to tell me all about the issue of climate change.

PERSON ANN IS INTERVIEWING: Climate change? Like what about it? Like, as far as the ozone layer and ice caps are melting, water level rising, rainforest going down, oxygen going down because of that? All of that kind of stuff?

ANN: Anything else?

PERSON ANN IS INTERVIEWING: Well, erosion all over the place. Umm, topsoils down into everywhere. Fertilizer poisoning.

ANN: Anything else that comes to mind related to climate change?

PERSON ANN IS INTERVIEWING: Climate change. Winters aren't like they used to be. Nothing's as severe. Not as much snow. Nothing like that.

When we were developing our approach, we did lots of mental model interviews like these on several different topics, and Ann developed a careful strategy to record (code) all the correct, incorrect, and extraneous concepts that people brought up. After doing 10–15 interviews on some topic like climate change, we found we were often not hearing any new concepts when we ran another interview, so there was not much point to running a lot more time-consuming interviews. On the other hand, some of the concepts that people had brought up in the interviews we'd run were probably the unique views of just that one individual. The point of Step 3 in the process we designed is to weed out these unique concepts[2] so that we could focus on the few concepts that are most important to address in the communication.

We conducted such a follow-up survey among Pittsburgh residents in 1992 and published a pair of papers. The first one described the mental model studies, and the second one reported on the survey. We found that people viewed global warming as bad and very likely. Many told us they believed it was already occurring. They were confused about what was causing it, and they were also confused about the difference between weather and climate. Climate, of course, is the average weather that occurs over time in a specific region. While the primary cause of global warming is burning coal, oil, and gas, which puts carbon dioxide into the atmosphere, many of the people we interviewed or who completed our survey did not know this. They suggested a variety of causes that are *not* important contributors to global climate change – things like aerosol spray cans, the hole in the ozone layer, and general air pollution. Roughly 10% of people thought nuclear power (which releases no carbon dioxide to the atmosphere) and the space program (which some argued punches holes in the atmosphere and lets in heat, which is not true) contributed to climate change. Most peoples' mental model was that pretty much anything that is bad for the environment causes climate change. The fact that few people realized that the big problem is burning fossil fuel and emitting carbon dioxide to the atmosphere meant that they were really not in a position to judge what needs to be done to avoid climate change. On the other hand, people did a pretty good job of anticipating what many of the ecological consequences of climate change are likely to be.

Based on these results, I worked with Tom Smuts and several other colleagues to develop a risk communication about climate change for the general public. This was a few years before pretty much everyone had access to the Internet, so we developed something we called "paper hypertext." The main brochure started with a two-page spread that hit the key points (and listed and corrected important misconceptions). Three subsequent two-page spreads gave more details about *What is climate change?*; *If climate changes, what might happen?*; and *What can be done about climate change?* For each of these sections, there was also a pouch containing a smaller brochure that provided additional details. You can find the text of all this in the first Appendix (pp. 185–238) of our book *Risk Communication: A Mental Model Approach*, which is referenced at the end of this chapter.

Seventeen years after we ran our first surveys on public understanding of climate change, in 2009, we used the exact same questionnaire to conduct another study. We found that while some of the incorrect beliefs persisted, there had been a big improvement in peoples' basic understanding. Most people had figured out that adding carbon dioxide to the atmosphere by burning fossil fuels was the primary cause of climate change. We concluded that there had been real progress in public understanding of the first two facts I listed at the beginning of this chapter. However, as I gave general talks about climate issues to non-technical audiences, I began to conclude that there was still deep public misunderstanding about that third fact: "Once carbon dioxide gets into the atmosphere, much of it remains there for hundreds of years."

In our earlier work, we had not thought to include questions about how long $CO_2$ remains in the atmosphere once it gets there. As Figure 5.1 shows, the answer is not just a single number ($CO_2$ slowly gets absorbed by the land, the oceans, etc.), but much of it remains in the atmosphere for hundreds of years. Indeed, as you read this, some of the $CO_2$ in the air you are breathing got put there by factories in England and other countries back in the late 1700s during the Industrial Revolution!

To explore what people think about how long $CO_2$ remains in the atmosphere, in 2016–17, PhD student Rachel Dryden worked with me, Ann, and Wändi to develop a survey. Rachel mailed her survey to people all across Allegheny County (the Pittsburgh region), and Ann used the two key questions in an online national

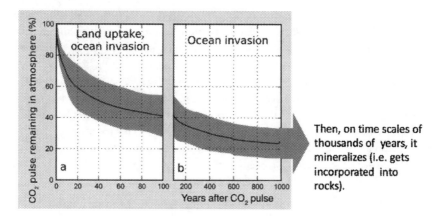

**FIGURE 5.1** When a bit of $CO_2$ (a "pulse") gets added to the atmosphere, this figure shows how the amount of $CO_2$ decays over time. Notice that even after several hundred years, roughly 1/3 of what first got added is still in the atmosphere. (This figure modified from IPCC 2013 AR5 WG1.)

IPCC, 2013: Climate Change 2013: The Physical Science Basis. Contribution of Working Group I to the Fifth Assessment Report of the Intergovernmental Panel on Climate Change, Box 6.1, Figure 1.

computer survey she ran using a system called MTurk. What we found is shown in Figure 5.2.

As I had feared, the results were alarming. Most people think that once $CO_2$ enters the atmosphere, it stays there just about as long as regular air pollution. Why do I say I find this alarming? Because it is consistent with people also thinking "I don't know if this climate change stuff is real or how big a problem it will be, but if it ever gets serious, we'll just fix it by stopping emissions—just like we cleaned up air pollution in places like LA and Pittsburgh." *BUT that won't be possible.* We could clean up air pollution because it only stays in the atmosphere for a few hours or days, so when we stop emissions, it quickly disappears. If we stop emissions of $CO_2$, much of what we have already put into the atmosphere will still be there and continue to cause warming and climate change for hundreds of years.

The negotiations that lead to the climate agreement in Paris in 2015 concluded that the nations of the world should hold the average warming to well under 2°C (3.6°F). However, compared with the amount that has already been added, there is no room in the atmosphere to add much more if we want to hit the Paris target. This means that unless the nations of the world dramatically reduce their emissions over the next two or three decades, average warming will soar well above 2°C. As I recently explained in a talk I gave at Tufts University, if that happens, we'll be "toast," or perhaps more accurately "burnt toast."

In addition to warming, another impact of climate change is an increase in the frequency and severity of extreme events such as storms. This doesn't mean that

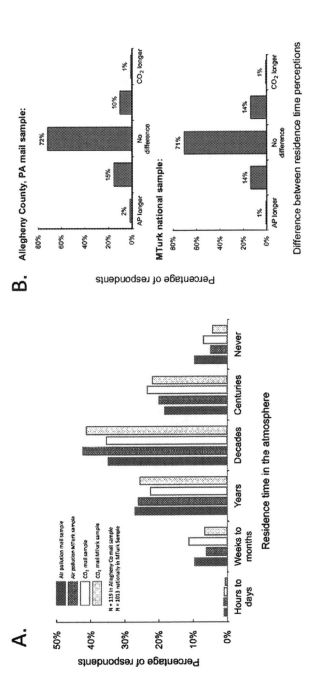

**FIGURE 5.2 A.** The diagram on the left shows the average results for how long people thought air pollution and $CO_2$ remain in the atmosphere (the residence time). Most people think the answer is about the same for both, and they think it is a few decades. *That is not correct.* Once emissions stop, air pollution only stays in the atmosphere for hours or days. Once $CO_2$ enters the atmosphere, much of it stays there for centuries. **B.** In contrast to the diagram on the left, which shows average results for all the people who answered our questions, the diagram on the right shows individual-by-individual responses and confirms that most people think both $CO_2$ and air pollution stay in the atmosphere about the same amount of time. Figures are from Dryden et al., 2018.

Dryden, R., Morgan, M. G., Bostrom, A., & Bruine de Bruin, W. (2018). Public perceptions of how long air pollution and carbon dioxide remain in the atmosphere. *Risk Analysis, 38*(3), 525–534.

climate change *causes* these events. Rather, it increases the probability that they will occur. The science of figuring out whether and how much climate change contributed to an extreme event is called "climate attribution." In another part of her work, Rachel Dryden conducted a series of studies to learn how best to explain these issues to members of the general public.[3]

She did this by using spinner boards (Figure 5.3) in which the odds of an event occurring are equal to the chance that a spinner will land on the colored segment of the spinner board. So for example, if there was a 1 in 100 chance of the event occurring in a year, the width of the colored section would be 360°/100=3.6° wide. She explained the fact that a "100-year flood" could happen two years in a row by noting that, while it was most unlikely, there is a slim chance that the spinner could land twice in a row on the narrow colored segment. Once someone had played with the spinner board for a few moments, Rachel asked: "If climate change makes it more likely that an extreme event will occur, how would the spinner board change, and what would this difference mean?" The answer, of course, is that the colored wedge would get wider.

In interviews Rachel conducted with members of the public, every single person correctly figured this out without seeing Figure 5.3 B.

Then she asked people to imagine that they were going to use the two spinner boards in Figure 5.3 to explain the effect of climate change on the likelihood of an

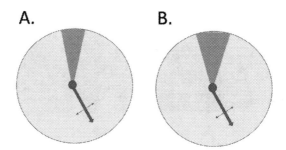

**FIGURE 5.3  A.** Basic spinner board used to explain the likelihood that an extreme event will occur in the course of some specified period of time, such as the next year. **B.** Modified spinner board showing how climate change might modify the board. In our study, all respondents were able to describe the modified board before seeing it, and then, when given the second board, correctly use the pair of boards to explain the effect of climate change on the frequency of an extreme event to a friend or neighbor. This figure shows the spinner boards used in this study. It is not intended to represent any actual expected. Source: Dryden and Morgan (2020).

Dryden, R., & Morgan, M. G. (2020). A simple strategy to communicate about climate attribution. *Bulletin of the American Meteorological Society, 101*(6), E949–E953. © American Meteorological Society. Used with permission.

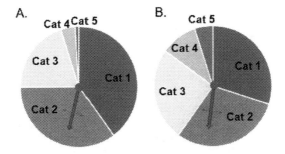

**FIGURE 5.4 A.** Basic spinner board used to explain that the distribution of peak hurricane strengths could change in the future. **B.** Modified spinner board showing how climate change might increase the probability that a storm would be more intense. As in Figure 5.3, in our study, all respondents were able to describe the modified board before seeing it, and then when given the second board, correctly used the pair of boards to explain the effect of climate change on the intensity of a hurricane to a friend or neighbor. This figure shows the spinner boards used in this study. It is not intended to represent any actual expected change in the frequency of any specific extreme events. Source: Dryden and Morgan (2020).

Dryden, R., & Morgan, M. G. (2020). A simple strategy to communicate about climate attribution. *Bulletin of the American Meteorological Society, 101(6),* E949–E953. © American Meteorological Society. Used with permission.

extreme event to a neighbor or friend. Once again, every single respondent gave a correct explanation.

Turning her focus to hurricanes, she showed people the spinner board in Figure 5.4 A, which is divided in proportion to the occurrence of the 1–5 Saffir–Simpson hurricane intensity values (i.e., Cat. 1 to Cat. 5 storms). Again, she asked respondents to explain the spinner board for a specific storm event and asked how the board would change if climate change contributes to making hurricanes stronger. Their explanations were again correct. They also gave correct explanations when they were asked how they would explain the two boards to a neighbor or friend.

In a short paper we published on this work in the *Bulletin of the American Meteorological Society,* we argued that:

> The use of spinner boards appears to be an easily understood and highly effective strategy to explain issues of climate attribution to the general public. Spinner boards could be readily adapted for use by TV weather forecasters and in news coverage of hurricanes and other extreme weather events… [such an approach could] provide weather forecasters and news reporters with a vehicle to provide compelling visual explanations.

## Technical Details for Chapter 5

Here are several papers we wrote about our work on risk communication:

Bostrom, A., Fischhoff, B., & Morgan, M. G. (1992). Characterizing mental models of hazardous processes: A methodology and an application to radon. *Journal of Social Issues, 48*(4), 85–100.

Morgan, M. G. Fischhoff, B., Bostrom, A., Lave, L., & Atman, C. J. (1992). Communicating risk to the public. *Environmental Science & Technology, 26*(11), 2048–2056.

Atman, C. J., Bostrom, A., Fischhoff, B., & Morgan, M. G. (1994). Designing risk communications: Completing and correcting mental models of hazardous processes, Part I. *Risk Analysis, 14*(5), 779–788. Also reprinted in Gerrard, S., Turner, R. K., & Bateman, I. (eds.). (2001). *Environmental Risk Planning and Management*, Chapter 18, Edward Elgar Publishers, 251–260.

Bostrom, A., Atman, C. J., Fischhoff, B., & Morgan, M. G. (1994). Evaluating risk communications: Completing and correcting mental models of hazardous processes, Part II. *Risk Analysis, 14*(5), 789–798. Also reprinted in Gerrard, S., Turner, R. K., & Bateman, I. (eds.) (2001). *Environmental Risk Planning and Management*, Chapter 19, Edward Elgar Publishers, 261–270.

Here is a book we wrote summarizing our work on risk communication:

Morgan, M. G., Fischhoff, B., Bostrom, A., & Atman, C. (2002). *Risk Communication: A mental models approach.* Cambridge University Press, New York, 351pp.

Here are the first two papers we published on public understanding of climate change. The first one reports on our mental model studies, and the second one reports on the results of our survey:

Bostrom, A., Morgan, M. G., Fischhoff, B., & Read, D. (1994). What do people know about global climate change? 1. Mental models. *Risk Analysis, 14*(6), 959–970.

**Abstract:** A set of exploratory studies and mental model interviews was conducted in order to characterize public understanding of climate change. In general, respondents regarded global warming as both bad and highly likely. Many believed that warming has already occurred. They tended to confuse stratospheric ozone depletion with the greenhouse effect and weather with climate. Automobile use, heat and emissions from industrial processes, aerosol spray cans, and pollution in general were frequently perceived as primary causes of global warming. Additionally, the "greenhouse effect" was often interpreted literally as the cause of a hot and steamy climate. The effects attributed to climate change often included increased skin cancer and changed agricultural yields. The mitigation and control strategies proposed by interviewees typically focused on general pollution control, with few specific links to carbon dioxide and energy use. Respondents appeared to be relatively unfamiliar with such regulatory developments as the ban on CFCs for nonessential uses. These beliefs must be considered by those designing risk communications or presenting climate-related policies to the public.

Read, D., Bostrom, A., Morgan, M. G., Fischhoff, B., & Smuts, T. (1994). What do people know about global climate change? 2. Survey studies of educated laypeople. *Risk Analysis, 14*(6), 971–982.

**Abstract:** Drawing on results from earlier studies that used open-ended interviews, a questionnaire was developed to examine laypeople's knowledge about the possible causes and effects of global warming, as well as the likely efficacy of possible interventions. It was administered to two well-educated opportunity samples of laypeople. Subjects had a poor appreciation of the facts that (1) if significant global warming occurs, it will be primarily the result of an increase in the concentration of carbon dioxide in the earth's atmosphere, and (2) the single most important source of additional carbon dioxide is the combustion of fossil fuels, most notably coal and oil. In addition, their understanding of the climate issue was encumbered with secondary, irrelevant, and incorrect beliefs. Of these, the two most critical are confusion with the problems of stratospheric ozone and difficulty in differentiating between causes and actions specific to climate and more general good environmental practice.

Roughly 17 years later, we conducted another survey using exactly the same detailed questionnaire about climate change. Here is the paper that reports the differences in public understanding that we found:

Reynolds, T. W., Bostrom, A., Read, D., & Morgan, M. G. (2010). Now what do people know about global climate change? Survey studies of educated laypeople. *Risk Analysis, 30*(10), 1520–1538.

**Abstract:** In 1992, a mental-models-based survey in Pittsburgh, Pennsylvania, revealed that educated laypeople often conflated global climate change and stratospheric ozone depletion, and appeared relatively unaware of the role of anthropogenic carbon dioxide emissions in global warming. This study compares those survey results with 2009 data from a sample of similarly well-educated laypeople responding to the same survey instrument. Not surprisingly, following a decade of explosive attention to climate change in politics and in the mainstream media, survey respondents in 2009 showed higher awareness and comprehension of some climate change causes. Most notably, unlike those in 1992, 2009 respondents rarely mentioned ozone depletion as a cause of global warming. They were also far more likely to correctly volunteer energy use as a major cause of climate change; many in 2009 also cited natural processes and historical climatic cycles as key causes. When asked how to address the problem of climate change, while respondents in 1992 were unable to differentiate between general "good environmental practices" and actions specific to addressing climate change, respondents in 2009 have begun to appreciate the differences. Despite this, many individuals in 2009 still had incorrect beliefs about climate change, and still did not appear to fully appreciate key facts such as that global warming is primarily due to increased concentrations of carbon dioxide in the atmosphere, and the single most important source of this carbon dioxide is the combustion of fossil fuels.

Here is the paper in which we explored what people know about how long carbon dioxide remains in the atmosphere:

Dryden, R., Morgan, M. G., Bostrom, A., & Bruine de Bruin, W. (2018). Public perceptions of how long air pollution and carbon dioxide remain in the atmosphere. *Risk Analysis, 38*(3), 525–534.

**Abstract:** The atmospheric residence time of carbon dioxide is hundreds of years, many orders of magnitude longer than that of common air pollution, which is typically hours to a few days. However, randomly selected respondents in a mail survey in Allegheny County, PA ($N = 119$) and in a national survey conducted with MTurk ($N = 1,013$) judged the two to be identical (in decades), considerably overestimating the residence time of air pollution and drastically underestimating that of carbon dioxide. Moreover, while many respondents believed that action is needed today to avoid climate change (regardless of cause), roughly a quarter held the view that if climate change is real and serious, we will be able to stop it in the future when it happens, just as we did with common air pollution. In addition to assessing respondents' understanding of how long carbon dioxide and common air pollution stay in the atmosphere, we also explored the extent to which people correctly identified causes of climate change and how their beliefs affect support for action. With climate change at the forefront of politics and mainstream media, informing discussions of policy is increasingly important. Confusion about the causes and consequences of climate change, and especially about carbon dioxide's long atmospheric residence time, could have profound implications for sustained support of policies to achieve reductions in carbon dioxide emissions and other greenhouse gases.

Here is the paper on helping members of the general public to understand "climate attribution" using spinner boards:

Dryden, R., & Morgan, M. G. (2020). A simple strategy to communicate about climate attribution. *Bulletin of the American Meteorological Society, 101*(6), E949–E953. And in the print version: Using Spinner boards to explain climate attribution, *BAMS* 102 (1), 27–29, January 2021.

**Abstract:** Hurricane Harvey and other recent weather extremes stimulated extensive public discourse about the role of anthropogenic climate change in amplifying, or otherwise modifying, such events. In tandem, the scientific community has made considerable progress on statistical "climate attribution." However, explaining these statistical methods to the public has posed challenges. Using appropriately designed "spinner boards," we find that members of the general public are readily able to understand basic concepts of climate attribution and explain those concepts to others. This includes both understanding and explaining the way in which the probability of an extreme weather event may increase as a result of climate change and explaining how the intensity of hurricanes can be increased. If properly developed and used by TV weather forecasters and news reporters, this method holds the potential to significantly improve public understanding of climate attribution.

### Some notable papers by authors not affiliated with our Centers:

Budescu, D.V., & Broomell, S. B. (2012). Effective communication of uncertainty in the IPCC reports. *Climatic Change, 113*(2), 181–200.

National Academies of Sciences, Engineering, and Medicine. (2017). *Communicating Science Effectively: A research agenda*. National Academies Press.

Pidgeon, N., & Fischhoff, B. (2013). The role of social and decision sciences in communicating uncertain climate risks. In *Effective Risk Communication* (pp. 345–358). Routledge.

Spiegelhalter, D. (2017). Risk and uncertainty communication. *Annual Review of Statistics and Its Application, 4*, 31–60.

van der Bles, A. M., van der Linden, S., Freeman, A. L. J., Mitchell, J., Galvao, A. B., Zaval, L., & Spiegelhalter, D. J. (2019). Communicating uncertainty about facts, numbers and science. *Royal Society Open Science, 6*(5), 181870.

van der Bles, A. M., van der Linden, S., Freeman, A. L., J., & Spiegelhalter, D. J. (2020). The effects of communicating uncertainty on public trust in facts and numbers. *Proceedings of the National Academy of Sciences, 117*(14), 7672–7683.

Yale Program on Climate Change Communication is a leading program conducting scientific studies on public opinion and behavior and provides support for climate-related decision making. Details are at: https://climatecommunication.yale.edu/

## Notes

1 We recruited a group of "well-educated lay people." By that, we basically meant people who had paid attention in high school.
2 In technical terms, the point of the survey in Step 3 is to gain "statistical power" in identifying critical concepts.
3 Rachel also used a strategy called signal detection theory to see how well people could judge whether specific time sequences of hurricane events were unusual. This work is described in Dryden, R., Morgan, M. G., & Broomell, S. (2020). Lay detection of unusual patterns in the frequency of hurricanes. *Weather, Climate, and Society, 12*(3), 597–609.

# 6

# SOCIAL/ECOLOGICAL DIMENSIONS OF CLIMATE CHANGE

*Hadi Dowlatabadi, Granger Morgan, Mike Griffin, and Tim McDaniels*

## Background on Socio-Ecological Systems

### Hadi Dowlatabadi

Any discussion of the impacts of climate change should consider interactions between social and ecological systems. Humans derive many benefits and resources from ecosystems and from resource extraction and harvesting. Meanwhile, human disturbances and attempted management along with underlying climate and other conditions determine what ecosystems exist.

The benefits that humans derive from ecosystems are both material and spiritual. Beginning with our very earliest discussions about integrated assessment of climate change, in January of 1991, Lester Lave emphasized the importance of what he, as an economist, described as the "non-market" impacts of climate change on natural ecosystems. Granger chimed in with his favorite example of how sad he would be if climate change meant that white pine, sugar maples, and birch trees were replaced by yellow pine and oak trees in New England.

Developing a credible approach to how ecosystems should be included in integrated assessments was (and remains) very challenging. Here are three fundamental challenges and questions that I see:

- Most natural ecosystems change much more slowly than human actions – therefore, what is prevalent on the "natural" undisturbed landscape is not shaped by the recent past (or even the climate record of the past few hundred years). Given the rapid change in environmental conditions and the varied response rates of different elements of ecosystems, we have no idea what the "equilibrium conditions" would be. Therefore, projecting what will be prevalent in response to climate change has to be based on untestable assumptions.

DOI: 10.4324/9781003330226-6

- Direct human action, such as encroachment, is changing ecosystems much faster than climate is changing. How should we model these or attempts to intervene to protect nature?
- The general public values what appears to be healthy ecosystems.[1] Experts, on the other hand, know and value each species and their interactions and have a much more sophisticated, nuanced sense of ecosystem health and services. Media coverage of expert opinions therefore often shapes and reshapes public perceptions of the state of ecosystems, their peril, their importance, and any vital services they provide.

At the time of developing ICAM-1, the field of ecosystem services had not yet been born. The concepts of natural capital and Green GDP had been explored at Resources for the Future (RFF) since the 1960s, but no one had a foolproof way of representing damage to ecosystems in a model for decision-making. Yet we were not willing to develop a framework that ignored their importance. Therefore, in the process of developing ICAM-1, we chose to represent damage to ecosystems under the banner of non-market damages and afforestation as a means of reducing atmospheric $CO_2$.

Lester Lave suggested we use insights from studies of how the public valued visits to national parks, or simply knowing that they are there. We used these values to inform a wide range of values for non-market damages being caused to ecosystems by climate change. We also wondered about the nature of psychological adjustments to a changing landscape. These could help to offset the sense of loss if one green landscape was replaced with another; or loss could be amplified, if there were extinctions (local or global) reducing the chance of seeing *specific* ecosystems or species. For these reasons, we used a range of values for damages from one that diminished over time to one that was amplified with time.

We also modeled the option of 2 billion tons per year of new forests (afforestation) as a potential Negative Emissions Technology to capture carbon dioxide from the atmosphere and lock it up, for at least a while, in wood and soil.

In this chapter, we begin by providing an overview of studies our group did with the aim of understanding different aspects of the socio-ecological systems within the scope of our integrated assessment of climate change. Later, we expanded these studies with more detailed examinations of: how ecosystems may be impacted by "solutions to climate change" (e.g., biofuels production); the history of policies to protect specific aspects of ecosystems; and subjective perceptions of ecologists about planned interventions to maintain specific landscapes and species in a given location.

Our studies span a 30-year period and were both rewarding in what they revealed and in showing just how much we yet need to consider when thinking about the dynamics of socio-ecological systems. The rest of this chapter is devoted to providing an overview of this work and covers the following topics:

- Modeling the ecosystem impacts of climate change;
- Asking experts how they think forests will respond to climate change;

- Growing corn to make biofuel;
- Limiting habitat fragmentation from gas well development;
- Managing soil carbon;
- Drivers and dynamics of environmental policy change;
- Decision aids for resource managers.

## Modeling the Ecosystem Impacts of Climate Change

### Hadi Dowlatabadi

In the early 1990s, a number of models were projecting very large-scale land cover change in response to business-as-usual (BAU) climate change scenarios. The majority of these papers relied on the concept of an optimal biome for each climate zone. The best-known example of this approach is the Holdridge zones (Figure 6.1). This and its more sophisticated variants were used in early climate impact models to estimate impacts of climate change on ecosystems and land cover. These exercises first established what should ideally be the land cover of a region and how this ideal would change under projected climate change. These models projected huge changes in land cover and vast quantities of $CO_2$ released in transition from one type of land cover (e.g., temperate rainforest) to a successor land cover consistent with the new local climate (e.g., temperate hardwood).

This approach reproduced how climate modelers calculate change in climate due to buildup of greenhouse gases in the atmosphere. Early climate models were not good at reproducing today's climate in specific regions. So the models were used to calculate the difference between what a region's climate "should" have been and what it "could become" under a given greenhouse gas emissions scenario.

Similarly, land cover modelers did not question whether the Holdridge zone models of "optimal land cover" has validity on the ground – it does not. They ran their models using current climate (as it should be) to get land cover as it should be, and then they modeled land cover as it could be using climate models of what it could be and arrived at very large shifts in land cover. Of course, as one type of land cover was replaced by another, the $CO_2$ in trees and soils would be released and new species would grow in their place. During this transition period there were very large releases of $CO_2$ from land covers. This was touted as strong evidence for the positive feedback effect inherent in the earth's biosphere. Developers of such models argued that one can measure the potential change in land cover by comparing the two fictions of "optimal" land cover pre- and post-projected climate change. This is not a challenge limited to modeling ecosystems; models of the economy and energy systems are no better at projections of non-marginal change phenomena over time horizons longer than a few years.

When Elena Shevliakova joined the group, in keeping with our mantra of faithfully reflecting all scientific uncertainties, we set about the task of building a probabilistic model of land cover. We began by partnering with David Weinstein at Cornell, who had detailed field data on daily weather and Net Primary Productivity (NPP)[2] in Red Spruce that his team studied to understand impacts of exposure to ozone and nutrient stress. Our re-analysis of their data also revealed that cloudiness had a strong effect on NPP. This and further work, with the help of Milind Kandlikar and Mitch Small, led to Elena's PhD thesis.

The first paper I'll describe, from her thesis, is titled, "A probabilistic model of ecosystem prevalence." The full citation and abstract for this and the other papers described can be found at the end of the chapter.

Using empirical data on land cover at a resolution of half a degree of latitude and half a degree of longitude (and 17 land cover classifications), we trained an unstructured neural network model to predict land cover based on eight underlying variables. Four of the variables would be influenced by climate change: maximum temperature, average temperature, minimum temperature, and average precipitation; the other four were independent of climate change: soil type, latitude, slope, and elevation. While by today's standards a $0.5° \times 0.5°$ grid is very coarse, at the time, this was the finest resolution at which we also had reliable data on the independent variables.

As noted earlier, at the time, land cover "models" postulated an optimum cover for given conditions as determined by the Holdridge zone, as shown in Figure 6.1.

Before deriving our neural net model, we simply mapped the land cover prevalence data on a triangular grid with the same axes and saw that the actual range of each ecosystem extended far beyond the "cells" into which they are confined in the idealized frameworks. This observation alone has four implications:

a) That current land covers are more robust to changes in climate than predicted by models that rely on the differences between hypothetical optimal land covers under different climates;

b) That at any given time, the land cover we observe is not the optimum one unless there has been no disturbances for centuries. This is because it takes centuries for a stable land cover to be established;

c) That if a particular region is stripped of its land cover, there would be many competing land covers capable of occupying that space; and

d) That estimating succession in land cover has to consider the relative probability of neighboring land covers (assuming natural processes of succession rather than ecological engineering) while noting that ecological engineering using a more distant but higher probability of prevalence land cover may be more robust to climate change in the long run.

Elena's thesis shows that when we put together all the actual observations of land cover from all regions of the globe, the same land covers can be found across a much wider range of climate conditions than suggested in the Holdridge model, which pigeonholes land covers into narrow climate ranges. Furthermore, many

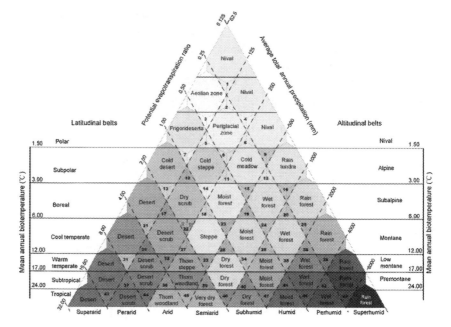

**FIGURE 6.1** Holdridge mapping of land cover according to latitude, altitude, and humidity.

Source: www.sciencedirect.com/science/article/abs/pii/S030438001200405X

Fan, Z. M., Li, J., & Yue, T. X. (2013). Land-cover changes of biome transition zones in Loess Plateau of China. *Ecological modelling, 252,* 129–140.

different land covers are found to occupy the same climate conditions. These dual observations from nature have two important implications: (a) the current land cover is more robust to climate change than models predict; and (b) if we were to start from barren lands, many different ecosystems can successfully occupy any region.

It is interesting to note that, on the ground, we do not see many competing land cover types vying for dominance in a given region. We hypothesize that the reason this is observed has to do with mechanisms of establishment and distance limits over which the various land cover types can easily disperse. This realization suggested to us that the current combinations of plants (and other living constituents) that define each land cover type may only be the result of the particulars of the regional climate (and other disturbance) regimes of the past two or so centuries. These assemblies could be unstable under big disturbances, so confident projections of land cover under climate change would need to address the issue of how the current bundles of biota defining a fairly stable ecosystem would disentangle due to the different sensitivity of each of the species that make them up. To the best of our knowledge, this unbundling and re-bundling of the different trees, plants, insects, fungi, etc. in ecosystems has not been directly modeled, as it requires the careful characterization of symbiotic services provided among the constituents of each ecosystem.

## Asking Experts How They Think Forests Will Respond to Climate Change

### Granger Morgan

In light of the large uncertainty that Hadi was finding in how ecosystems might change in response to climate change, we teamed up with ecologist Lou Pitelka to develop and conduct an expert elicitation with a group of 11 leading forest experts. Details on doing expert elicitations are discussed in Chapter 4. Our focus in this study was on tropical and far northern (boreal) forests. The results are summarized in a paper titled, "Elicitation of Expert Judgments of Climate Change Impacts on Forest Ecosystems."

In addition to being interested in the results in their own right, we knew that there were several experts building models of how forests would change and we wanted to see if our experts would give us results that were largely in agreement with what those models were predicting.

We asked experts to assume that the amount of carbon dioxide in the atmosphere had doubled and then stayed constant. We translated that for them in terms of the amount of warming in summer and winter and the change in precipitation in both regions. We then asked experts to estimate the change in above-ground and below-ground (roots) biomass in minimally disturbed forests after the new climate had been established for 500 years. There was general agreement that the changes would be larger in northern forests than in tropical forests. In Figure 4.2 we showed results for northern forests in the form of box plots.[3] Figure 6.2 shows similar results for tropical forest. Notice that in some cases, the experts did not even agree about whether the change would be an increase (values bigger than 1) or a decrease (values less than 1) in biomass.

We also asked experts to make judgments about several other things, such as the rate of northward migration of the boundary between boreal forest and tundra and the boundary between mixed northern forest and boreal forest. Estimates of migration rates in northern forests displayed a range of more than four orders of magnitude (i.e., 10,000 times different).

We asked a different set of five experts for judgments about impacts on biodiversity and extinction rates. These estimates also showed significant variation between experts.

A series of questions about research needs found consensus on the importance of expanding observational and experimental work on ecosystem processes and expanding regional and larger-scale observational, monitoring, and modeling studies.

I must say it is disappointing that despite the rather high levels of disagreements among the experts we interviewed, groups building ecosystem-impact models for

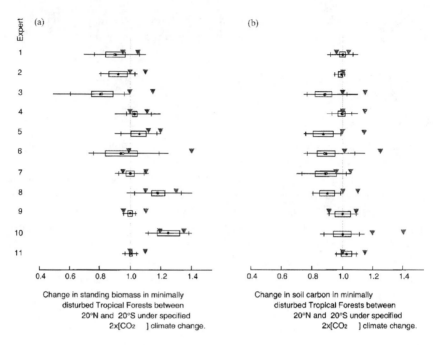

**FIGURE 6.2** Box plots summarizing elicited expert subjective probability distributions of (a) standing biomass and (b) soil carbon in minimally disturbed tropical forests at least 500 years after the warming that would occur if carbon dioxide in the atmosphere is doubled. The shaded triangles indicate the estimated range of response for a doubling of $CO_2$ alone without any accompanying climate change. Results for tropical forests involved much smaller changes than those for northern forests (Figure 4.2) but just as in that case, there was not complete agreement about the sign of the change.

Source: Morgan, Pitelka, & Shevliakova, 2001.

Morgan, M. G., Pitelka, L. F., & Shevliakova, E. (2001). Elicitation of expert judgments of climate change impacts on forest ecosystems. *Climatic Change, 49*(3), 279–307.

use in climate assessments did not significantly change their results or expand their discussion of the uncertainty in their results!

## Growing Corn to Make Biofuel

### Mike Griffin

In 2007, Congress passed the Energy Independence and Security Act which, as the name suggests, was intended "to move the United States toward greater energy independence." One of the several things this act called for was "to increase the production of clean renewable fuels." More specifically, it called for making 36 billion gallons of biofuel by 2022. Fifteen billion gallons of that was to be ethanol made from corn and sometimes also the leaves, stalks, and cobs of corn. The remaining 21 billion was to be "advanced biofuel." Growing corn to make fuel for cars and trucks is obviously popular with Midwestern corn farmers, but is it good for the environment and does growing it, or other crops like some grasses, really help to address the problem of climate change? My colleague Scott Matthews and I had our doubts.

When we started to work on this question, others had already done analysis that identified some of the environmental problems of using large areas to grow crops to make fuel. Most of those crops need to be fertilized with nitrogen fertilizer. With two graduate students, we decided to use a system-wide approach to take a closer look at the consequences of using all that fertilizer. Much of the fertilizer that is applied to a crop doesn't end up in the plants – it runs off in the water. In the Midwest, that water makes its way to the Mississippi River and ends up in the Gulf of Mexico.

You may have heard about the so-called "dead zone" in the Gulf where there is so little oxygen in the water that almost no life exists because of "hypoxia" (low oxygen in living tissues). This dead (or anoxic) zone exists because all the nitrogen fertilizer in the water causes algae blooms that suck up all the oxygen in water. Figure 6.3 shows how big the dead zone was in July 2017.

To conduct our analysis, we constructed several different scenarios of how biofuel development might proceed – just corn, the entire corn plant (corn stover), switch grass, soybeans to make diesel fuel, etc. We also explored different run-off management strategies like creating "buffer strips" of other plants. We added uncertainty to all this and conducted a simulation.

The U.S. EPA is very concerned about the dead zone and wants to reduce its size dramatically in the future. While several of the scenarios we studied led to modest improvements and some reduced run-off more than just growing corn the way we have in the past, none came at all close to meeting the EPA target. To do that, the country will need to mount an aggressive strategy to manage fertilizer run-off all across the Midwest. One clear lesson from all this is that it is really important to do careful analysis before embarking on big projects that "sound green." Details on this work can be found in the paper we published in 2009 in the scientific journal *Environmental Science & Technology* titled, "Impact of biofuel crop production on the formation of hypoxia in the Gulf of Mexico." The full citation and the abstract are listed at the end of the chapter.

## Limiting Habitat Fragmentation from Gas Well Development

### Mike Griffin

Here in Pennsylvania where Carnegie Mellon is located, we've been in the heart of the fracking boom that has led to a dramatic increase in the production of natural gas. That has had a couple big benefits – it has switched a lot of electricity production from coal to natural gas, which produces only about half as much carbon dioxide. It has also helped to keep prices low and make the country more energy-independent.

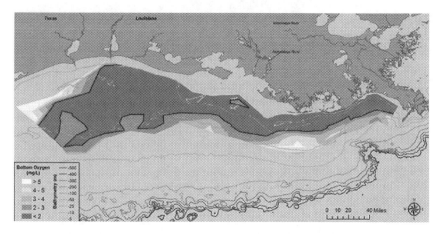

**FIGURE 6.3** A map produced by the U.S. National Oceanic and Atmospheric Administration of the dead zone in the Gulf of Mexico, which covered almost 8800 square miles in July of 2017.

Source: NOAA and N. Rabalais, LSU/LUMCON.

Forest covers almost 65% of Pennsylvania, with much of it in the central and northern parts of the state where there are large, uninterrupted patches of core forest. I've done lots of work in ecology, and as Scott and I watched the gas developers cutting down paths through the woods to gain access to drilling sites and lay out the routes for gas pipelines, we began to suspect that there were ways to do this that could be much less disruptive in terms of "habitat fragmentation." The pattern of habitat loss is often more important than the quantity of that loss. You may not have thought much about it, but when a large, uninterrupted section of a forest is broken up, many sensitive species are unable or unwilling to cross between the different forest patches. Another impact of forest disruption is the creation of forest edges. These edges can affect the physical and biological conditions at an ecosystem's boundary and can increase predation, change lighting and humidity, and increase the presence of invasive species. For example, songbirds – the Ness near edges and openings – are less likely to be successful raising young then nesting if they can nest in interior forests. The ongoing fragmentation of Pennsylvania's forests poses risks to the state's rich biodiversity.

Our study projected the structure of future alternative pathways for Marcellus shale development and quantified the potential ecological impact of future drilling using a core forest region of Bradford County, PA as our study region. Our modeling found that future development could cause the level of fragmentation in the study area to more than double throughout the lifetime of gas development. Specifically, gathering lines that bring the gas from separate wells to a central location turned out to be responsible for approximately 94% of the increased fragmentation in the core forest study region.

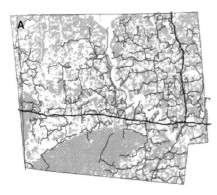
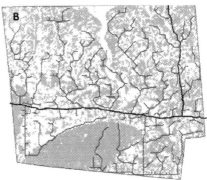

**FIGURE 6.4** The figure on the left (A) shows the gathering line networks that have been built up in our study area of Bradford County, PA through 2012 to carry gas to the main pipeline in black. The figure on the right (B) shows the much lower level of disruption in the form of habitation fragmentation that would have occurred if our siting proposal had been in use.

Source: Abrahams, Griffin, & Matthews, 2015.

Abrahams, L. S., Griffin, W. M., & Matthews, H. S. (2015). Assessment of policies to reduce core forest fragmentation from Marcellus shale development in Pennsylvania. *Ecological Indicators*, *52*, 153–160.

But there is good news. We found that by requiring gathering lines to follow pre-existing road routes in forested regions, shale resources can be developed to their full potential while essentially preventing any further fragmentation from occurring across the core forested landscape of the county we studied. Figure 6.4 shows how the disruption in our study area would have changed if existing lines had been located using our method.

If we estimate that the ultimate recovery of gas from our study region will be between 1 and 3 billion cubic feet per well, the policy proposal we developed could be implemented for a minimal incremental economic investment of approximately $0.005–$0.02 per million cubic feet of natural gas produced over the more traditional placement of gathering lines.

Details on this work can be found in the paper we published in 2015 in the scientific journal *Ecological Indicators* titled, "Assessment of policies to reduce core forest fragmentation from Marcellus shale development in Pennsylvania." The full citation and the abstract are listed at the end of the chapter.

## Managing Soil Carbon

### Granger Morgan

As Lester Lave and others in our group began to talk more and more about the possibility of gathering up crop waste and turning it into biofuel, I began to get concerned about whether problems might arise from removing all that material that would otherwise rot down and return to the soil. To get us educated, we reached out to Ratan Lal, a distinguished soil physicist at Ohio State.[4]

Ratan explained to us that what matters is the amount of carbon in the soil to which critical nutrients get attached. Restoring soil carbon is essential to improving soil quality, sustaining and improving food production, maintaining clean water, and reducing increases in atmospheric $CO_2$.

We found this so interesting and important that we decided we should combine Ratan's scientific expertise with our policy expertise and write a general "policy forum" piece for the scientific journal *Science*. In that short piece, we explained that "leaving crop residues after harvest increases the carbon content of soil and controls erosion, but the benefits are lost if the biomass is plowed under, because microorganisms quickly degrade the residue carbon into carbon dioxide. Essential nutrients that adhere to soil organic carbon disappear with its depletion. Thus, farmers require more fertilizer irrigation and pesticides to preserve yield."

We noted that short-sighted farming practices have resulted in the loss of an estimated $4\pm1$ billion tons of carbon from soils of the United States and $78\pm12$ billion tons from the world's soils, a large fraction of which ended up in the atmosphere. Soil carbon loss has come principally from plowing that turns over the soil, making it susceptible to accelerated erosion. This is exemplified by the Dust Bowl era in the United States and is a serious issue in most developing countries.

We concluded our policy forum piece by arguing for the benefits of "no-till" agriculture. In no-till agriculture, seeds are planted without turning the soil with a plow. This reduces the loss of soil organic carbon while conserving soil water and inhibiting weeds. We argued that implementing a program to increase soil organic carbon "requires that governments mandate no-till agriculture or provide financial incentives to farmers." This may be much easier to accomplish in places like the U.S. than in developing nations.

## Drivers and Dynamics of Environmental Policy Change

### Hadi Dowlatabadi

Every year, the Ministry of Forests in British Columbia announces the "annual allowable cut." Shannon Hagerman and I wanted to know how this strange "policy" had come about. What did this mean under the ministry's much ballyhooed multi-use and collaborative forest management rubric? Was there, in fact, any difference between the annual allowable cut before and after this newly enlightened policy regime?

In the paper "Observations on Drivers and Dynamics of Environmental Policy Change: Insights from 150 Years of Forest Management in British Columbia," Shannon and I looked at the history of major shifts in forest "management" policy over time, along with noting lumber production and other metrics of actual forest land cover change/protection. We found continuity in forestry trends that persisted across major shifts in policy, from conservation to sustainable yield and now to multi-use multi-stakeholder. In a sense, the things that could be directly observed, such as how much wood was cut, show that the policymakers were rejigging their pseudo-management story to match what was happening on the land.

We found that factors like technology and markets played major roles and the government played catch-up policy. We found that at every phase of these policy shifts, there was a significant uncertainty about the state of the forests, the markets for forest products, and so on. Yet these uncertainties did not act as a barrier to the adoption of new management paradigms at any time – likely because the paradigms did not have any meaning in and of themselves. Given the uncertainty that climate change brings to forest health, markets for forest products, and more, we would have hoped that management regimes would be forward-looking, responsive to uncertainties, and designed to ensure robust outcomes for this socio-ecological sector. Sadly, we did not find that to be the case. Therefore, we very much doubt that, absent substantial institutional change, the government is capable of ameliorating the impacts of climate change on forests and the forestry sector in British Columbia.

When thinking about the impacts of climate change on ecosystems, there is always someone touting the value in "migration corridors" and "contiguous land covers," etc. Shannon Hagerman and I wondered if ecologists were thinking in terms of triage or assisting the most vulnerable when thinking about how we can help ecosystems adapt to climate change. Our findings were published in 2010 in a paper titled, "Expert views on biodiversity conservation in an era of climate change."

Adapting conservation policy to the impacts of climate change has emerged as a central and unresolved challenge. In this paper, we report on 21 in-depth interviews with experts in biodiversity and climate change in which we asked for their views of the implications of climate change for conservation policy. We found a diversity of views across a set of topics that included: changing conservation objectives, conservation triage and its criteria, increased management interventions in protected areas, the role of uncertainty in decision-making, and evolving standards of conservation success. Notably, our findings reveal active consideration among experts of some more controversial elements of policy adaptation (including the role of disturbance in facilitating species transitions and changing standards of conservation success). Yet at the time, there was very little on these topics in the published literature.

As a follow-up to the research above, Shannon attended the annual meeting of the World Conservation Congress to see whether and how the burgeoning ideas raised in the expert elicitations would be received and discussed in the key gathering of their peers.

The final paper I'll describe is titled, "Climate Change Impacts, Conservation and Protected Values: Understanding Promotion, Ambivalence and Resistance to Policy Change at the World Conservation Congress." The impacts of climate change imply substantive changes to current conservation policy frameworks. Debating and formulating the details of these changes was central to the agenda of the Fourth World Conservation Congress (WCC) of the International Union for Conservation of Nature (IUCN). In this paper, we document the promotion of, and resistance to, various proposals related to revising conservation policy given climate impacts as they unfolded at this key policy-setting event. Our analysis finds that during one-on-one interviews, many experts acknowledged the need for new policy means (including increased interventions) and revised policy objectives given anticipations of habitat and species loss. However, this same pattern and the implied willingness to consider more controversial strategies were less evident at public-speaking events at the WCC. Rather, active avoidance of contentious topics was observed in public settings. This resulted in the reinforcement (not revision) of conventional policy means and objectives at this meeting. We suggest that this observation can at least partly be explained by the fact that the difficult trade-offs (species for species or land base for land base) implied by nascent proposals severely violate prevailing value-based conservation commitments, and so understandable resistance to change is observed.

## Decision Aids for Resource Managers

### Tim McDaniels

Two fundamental themes that have guided research efforts throughout the work of our Centers are:

- Decision-making under irreducible uncertainty for climate, energy, and environmental issues;
- The importance of engaging real-world stakeholders in significant decisions related to these themes.

Linking these two themes faces great barriers because of the complexity associated with global change issues and recognition that standard tools of policy analysis are often not up to the task.

Climate and broader global change issues involve profound uncertainties, multiple objectives, very long-term consequences, deep ethical dimensions, and often conflicts with more pressing concerns such as human livelihoods. Even top experts do not understand the totality of these issues, so expecting citizens and civil society groups to be informed and deliberative in providing advice to regulatory bodies is demanding.

There is also the important question of where the best leverage for research lies in these vast themes. For the work I lead with others at the University of British Columbia, *adaptation to global change processes in natural resource management issues* seemed like a great opportunity. Forestry, fisheries, wilderness, and water resources decision processes all have to grapple with a changing climate, which makes all the pre-existing decision complexities that much tougher. The amount of uncertainty in all the future climate parameters increased and the risk of exceeding (unknown) thresholds in complex human and natural systems is heightened with unpredictable but potentially grave consequences. These are pressing issues in Western Canada and also in many other areas around the world.

My colleagues, students, and I were able to build on early experience in regulatory-focused work for BC Hydro, the province's electric utility, which has a largely hydroelectric generation infrastructure. In that process, Tim McDaniels and Robin Gregory were able to develop a decision analytic practice-oriented approach to working with advisory committees, linking managers, civil society groups, and technical specialists. This approach was replicated with great success in the utility's 23 managed watersheds and resolved many of their most pressing resource management questions through decision processes that largely led to consensus on future plans. Subsequently, additional decision analytic small-group processes were conducted, again for regulatory purposes, for integrated resource planning at BC Gas.

With support from the Centers, we developed more research-oriented small-group experiments that explored whether the successes we had in longer processes could be achieved in shorter time periods, with more limited information, in work-shop settings. That work allowed us to demonstrate that some of the previously documented obstacles to preference elicitation can be overcome and eliminated with a decision framing focused on the pros and cons of management alternatives.

We built on these experimental insights to work more directly on a major climate-related forest land management question that looms large in British Columbia: how best to do forest management after the province's catastrophic pine beetle infestation had passed and the province was left with vast areas of formerly forested, now largely denuded lands. We employed our small-group approach, working with Tamsin Mills of the University of British Columbia and Dan Ohlson of Compass Resource Management, with an advisory group of forestry specialists in British Columbia. The result was an approach that pointed to a new robust alternative for regional forest land management. This paper won an award from the Society of Risk Analysis.

Turning to fisheries issues, we tackled three questions of great interest in Western Canada and elsewhere around the world:

1. Governance of multilevel resource management questions such as aquaculture, which involves corporate actors and regulators that have implications from global to local levels;

2. The potential risks of warming rivers due to climate change on sockeye salmon given their complex multistage life cycle in rivers, lakes, and the ocean; and

3. The consequences of the loss of traditional foods and cultural practices for First Nations people who have fished for sockeye salmon for millennia on specific sites along the Fraser River in British Columbia.

For each of these projects, a number of graduate students at the University of British Columbia worked with the benefit of Center support to develop the instruments and conduct the interviews and workshops required for the studies. We did not strictly adhere to the decision analytic small-group paradigm in these projects but rather were informed by that set of methods and related new approaches. For the work on multi-scale management, we considered relevant objectives and alternatives for management decisions across scales from international to local, to investigate potential gaps and mismatches. For the work on sockeye salmon risks, we conducted a group expert elicitation among fisheries managers of how well sockeye may be able to survive or adapt under various climate scenarios. Finally, the project on traditional foods and cultural practices involved on-site interviews among traditional fishers conducted by Colleen Jacobs, a master's student at UBC. She is a member of the St'át'imc First Nation near Lilloett, BC, and her extended family has been traditional subsistence harvesters of salmon on the Fraser River throughout their history.

Turning again to decision analytic approaches for our final topic (among many) discussed here, we conducted research and wrote papers on two aspects of learning over time as a means to manage irreducible uncertainty for climate and global change issues. One of these projects was concerned with treating learning as an explicit objective in stakeholder decision processes. We developed a definition and framework for *the value of learning* as an explicit complement to the value of information concept in decision analysis. The second study, conducted with Ralph Keeney, focused on a framework for guiding mitigation and adaptation decisions through conscious efforts to learn over time in cycles of policy choices.

## Technical Details for Chapter 6

Here are three papers on climate impacts on ecosystems:

Siegel, E., Hadi Dowlatabadi, H., & Small, M. J. (1995). A probabilistic model of ecosystem prevalence. *Journal of Biogeography*, 875–879. https://doi.org/10.2307/2845988

**Abstract:** We have developed a probabilistic model of ecosystem response to climate change. This model differs from previous efforts in that it is statistically estimated using Olson ecosystem and climate data, yielding a joint multivariate probability of prevalence for each ecosystem, given climatic conditions. We expect this approach to permit simulation of inertia and competition which have, so far, been absent in transfer models of continental scale ecosystem response to global change. We simulate inertia through estimating the probability of an ecosystem persisting in situ beyond the prevailing climate conditions (and possibly beyond the limits usually prescribed in deterministic schema such as Holdridge Zones). Conversely, we can observe today that the same environmental conditions support different ecosystems in different locations. Thus, despite the probability that one ecotype will dominate others at a given point, others would have the possibility of establishing an early foothold. This is similar to stand models showing changes in biome composition. We believe these modifications to transfer models will play a significant role in the simulation of the transient response of ecosystems to climate change.

Dowlatabadi, H., Shevliakova, E., & Kandlikar, M. (1994). *Issues in evaluation of ecosystem change in response to global change* (No. CONF-9406429–1). Carnegie-Mellon Univ., Pittsburgh, PA (United States). Dept. of Engineering and Public Policy. Available online at: www.osti.gov/servlets/purl/465908

**Abstract:** Uncertainty analysis of our integrated climate assessment model has revealed the importance of obtaining better market and non-market impacts… Improving market and non-market damage assessments has necessitated advances in the theoretical and applied dimensions of the problem. The assessment of climate change impacts on ecosystems provides a severe test of the new ideas being put forward. This paper provides a brief overview of: i) the challenges inherent in modeling ecosystem dynamics; ii) the problem of selecting an appropriate metric of change; and, iii) the thorny issue of how to place a monetary value on market and nonmarket impacts. We focus on two central issues in estimation of impacts: i) before climate change, are the systems being impacted (both ecological and economic) in equilibrium? and ii) how quickly do ecological and related economic systems adapt to change? In addition, we attempt to be comprehensive in laying out the magnitude of the challenge ahead.

Morgan, M. G., Pitelka, L. F., & Shevliakova, E. (2001). Elicitation of expert judgments of climate change impacts on forest ecosystems. *Climatic Change, 49*(3), 279–307.

**Abstract:** Detailed interviews were conducted with 11 leading ecologists to obtain individual qualitative and quantitative estimates of the likely impact of a $2\times[CO_2]$ climate change on minimally disturbed forest ecosystems. Results display a much richer diversity of opinion than is apparent in qualitative consensus summaries, such as those of the IPCC. Experts attach different relative importance to key factors and processes such as soil nutrients, fire, $CO_2$ fertilization, competition, and plant-pest-predator interactions. Assumptions and uncertainties about future fire regimes are particularly crucial. Despite these differences, most of the experts believe that standing biomass in minimally disturbed Northern forests would increase and soil carbon would decrease. There is less agreement about impacts on carbon storage in tropical forests. Estimates of migration rates in northern forests displayed a range of more than four orders of magnitude. Estimates of extinction rates and dynamic response show significant variation between experts. A series of questions about research needs found consensus on the importance of expanding observational and experimental work on ecosystem processes and of expanding regional and larger-scale observational, monitoring and modeling studies. Results of the type reported here can be helpful in performing sensitivity analysis in integrated assessment models, as the basis for focused discussions of the state of current understanding and research needs, and, if repeated over time, as a quantitative measure of progress in this and other fields of global change research.

Here are papers on hypoxia in the Gulf of Mexico and on limiting habitat fragmentation caused by fracking:

Costello, C., Griffin, W. M., Landis, A. E., & Matthews, H. S. (2009). Impact of biofuel crop production on the formation of hypoxia in the Gulf of Mexico. *Environmental Science & Technology, 43*(20), 7985–7991.

**Abstract:** Many studies have compared corn-based ethanol to cellulosic ethanol on a per unit basis and have generally concluded that cellulosic ethanol will result in fewer environmental consequences, including nitrate ($NO_3^-$) output. This study takes a system-wide approach in considering the $NO_3^-$ output and the relative areal extent of hypoxia in the Northern Gulf of Mexico (NGOM) due to the introduction of additional crops for biofuel production. We stochastically estimate $NO_3^-$ loading to the NGOM and use these results to approximate the areal extent of hypoxia for scenarios that meet the Energy Independence and Security Act of 2007's biofuel goals for 2015 and 2022. Crops for ethanol include corn, corn stover, and switchgrass; all biodiesel is assumed to be from soybeans. Our results indicate that moving from corn to cellulosics for ethanol production may result in a 20-percent decrease (based on mean values) in $NO_3^-$ output from the Mississippi and Atchafalaya River Basin (MARB). This decrease will not meet the EPA target for hypoxic zone reduction. An aggressive nutrient management strategy will be needed to reach the 5000 km² areal extent of hypoxia in the NGOM goal set forth by the Mississippi River/Gulf of Mexico Watershed Nutrient Task Force even in the absence of biofuels, given current production to meet food, feed, and other industrial needs.

Abrahams, L. S., Griffin, W. M., & Matthews, H. S. (2015). Assessment of policies to reduce core forest fragmentation from Marcellus shale development in Pennsylvania. *Ecological Indicators, 52*, 153–160.

**Abstract:** Marcellus Shale development is occurring rapidly and relatively unconstrained across Pennsylvania (PA). Through 2013, over 7400 unconventional wells had been drilled in the Commonwealth. Well pads, access roads, and gathering lines fragment forestland resulting in irreversible alterations to the forest ecosystem. Changes in forest quantity, composition, and structural pattern can result in increased predation, brood parasitism, altered light, wind, and noise intensity, and spread of invasive species. These fragmentation effects pose a risk to PA's rich biodiversity. This study projects the structure of future alternative pathways for Marcellus shale development and quantifies the potential ecological impact of future drilling using a core forest region of Bradford County, PA. Modeling presented here suggests that future development could cause the level of fragmentation in the study area to more than double throughout the lifetime of gas development. Specifically, gathering lines are responsible for approximately 94% of the incremental fragmentation in the core forest study region. However, by requiring gathering lines to follow pre-existing road routes in forested regions, shale resources can be exploited to their full potential, while essentially preventing any further fragmentation from occurring across the core forested landscape of Bradford County. In the study region, assuming an estimated ultimate recovery (EUR) of 1–3 billion cubic feet (Bcf) per well, this policy could be implemented for a minimal incremental economic investment of approximately $0.005–$0.02 per Mcf of natural gas produced over the modeled traditional gathering line development.

This is the citation for our paper in *Science* on managing soil carbon:

Lal, R., Griffin, M., Apt, J., Lave, L., & Morgan, M. G. (2004). Managing soil carbon. *Science, 304* (5669), 393.

This is a one-page policy forum piece available online at: https://science.sciencemag.org/content/304/5669/393.full
See the main text of this chapter for a brief description of the arguments we made.

Here are three papers on drivers and dynamics of environmental policy change:

Hagerman, S. M., Dowlatabadi, H., & Satterfield, T. (2010). Observations on drivers and dynamics of environmental policy change: Insights from 150 years of forest management in British Columbia. *Ecology and Society, 15*(1).

**Abstract:** Human and ecological elements of resource management systems co-adapt over time. In this paper, we examine the drivers of change in forest management policy in British Columbia since 1850. We asked: How has a set of system attributes changed over time, and what drivers contributed to change when it occurred? We simultaneously examined a set of three propositions relating to drivers and dynamics of policy change. We find that factors contributing to the level of impacts, like technology, changed substantially over time and had dramatic impacts. In partial contrast, the institutions used to exercise control (patterns of agency and governance) remained the same until relatively recently. Other system attributes remained unchanged (e.g., the concept of ecosystems as stable entities that humans can manage and control). Substantive, decision-relevant uncertainties characterized all periods of management but did not act as a barrier to the adoption of new regimes at any time. Against this backdrop of constancy in some attributes, and change in others, a few exogenous drivers (e.g., technology, war, markets, legal decisions, ideas, and climate) triggered episodic reexamination of guidelines for resource management. The implications of these findings for future policy change in this system are discussed.

Hagerman, S., Dowlatabadi, H., Satterfield, T., & McDaniels, T. (2010). Expert views on biodiversity conservation in an era of climate change. *Global Environmental Change*, *20*(1), 192–207.

**Abstract:** Adapting conservation policy to the impacts of climate change has emerged as a central and unresolved challenge. In this paper, we report on the results of 21 in-depth interviews with biodiversity and climate change adaptation experts on their views of the implications of climate change for conservation policy. We find a diversity of views across a set of topics that included: changing conservation objectives, conservation triage and its criteria, increased management interventions in protected areas, the role of uncertainty in decision-making, and evolving standards of conservation success. Notably, our findings reveal active consideration among experts with some more controversial elements of policy adaptation (including the role of disturbance in facilitating species transitions, and changing standards of conservation success), despite a comparative silence on these topics in the published literature. Implications of these findings are discussed with respect to: (a) identifying future research and integration needs and (b) providing insight into the process of policy adaptation in the context of biodiversity conservation.

Hagerman, S., Satterfield, T., & Dowlatabadi, H. (2010). Climate change impacts, conservation and protected values: Understanding promotion, ambivalence and resistance to policy change at the World Conservation Congress. *Conservation and Society*, *8*(4), 298–311.

**Abstract:** The impacts of climate change imply substantive changes to current conservation policy frameworks. Debating and formulating the details of these changes was central to the agenda of the Fourth World Conservation Congress (WCC) of the International Union for Conservation of Nature (IUCN). In this paper, we document the promotion of, and resistance to, various proposals related to revising conservation policy given climate impacts as they unfolded at this key policy-setting event. Our analysis finds that, during one-on-one interviews, many experts acknowledged the need for new policy means (including increased interventions) and revised policy objectives given anticipations of habitat and species loss. However, this same pattern and the implied willingness to consider more controversial strategies were less evident at public speaking events at the WCC. Rather, active avoidance of contentious topics was observed in public settings. This resulted in the reinforcement (not revision) of conventional policy means and objectives at this meeting. We suggest that this observation can at least partly be explained by the fact that the difficult trade-offs (species for species or land base for land base) implied by nascent proposals severely violate prevailing value-based conservation commitments and so understandable resistance to change is observed.

Finally, here are four papers on supporting resource management decision-making by Tim McDaniels and his colleagues:

McDaniels, T., Mills, T., Ohlson, D., & Gregory, R. (2012). Exploring robust alternatives for climate adaptation in forest-land management through expert judgments. *Risk Analysis, 32*(12), 2098–2112.

**Abstract:** We develop and apply a judgment-based approach to selecting *robust* alternatives, which are defined here as *reasonably likely to achieve objectives, over a range of uncertainties*. The intent is to develop an approach that is more practical in terms of data and analysis requirements than current approaches, informed by the literature and experience with probability elicitation and judgmental forecasting. The context involves decisions about managing forest lands that have been severely affected by mountain pine beetles in British Columbia, a pest infestation that is climate-exacerbated. A forest management decision was developed as the basis for the context, objectives, and alternatives for land management actions, to frame and condition the judgments. A wide range of climate forecasts, taken to represent the 10–90% levels on cumulative distributions for future climate, were developed to condition judgments. An elicitation instrument was developed, tested, and revised to serve as the basis for eliciting probabilistic three-point distributions regarding the performance of selected alternatives, over a set of relevant objectives, in the short and long term. The elicitations were conducted in a workshop comprising 14 regional forest management specialists. We employed the concept of *stochastic dominance* to help identify robust alternatives. We used extensive sensitivity analysis to explore the patterns in the judgments, and also considered the preferred alternatives for each individual expert. The results show that two alternatives that are more flexible than the current policies are judged more likely to perform better than the current alternatives on average in terms of stochastic dominance. The results suggest judgmental approaches to robust decision making deserve greater attention and testing.

Jacob, C., McDaniels, T., & Hinch, S. (2010). Indigenous culture and adaptation to climate change: Sockeye salmon and the St'át'imc people. *Mitigation and Adaptation Strategies for Global Change, 15*(8), 859–876.

**Abstract:** This paper provides a culturally-informed understanding of the impacts of climate change on a highly important subsistence activity that has been practiced by First Nations of central British Columbia for thousands of years. The paper begins with a review of the science regarding sockeye salmon and climate change. It discusses harvest patterns, and how the timing of runs has changed. A survey was conducted by the first author regarding St'át'imc traditional fishing at a historic site on the Fraser River, in 2005. The results show that the impacts of climate change are apparent to those conducting traditional fishing practices, in terms of changed timing and abundance of salmon runs. These perceptions fit closely with the information available from scientists and management agencies. These changes are highly problematic for the St'át'imc, in that the preservation method (drying) is tied to seasonal weather patterns. The whole cultural setting, and the relevance of salmon for subsistence would be highly altered by climate change that leads to changes in the timing and abundance of sockeye salmon. The paper discusses mitigation and adaptation alternatives, but also indicates the scope of these seem limited, given the resource systems and the context of these activities.

McDaniels, T., & Gregory, R. (2004). Learning as an objective within structured decision processes for managing environmental risks. *Environmental Science & Technology, 38*(7), 1921–1926.

**Abstract:** Social learning through adaptive management holds the promise of providing the basis for better risk management over time. Yet the experience with fostering social learning through adaptive management initiatives has been mixed and would benefit from practical guidance for better implementation. This paper outlines a straightforward heuristic for fostering improved risk management decisions: specifying learning for current and future decisions as one of several explicit objectives for the decision at hand, drawing on notions of applied decision analysis. In keeping with recent guidance from two important U.S. advisory commissions, the paper first outlines a view of risk management as a policy-analytic decision process involving stakeholders. Then it develops the concept of the value of learning, which broadens the more familiar notion of the value of information. After that, the concepts and steps needed to treat learning as an explicit objective in a policy decision are reviewed. The next section outlines the advantages of viewing learning as an objective, including potential benefits from the viewpoint of stakeholders, the institutions involved, and for the decision process itself. A case-study example concerning water use for fisheries and hydroelectric power in British Columbia, Canada is presented to illustrate the development of learning as an objective in an applied risk-management context.

Keeney, R. L., & McDaniels, T. L. (2001). A framework to guide thinking and analysis regarding climate change policies. *Risk Analysis, 21*(6), 989–1000.

**Abstract:** The potential impacts from climate change, and climate change policies, are massive. Careful thinking about what we want climate change policies to achieve is a crucial first step for analysts to help governments make wise policy choices to address these concerns. This article presents an adaptive framework to help guide comparative analysis of climate change policies. The framework recognizes the *inability to forecast long-term impacts* (due in part to *path dependence*) as a constraint on the use of standard policy analysis, and stresses *learning over time* as a fundamental concern. The framework focuses on the objectives relevant for climate change policy in North America over the near term (e.g., the next 20 years). For planning and evaluating current climate policy alternatives, a combination of fundamental objectives for the near term and proxy objectives for characterizing the state of the climate problem and the ability to address it at the end of that term is suggested. Broad uses of the framework are discussed, along with some concrete examples. The framework is intended to provide a basis for policy analysis that explicitly considers the benefits of learning over time to improve climate change policies.

**Some notable writings by authors not affiliated with our Centers:**

Carpenter, S. R., et al. (2009). Science for managing ecosystem services: Beyond the millennium ecosystem assessment. *Proceedings of the National Academy of Sciences, 106*(5), 1305–1312.

Foley, J. A., et al. (2005). Global consequences of land use. *Science, 309*(5734), 570–574.

Gates, D. M. (1993). *Climate Change and Its Biological Consequences.* Sinauer Associates, 280pp.

IPCC working group II "assesses the vulnerability of socio-economic and natural systems to climate change, negative and positive consequences of climate change and options for adapting to it." Available online at: www.ipcc.ch

Liu, J., et al. (2013). Framing sustainability in a telecoupled world. *Ecology and Society*, *18*(2).

Manning, D. W., Rosemond, A. D., Benstead, J. P., Bumpers, P. M., & Kominoski, J. S. (2020). Transport of N and P in U.S. streams and rivers differs with land use and between dissolved and particulate forms. *Ecological Applications*, *30*(6), e02130.

Prentice, I. C., et al. (1992). Special paper: A global biome model based on plant physiology and dominance, soil properties and climate. *Journal of Biogeography*, 19(2), 117–134.

Rockström, J., et al. (2009). A safe operating space for humanity. *Nature*, *461*(7263), 472–475.

Ramankutty, N., & Foley, J. A. (1999). Estimating historical changes in global land cover: Croplands from 1700 to 1992. *Global Biogeochemical Cycles*, *13*(4), 997–1027.

U.S. National Academy of Sciences and The Royal Society. (2009). *Climate change and ecosystems*. The National Academies Press, 28pp.

U.S. National Assessment provides detailed regional assessments of the ecological on social consequences of climate change across the United States. Available online at: https://nca2014.globalchange.gov/

Van Meter, K. J., Van Cappellen, P., & Basu, N. B. (2018). Legacy nitrogen may prevent achievement of water quality goals in the Gulf of Mexico. *Science*, *360*(6387), 427–430.

Walker, B., et al. (2004). Resilience, adaptability and transformability in social–ecological systems. *Ecology and Society*, *9*(2).

## Notes

1 The general public looks at the lush green landscape across much of the Eastern U.S. and thinks it looks just great. Some forest experts look at that same landscape and miss Longfellow's lovely "spreading chestnut trees" that were almost completely wiped out across the North American landscape by a fungal blight in the first half of the 20th century. For details, see: https://forestpathology.org/canker/chestnut-blight/

2 NPP, often expressed in terms of grams of carbon per square meter per year (gC m$^{-2}$ y$^{-1}$), is a measure of the rate at which plants accumulate carbon per area of cover. NPP varies by region and species and biome.

3 Box plots are explained in Chapter 4.

4 Interested readers can find some of the international prizes Prof. Lal has received in recognition of his work on soil and sustainable agriculture at https://en.wikipedia.org/wiki/Rattan_Lal

# 7

# IMPACTS ON PUBLIC HEALTH

*Elizabeth Casman and Hadi Dowlatabadi*

When he was Vice President, Al Gore worked hard to promote serious government attention to reducing emissions of carbon dioxide and slowing climate change. In the latter years of the Clinton-Gore presidency, as efforts to promote effective climate policy encountered growing obstacles, Gore and others in his office talked at length about the possibility that global warming would have catastrophic consequences for public health.

By the early 1990s, scientific journals like *Climatic Change* were being flooded with studies that reported on results from models that related some aspect of climate to a serious impact on people's health. These included papers that counted deaths due to heat waves and studies that modeled the impact of warming on the life cycle of mosquitoes, amplifying their capacity to transmit diseases such as dengue and malaria. There were even studies of how stream flows would change, leading to more hospitable conditions for snails that complete the life cycle of flat-worm parasites that cause schistosomiasis.

Steve Schneider, who was the editor of *Climatic Change*, kept sending papers like these to Hadi Dowlatabadi, for review. For three reasons, Hadi found it very hard to give most of these papers a positive review: (a) none of them were looking at processes that lead to health impacts in a systematic fashion; (b) they all assumed that for decades into the future only climate would change and everything else would remain the same (a *ceteris paribus* assumption); and (c) none included a consideration of the possible good or bad health consequences of interventions that might be taken to mitigate climate change or adapt to it.

One paper Hadi was asked to review that he remembers especially well was about the impact of climate change on the occurrence of schistosom-iasis, or "snail fever," which occurs when people pick up a parasitic flatworm from wading in slow moving streams. This can cause people's urinary tract or intestines to become infected and results in serious abdominal pain, diarrhea,

DOI: 10.4324/9781003330226-7

bloody stool, or blood in the urine. The paper Hadi was asked to review used computer modeling results of climate change to see how rainfall patterns might change in 2100, and then it calculated stream flows in Sub-Saharan Africa and the impacts that these changes would have on the ecosystem of the freshwater snails. Changes in the prevalence of the snails, which are part of the life cycle of flatworms that cause the disease, were then used to project how many more humans would suffer from this disease.

Sounds great! But think about it. Why assume that nothing other than stream flows will change in the next 100+ years? Surely if there is a health problem for which there are known and effective interventions, a century is long enough to get some of these interventions put in place – *not* motivated by climate change impacts but by a genuine interest in reducing cases of schistosomiasis. If we are 100 years into the future, it is likely there will be programs in place to control the snails. Many irrigation ditches will probably be lined with concrete and fewer people will be wading through streams in their bare feet. The notion of doing a *ceteris paribus* study is fine for analyses of a small (i.e., marginal) change, but it is *not* appropriate for a problem over a century into the future, during which time lots of other things will probably also change. Of course, we do not know what changes will happen, but rather than completely ignore the possibility, a modeler should at least run through a number of stories of the form: if such-and-such happens, here are the likely consequences.[1]

After Hadi had recommended that quite a few papers like this one be rejected, Steve Schneider invited Hadi to write an editorial essay in which he argued that climate is not the only thing that is changing and that good impact assessments need to acknowledge this fact and deal with it. He argued, for example, that the reason tropical diseases, particularly malaria, are currently limited to the tropics is not mainly a matter of climate. Across many more northerly places that once had malaria, it is now gone. It was the actions of people and the resources they expended in response to malaria (contextual determinants) that determined its geographic range. He argued that relative to the human drivers of malaria transmission and suppression, the effects of changes to mosquito vectors and parasite metabolic rates in a warmed future would be small, so long as there were no widespread economic collapses or large movements of infected populations that overwhelmed the capacity to mount public health responses, as happened after the return of malaria among occupying Soviet forces in Afghanistan.

Hadi and Liz Casman discussed these issues and set out to challenge researchers in our Center to develop health impact studies that attempted to address the failings of the ever-proliferating and increasingly alarming "desk studies." It was then that Liz noted that no one contributing to the IPCC[2] 2nd Assessment Report chapter on health had any frontline experience with the topics covered.

In the remainder of this chapter, we summarize five of our studies in which we worked to adopt a more systematic approach to climate and health. Two are on malaria. One that led to a workshop and book on the contextual determinants of malaria brought together frontline experts with convening lead authors of the IPCC

Health chapter. Another, was a "desk study" using models comparing the impact of climate versus economics on malaria cases. A third study involved the relative risks to different populations when intervening to control a vector: in that case, mosquitoes and the West Nile Virus. A fourth project involved exploring effective detection and intervention for an outbreak of cryptosporidiosis following an event that overwhelms water supply. And finally, a study shows how mitigation of greenhouse gas (GHG) emissions at the cost of higher $PM_{2.5}$ is not a good tradeoff. Failing to adopt a more systematic approach has already cost many lives, as we detail below.

## The Contextual Determinants of Malaria

Malaria is called a "vector-borne disease" because it is transmitted by mosquitoes that serve as the "vector" for transmission. The link between vector-borne diseases and environmental conditions has long been known, and mathematical models have been developed that quantify the relationship between temperature and the rates of different life stages and behaviors of both the mosquitoes (the vectors) and the disease-causing organisms they carry. Mosquitoes are cold-blooded. In winter, they are dormant or dead. As temperatures warm, they become more active and complete their life cycle more quickly. When the temperature is higher still, they die off. Of course, up to a point, mosquitoes are smart enough to defy this logic. They find warmer niches when it is cold outdoors and cooler places when it is hot. But there are limits to their life cycle and survival.

If the disease involves an organism that completes its life cycle within the mosquito, this too is accelerated in warmer temperatures. So the time between a mosquito biting an infected human (or animal), maturation, and transmission to a new host becomes shorter. This can dramatically increase the potential for infection by the vector. In addition, warming can increase the suitable habitat for mosquitoes if, on balance, more areas are currently too cold and not as much area will be too hot to be hospitable to their life cycle. This type of logic is all based on assuming that enough standing water exists for laying eggs and that organisms that feed off mosquito eggs and larva are not impacted, etc.

Thus, climate model predictions logically lead to mosquitoes that have a greater geographic distribution and mature more rapidly, increasing the number of mosquito bites per person in an expanding population. These observations led some to claim that several important vector-borne tropical diseases would extend their range into currently temperate climatic zones as the world warms.

It is logical that changes in temperature and moisture would affect the number of mosquitoes; that parasite maturation within infected mosquitoes might be sped up; and that more mosquitoes would, absent some intervention, bite humans more frequently; and that some malaria-free areas currently considered incompatible with the mosquito life cycle, like highlands and mountain slopes, could become hospitable habitats. It is also logical to consider how an ecosystem responds to rising vector populations; whether people would undertake suppression and screening that had not been necessary in the past; and so on.

Ignoring *normal* system responses renders the vision of tropical diseases reconquering high-latitude industrial countries as less than compelling. These diseases ravaged high-latitude regions like the upper Midwest before effective interventions had been devised. Why would climate change lead to amnesia about effective vector control or isolation of infected populations?

We only need a short review of history to appreciate the role of humans in controlling malaria. In the 1830–60s when railways were spreading across British North America (present-day Canada), a significant fraction of railway workers suffered from malaria. Since then, Canada has experienced the fastest rate of warming of any OECD[3] country, and currently there are at least three efficient vectors of malaria in southern Canada. Yet the only malaria cases in the country are travelers who have returned from world regions where malaria is still poorly controlled. By the logic of the desk studies of climate change impacts, Canada should have many more malaria cases than it did in the mid-19th century.

Many readers may not be aware that malaria was only eradicated in Europe in the 1970s. Its elimination is not because we have experienced a climate that is less hospitable to mosquitoes. Its elimination is due to isolation of infected humans and driving down vector populations – initially by using DDT and now by other means. Once infected humans are isolated, further transmission of the disease becomes independent of vector populations.

The mosquitoes that carry malaria have adapted to a range of habitats as diverse as jungles, grasslands, cities, and deserts. Some prefer brackish water, others fresh water. Some prefer to lay eggs in running water, others in stagnant. Some lay eggs in water-storage vessels in houses or in puddles made in tire tracks, while others prefer forest swamps. Some gravitate to clearings in dense jungles; others, above the Arctic Circle, overwinter in human dwellings.

Seasonal changes in temperature, humidity, and precipitation are known to affect the anopheline life cycle. And mosquitoes have coping strategies: some hibernate in winter; some chill in a cool place during hot spells; however, many are killed off by temperature extremes.

Local climate also influences the life history of the malaria parasite, particularly those phases of the life cycle that take place in mosquitoes. This is key to the climate change/malaria spread argument. Certain biological reactions vary with temperature and are sped up by increases in temperature. For example, for a mosquito to become infectious after biting an infected human, it must live longer than the time it takes the parasite to complete its maturation (8 days for *Plasmodium vivax* and 10–12 days for *P. falciparum*). Temperature-induced speeding up of this part of the parasite life cycle would increase the odds of producing new infectious mosquitoes during human-to-mosquito transmission. As explained above, mosquitoes actively seek micro-habitats with temperature and moisture they find more comfortable. In cooler temperatures, they seek warmer niches. In hot climates, they find cooler spaces. So while climate change scenarios project temperature rises by many degrees, the habitats where mosquitoes rest will likely have a more modest temperature swing.

One of the most important conceptual breakthroughs in understanding the dynamics of malaria transmission was Macdonald's[4] calculation of the basic reproduction rate $R_0$, the number of infected persons resulting from the introduction of a single infected human case into a susceptible population. This is a very important tool for epidemiologists. When the value of $R_0$ is greater than 1, an epidemic persists. When the value of $R_0$ falls below 1, the epidemic dies out. This parameter is a function of the ratio of mosquitoes to people, the number of mosquito bites per person per day, the efficiency of mosquito-to-human infection transmission per bite, the efficiency of human-to-mosquito infection transmission per bite, the time it takes for parasites to complete their metamorphosis inside the mosquitoes, the rate that infected humans are cleared of infection, and the probability of a mosquito surviving a day. Many of these factors are functions of temperature.

$R_0$ can be used to estimate the impact of malaria-control factors such as the reduction of mosquito numbers, the treatment of human cases (reducing the reservoir of malaria parasites in humans so uninfected mosquitoes won't pick up malaria), or barrier measures like bed-nets. If the malaria transmission is interrupted long enough that all mosquitoes and humans are malaria-free, malaria is eradicated. This is accomplished by implementing the first two interventions but not the third, which is for "control" but not eradication.

The $R_0$ equation also allows people to isolate the effect of temperature increases on malaria dynamics by adjusting the daily survival probability of the mosquito and the incubation period for the parasite for temperature. Maps of changes in $R_0$ in the warmer future are often mistakenly interpreted as meaning that malaria would become established in areas where it doesn't currently exist.

Endemicity describes the severity of disease transmission in a given area. Malaria is considered endemic when inter-annual incidence is more or less constant. Hyper-endemic transmission occurs when most people are exposed all the time. Epidemic malaria describes a sharp increase in transmission. A useful distinction in malaria transmission intensity is between stable and unstable transmission. Places with intense stable transmission have been described as hyper-endemic. The infection rate is so high that most adults have at least some degree of immunity to malaria. Infected adults in hyper-endemic areas will have symptoms of malaria (malaria causes severe anemia and fevers) but are not generally killed by it. In these areas, fatalities are concentrated among those with weaker immune systems: children under 5 years of age, pregnant women, and individuals whose immune function is impaired by disease or medicine (e.g., to fight cancer or help transplant recipients). With unstable malaria, transmission varies between years, and immunity is low. Fatalities are seen across the age spectrum and genders.

It occurred to us that while elevated temperatures could affect the dynamics of malaria transmission, it may not be the only, or even the most important, factor in the return of malaria to previously malaria-free areas.

The most important factor in the global retreat of malaria had been the widespread use of the insecticide DDT to control mosquitoes, coupled with the aggressive treatment of human cases. In some areas, however, cases of malaria fell sharply even before DDT was invented. For example, in the United States, where malaria closely followed European colonization, the disease effectively died out everywhere except the Tennessee River Valley after the Civil War with the introduction of window screens. In Israel and Iran, campaigns to pour kerosene on standing water eliminated malaria in treated areas.

More evidence of the importance of non-climatic variables on malaria endemicity is the modern return of malaria to places where it had been eradicated. An example of this was the return of malaria in Azerbaijan, Tajikistan, and other Central European countries after the fall of the Soviet Union because malaria surveillance and response activities were suspended but importation of the disease continued. Clearly, human activities and the resources devoted to suppressing malaria were too important to leave out of the future projections of changes in malaria in a warmed climate.

In 2000, we convened a workshop to discuss the things that contributed to *the determinants* of the different kinds of malaria around the world, especially with respect to climate change. Our objective was to better identify the major drivers of malaria endemicity. Participants from HDGC included Elizabeth Casman, Hadi Dowlatabadi, Baruch Fischhoff, Lester Lave, and Granger Morgan. We held the workshop in Lausanne, Switzerland because it is near the headquarters of the World Health Organization (WHO) and we wanted to involve as many WHO malaria area experts as could join us. We invited leading experts in malariology, entomology, epidemiology, demographics, climate science, economics, tropical medicine, and public health from around the world (Table 7.1). A glance at their names shows that

**TABLE 7.1** The malaria experts who participated in our workshop in Switzerland. Participants from our climate center included Elizabeth Casman, Hadi Dowlatabadi, Baruch Fischhoff, Lester Lave, and M. Granger Morgan.

| Name | Affiliation |
| --- | --- |
| Reid Basher | International Research Institute for Climate Prediction |
| Andrei Beljaev | Tropical and Parasitic Diseases, Russian Academy of Postgraduate Medical Training |
| Martin Birley | International Health Impact Assessment, University of Liverpool |
| Robert Bos | World Health Organization, United Nations Environment Program |
| David Bradley | Department of Infectious and Tropical Diseases, London School of Hygiene and Tropical Medicine |
| Jonathan St. H. Cox | London School of Hygiene and Tropical Medicine |
| Robert Desowitz | Tropical Medicine and Medical Microbiology, University of Hawaii; University of North Carolina |
| Ilya Fischhoff | Graduate Student, Ecology and Evolutionary Biology, Princeton University |
| Dana Focks | Center for Medical, Agricultural, and Veterinary Entomology, USDA; WHO |

**TABLE 7.1** Cont.

| Name | Affiliation |
| --- | --- |
| Duane Gubler | Vector-Borne Infectious Diseases, Centers for Disease Control and Prevention (CDC) |
| Renato d'A. Gusmão | Communicable Diseases Control, Pan American Health Organization; WHO |
| Chev Kidson | Tropical Medicine, Mahidol University, Bangkok; Chinese Academy of Preventive Medicine, Beijing |
| Anatole Kondrachine | Malaria Control, WHO |
| R. Sari Kovats | Epidemiology and Population Health; London School of Hygiene and Tropical Medicine |
| Socrates Litsios | Control of Tropical Diseases Program, World Health Organization |
| Janice Longstreth | Institute for Global Risk Research |
| Wolfgang Lutz | Population Project, International Institute for Applied Systems Analysis |
| Pim Martens | International Centre for Integrative Studies, Maastricht University |
| Tony McMichael | National Centre for Epidemiology and Population Health, Australian National University, Canberra |
| Jean Mouchet | Microbiology, Parasitology, and Medical Entomology Division, Institut de Recherches pour le Développement |
| Paul Reiter | Entomology Section, CDC |
| Donald Roberts | Center for Applications of Remote Sensing and GIS in Public Health, Uniformed Services University |
| Guido Sabatinelli | Roll Back Malaria, WHO, Eastern Mediterranean Region and European Region |
| Allan Schapira | WHO Regional Office for the Western Pacific |
| Michael Schlesinger | Atmospheric Sciences, Climate Research Group, University of Illinois at Urbana-Champaign |
| Vinod Prakash Sharma | WHO South-East Asia Regional Office |
| Kenneth Strzepek | International Water Management Institute, University of Colorado, Boulder |
| Tang Lin-Hua | Institute of Parasitic Diseases, Chinese Academy of Preventive Medicine |
| Mark Wilson | Epidemiology, School of Public Health, University of Michigan |

Drawn by authors.

this roster included IPCC authors as well as experts with on-the-ground experience in the fight against malaria in different regions around the world.

We commissioned a series of papers on malaria in different regions of the world to gain insights into the socio-ecological contexts of their prevalence and dynamics. The workshop participants held discussions on what conclusions about the impact of global warming on malaria could and could not be supported by existing knowledge of malaria dynamics. We summarized the results in a book, *The Contextual Determinants of Malaria* (2002).[5] The book includes chapters covering all the regional types of malaria. It described a diverse set of contextual determinants of malaria, frequently involving poverty, limited access to health services, population movement, inadequately funded or negligent public health systems, corruption, marginalized populations, or water management methods that increase vector habitat near human habitation. A picture emerged of malaria as a disease of poverty complicated by ecological factors, primitive infrastructure, and governmental failures.

Climate plays an underlying role in determining a region's ecology and associated malaria potential, and climate variability influences the intensity of transmission. Climate extremes disallow malaria endemicity but not malaria transmission in very cold and very dry areas where parasite and mosquito survival are brief or nonexistent. Some malaria vectors and parasites have adapted to inhospitable climates by colonizing amenable microhabitats or by taking advantage of resting states. Therefore, climate limits the epidemiological type of malaria transmission in a given location, modulates the seasonality of transmission in some areas, and prevents transmission in others. However, climate does not *determine* whether malaria is transmitted. That is controlled by human behaviors and responses to malaria. Thus, a climate that is amenable to malaria transmission is a necessary condition, but *not* a sufficient one.

Entomologists at our workshop argued that many of these areas were already amenable to malaria transmission, just as when it was transmitted worldwide. The competent mosquitoes have inhabited temperate climates when malaria was endemic, and they continue to inhabit the same zones now that malaria has been eradicated. What had changed was an interruption of the transmission cycle. The human population was no longer highly infected, so when a mosquito bit a person, the odds were high that the mosquito would not pick up the malaria parasite. When cases of human malaria were imported from abroad and began to spread locally, the cases were cured and the mosquitoes were poisoned.

Malaria is now geographically a tropical disease, but as we explained above, that was not always the case. It has been transmitted as far north as Canada and Siberia in recent centuries and was endemic in temperate parts of Europe and North America. Just as cold temperatures didn't circumscribe malaria endemicity in the past, cold temperatures are currently not responsible for the absence of malaria. The current localization of malaria to tropical areas may contribute to the perception that warmer temperature largely determines malaria risk, but it masks the stronger correlation of malaria with poverty and under-resourced or corrupt governance. Analogously, a future warming of temperate zones does not remove some barrier to malaria, because no such barrier exists. Warming could, however, accelerate the transmission in temperate regions if unstable transmission were re-established. Potential is not the same as expected transmission because the latter considers the actions of humans to oppose transmission.

By 2050, the non-climate-related determinants are expected to change but not necessarily in a predictable way on the local scale. If current trends persist, population, travel (both voluntary and coerced), and urbanization will increase. Public health anti-malaria activities are hard to predict because they are tied to other difficult-to-forecast local and global conditions such as incidence of armed conflict; local cooperation; technological breakthroughs in vector and parasite control, treatment, and prevention; and the prevailing local and global economic situation.

While most climate models don't currently predict climate change at the scale relevant to mosquito habitat, some generalizations about the future climate are possible. On a global scale, there will be perhaps several degrees of warming, on average. More hot days, fewer cold days, higher maximum temperatures and higher minimum temperatures, longer and earlier warm seasons, and more intense precipitation have already been observed. Larger year-to-year variation (droughts and floods) and more intense precipitation are likely over some areas. Increased wind speeds and rainfall from hurricanes and more intense precipitation from storms are not incompatible with forecasts of increased drought, as we have recently seen in the northern hemisphere. Episodes of destabilization of the jet stream by large areas of warm ocean (popularly known as a polar vortex effect) should increase in frequency as the oceans absorb more heat. Changes in the Asian monsoon and the intensification of ENSO-related droughts and deluges are also expected.[6]

But what will be the contribution of climate change relative to the other determinants of malaria? The answer depends on the geographic location.

For the most part, in Sub-Saharan Africa, the home of 90% of the world's malaria cases and fatalities, climate change is not expected to have much of an effect on malaria incidence. The main mosquito vectors in this region are efficient and widespread, and there are large areas of stable malaria where essentially everyone is repeatedly infected from infancy to death. In such areas, public health efforts are focused on saving lives, protecting the most vulnerable in the population (children and pregnant women), and case management. Malaria control has remained mostly outside the national health systems, which, coupled with inadequate transportation and health infrastructure, competing national interests, poverty, and government instability, contributes to slow progress in this region. Exceptions, with significant malaria control efforts, include Madagascar, South Africa, and Zimbabwe. On a broad regional scale, however, temperature and rainfall better predict malaria incidence in Sub-Saharan Africa than in any other region. This is because the non-climatic determinants are more important in the other regions. However, the impacts will be greatest in the relatively small areas of unstable transmission.

The concept of stable and unstable malaria is relevant to the question of whether climate change will alter the endemicity of malaria in Sub-Saharan Africa. Climate projections for Africa, interpreted through the lens of $R_0$ calculations, indicate that warming will not change hyper-endemicity. Marginal intensification of already high re-infection rates doesn't change the binary classification ($R_0$ greater than or less than 1). One exception could be at the margins of the Sahara, where future desert expansion could decrease mosquito habitat and locally reduce malaria transmission.

It is known that ENSO events influence the timing and intensity of transmission of unstable malaria. The predicted intensification of these and other coupled ocean-atmosphere oscillations potentially could impact malaria and other mosquito-borne diseases.

Air and water pollution are thought to be inhibitory to mosquito proliferation, and in fact, they do offer protection from malaria transmission in some highly polluted megacities in certain developing countries. Since population growth and the expansion of megacities in the southern hemisphere is part of most future scenarios, there may be some protection from malaria, though bought at other health and environmental costs.

Pollution does not currently protect the megacities of India from malaria. There, cities are ringed by extensive slums. Water used to be provided every few blocks by standpipes. Now, it is piped but intermittently available. This has led to water storage at home (whereas water in vessels filled from standpipes would be emptied on a daily basis). An exceptionally competent malaria vector, *Anopheles stephensi*, which lays eggs in the forks of trees and puddles, has now moved indoors, finding the water storage vessels a perfect and abundant habitat. This has changed malaria transmission in slums from seasonal to year-round. Near-term malaria control in India will have to include remedies for the water supply problems of the mega-slums.

Some areas in Asia and South and Central America are caught in a loop of partial malaria control followed by resurgence. These are regions where epidemics, drug resistance, and pesticide resistance are major problems. During the global anti-malaria campaigns of the mid-20th century, malaria was temporarily eradicated, only to be reintroduced when the control program ended. This pattern was exacerbated in places where malaria control was not extended to marginalized populations. Examples of this include mountain tribes in Peru and indigenous groups in India. Eventually these residual foci of endemicity re-inoculated the surrounding countries.

Another worrisome pattern that could intensify in the future is the mixing of infected and non-infected human populations in the presence of competent mosquito vectors coincident with the absence of vector control programs and medical support. This was observed in the mountainous tea plantations of Kenya, where malaria was unknown until the importation of workers from neighboring malarious lowlands. These workers quickly infected the local mosquitoes and malaria transmission became established.

In developing countries, the arrival of malaria has also been tied to labor movements, such as the case in eastern Turkey. Infected workers were imported to build a hydroelectric and flood control dam. The process of excavation created large amounts of vector habitat. The laborers camped at the worksite. The results were predictable; now malaria is endemic there. A variant on this pattern occurred when uninfected miners from south Asia flocked to the jungles of Cambodia to work in ruby mines where they contracted malaria, which they brought home with them and spread to urban areas. Human travel for economic gain, to escape conflict, and perhaps as climate refugees, is projected to increase with population growth. Without vigilance, the ground won from decades of malaria control in resource-poor regions could be lost.

Climate greatly affects the success of malaria control programs. Vector control is easier in temperate zones, where the number of competent vector species and the number of different environmental niches they occupy are often lower than in the tropics. Malaria was eradicated in countries that mounted and maintained the necessary medical and environmental interventions. It never really started in the hyper-endemic regions of Africa. Considering the magnitude of the non-climate drivers, if they still operate in the warmed future, the extrapolation is that they still would dominate the impact of changes in malaria transmission due solely to climate change. We don't dispute that climate change will alter malaria dynamics, but we think it would account for a relatively small fraction of the change in the global malaria burden.[7]

## Climate Change Compared with Policy Impacts on Malaria

A key insight from our malaria workshop involved the important role played by human factors in the control of malaria. In 1999, while Richard Tol was a visiting scholar in EPP at Carnegie Mellon, he pointed out that in tropical regions where GDP/capita was above US$3500, there was no endemic malaria. We took this as the threshold of economic development that would allow effective intervention. Richard and Hadi then decided to characterize the relative importance of climate change and climate policy on incidence of malaria in the future.

Using Tol's FUND integrated assessment model, we analyzed: (a) the role of demographics versus climate change in potential malaria cases; (b) the impact of economic growth (or pro-health policies) on malaria cases; and (c) the impact of climate policy on malaria cases.

Figure 7.1 displays the relative importance of population growth, climate change, and economic growth in projections of malaria cases. We can see that given malaria cases and modeled dynamics, absent population growth, total additional mortality due to climate change would almost quadruple from around 70,000 per year to 270,000 by 2100. This mostly arises from expansion in the geographic reach of malaria. With the addition of population growth in the projections, the total deaths from malaria are projected to balloon to 700,000 per year – an overall increase of 10x from today's figures. However, if one considers economic development and efforts to control malaria in the same pattern as observed today, the burden of malaria never rises above 180,000 and begins to fall after 2050 to near zero by 2090. These model outputs serve to emphasize two issues: (a) analyses that hold everything else unchanged over long time-horizons deliver the same type of erroneous message as bedeviled Malthus and the Club of Rome[8] and (b) the power of intervention can easily overwhelm the force of malaria – given enough attention

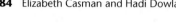

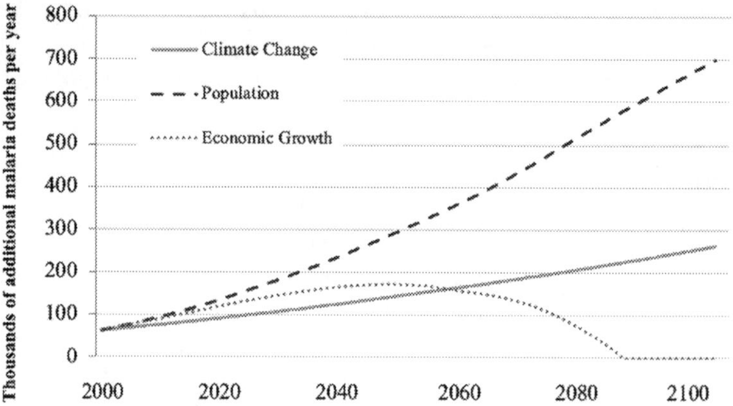

FIGURE 7.1 Additional deaths due to malaria projected from 2000 to 2100 using a baseline emissions scenario, Richard Tol's FUND integrated assessment model, and an assumption that once GDP/capita rises above US$3500 there is effective control of malaria. Figure from Tol and Dowlatabadi, 2001.

Tol, R. S., & Dowlatabadi, H. (2001). Vector-borne diseases, development & climate change. *Integrated Assessment, 2*(4), 173–181.

and resources. In fact, since these projections made in late 1999, the various efforts to control malaria have been making significant inroads into malaria control even where current resources are far below the economic development threshold used in these projections.

The second part of the analysis, which we did using FUND, explored how climate policy may slow down the economy of large GHG emitters and how this slowing down impacts raw material imports from developing countries. That analysis shows that raw material exports are a major contributor to the economic growth that can so effectively limit malaria. While a Kyoto type agreement by large, industrialized countries may be motivated by altruism and concern about climate change impacts, the net impact of such agreements could well be to exacerbate the challenges face by developing countries by retarding their economic growth. Most developing countries rely heavily on raw material exports. Any reduction in demand for these exports has a much higher economic impact on developing country economies than on OECD importers. This is reflected in the modeling results shown in Figure 7.2. This figure shows that, absent any countermeasures, GHG mitigation strategies in OECD countries that reduce demand for raw materials may exacerbate impacts of climate change in the most vulnerable regions of the world.

## Controlling the Spread of West Nile Virus

West Nile encephalitis is an infection of the brain that was first identified in Uganda in 1937. It is caused when a person is infected by a virus that is common

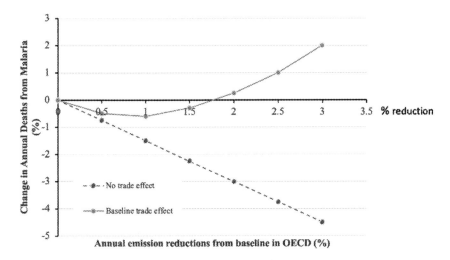

**FIGURE 7.2** The horizontal axis in this graph shows different levels of GHG mitigation in OECD countries. The economic model behind this analysis links mitigation effort to reduction in output, and reduction in output reduces imports of raw materials. Exports of raw materials are key to economic development of many lower-income countries. The reduction in exports arising from the OECD mitigation effort slows down their development and that in turn reduces the effort to control malaria. Hence, ignoring trade effects, mitigation by OECD reduces malaria cases, but if trade effects are included (of the type modeled here), the net impact would be to increase malaria cases. Figure from Tol and Dowlatabadi, 2001.

Tol, R. S., & Dowlatabadi, H. (2001). Vector-borne diseases, development & climate change. *Integrated Assessment*, *2*(4), 173–181.

in Africa and West Asia and the Middle East. It was first reported in New York City in 1999. This old-world virus had a devastating effect in the New World, where local immune systems were overwhelmed in crows, ravens and similar birds (corvids), horses, and more elderly human populations. By 2003, the virus had spread from east to west with incidences in 44 states and 5 provinces in the U.S. and Canada, respectively. During 2002, the outbreak in North America was particularly alarming. The human case count was over 4000 with 284 attributed deaths. Blood screenings in NY and other locations showed that a far larger population was exposed to the virus but with minimal to no health impacts. By 2018, the disease was present throughout the lower 48 states and Alaska with 2544 cases leading to 137 deaths. In Canada, there were a total of 200 cases in 2018. It seems that the dynamics of exposure to and transmission of the virus have led to its evolution into a less deadly form (or better equipped immune systems) as WNV transitioned from an emergent disease to one that is a continuing presence in the U.S. and Canada.

In 2005, Negar Elmieh along with the two of us thought it would be interesting to characterize the risks from interventions to control this emerging disease as compared with the impacts of the disease itself. A model called EMSHI was developed in the Analytica® software environment to characterize the exposure of different sub-populations to residual malathion fogging in urban areas.

We assumed ultra-low volume spraying of malathion in urban settings to estimate exposure to this broad-spectrum organophosphate, which has been licensed for use in adult mosquito control since 1953. We used field data on half-life of malathion in dry, moist, and wet locations. Malathion has a half-life of 24 hours in dry conditions but 10–20 days in moist and wet conditions. We calculated exposure via different pathways, building models of: inhalation, transdermal (through contact with surfaces coated with the pesticide), and ingestion (of foods/objects coated with it). We then modeled realistic behavior of children and adults during the spray period and the subsequent period the pesticide is expected to be in the environment before decomposition.

Using exposure limits for margin of exposure to malathion with no observed adverse effects (NOAEL), we estimated relative exposures of children and adults. Negar was imbedded in the public authority in British Columbia and was privy to how fogging was perceived as different publics. Politicians viewed it as a demonstration of active intervention. Environmentalists questioned its impacts. What we could offer was an analysis of whether fogging led to exceedance of harmless exposure – see Table 7.2.

We include this research here because it provides an illustration of how responses to any emergent challenge have distributional consequences and can naturally evolve into a norm that no longer commands extraordinary interventions.

Visions of pandemic and incurable encephalitis due to infection with the West Nile Virus (and many other vector-borne diseases) are alarming to the public. The level of alarm can be such that politicians feel compelled to demonstrate their intervention to ameliorate said risk. However, through the intervention, there can be harm to sub-populations who would otherwise be unaffected.

**TABLE 7.2** Percent exceedance in exposure to malathion based on standards from the Pest Management Regulating Agency of Canada.

| Exposure Pathway | NOAEL (mg/kg day) | | Target MOE | % Exceedance |
|---|---|---|---|---|
| Dermal | Adults | 25 | 300 | 0 |
| | Children | 5 | 1000 | 0.09 |
| Ingestion | Adults | 25 | 300 | 0 |
| | Children | 5 | 1000 | 0 |
| Inhalation | Adults | 25 | 1000 | 0 |
| | Children | 5 | 1000 | 2.76 |
| Overall | Adults | 25 | 300 | 0 |
| | Children | 5 | 1000 | 3.65 |

Elmieh, N., Dowlatabadi, H., & Casman, E. (2006). "Role of lifestyle characteristics and risk perceptions in preventing West Nile virus disease," APPH Annual Meeting, 2006.

## Cryptosporidiosis

Not all (and probably not even the most important) health effects resulting from climate change will be due to vector-borne diseases. A cursory review of the current health effects of climate-related events in the U.S. shows that the most important causes of death are cold and heat exposures, in that order. Flooding, wind, and wildfire related deaths are much less frequent in developed countries, though flooding and hurricanes (tropical cyclones) result in a large death toll in developing nations every year. Such problems are expected to increase in a warmed future, as are phenomena that weren't even on the radar in 2000, such as smoke-related morbidity from prolonged fire seasons.

Where do water-borne infectious diseases rank on this list? Climate change is expected to degrade water quality, periodically increasing suspended solids and algae in source water for treatment plants. Such conditions require rapid adaptation of drinking-water treatment protocols. Large water-borne infectious disease outbreaks in the past have been associated with failures in water treatment preceded by changes in source water quality. There is reason to suspect that such vulnerabilities will persist in the near future, especially at smaller, less-instrumented treatment plants.

To consider how the risk of water-borne infectious disease in the U.S. might change with global warming, we conducted a thought experiment, systematically analyzing the major drivers of cryptosporidiosis, a water-borne parasitic diarrheal disease, and considering how each driver might change in a future warmed climate. In this qualitative analysis, we looked at scenarios of socioeconomic change such as population growth and density changes, watershed protection advances, water infrastructure aging and replacement, improved drinking water treatment technologies, changes in the regulatory environment, medical advances, and wealth under the following climate scenarios: wet weather, sea level rise, dry weather, lowered water tables, longer recreational swimming season, increased use of wastewater in agriculture, and increased weather-related disaster frequency. Results are summarized in Table 7.3.

Increased temperature, changes in river flow, and increased water pollution could increase cryptosporidiosis in the U.S. More frequent algal blooms and heavier sediment loads provide more opportunity for failure to nimbly make the needed adjustments in water treatment and, until adaptation has been accomplished, will precipitate more water contamination emergencies. However, by itself, climate change is likely to be a poor predictor of water-borne cryptosporidiosis in countries with high standards of living. There, the effects of climate will probably be countered by the influences of public health investment, water treatment technologies, and drinking water regulations. Climate-related effects could be significant if compensating behaviors are not adopted, which we do not anticipate, barring some major socioeconomic failure. Therefore, we concluded that climate change will have little effect on cryptosporidiosis transmission in the U.S. if the U.S. invests the resources needed to maintain its current commitment to public health. Our

TABLE 7.3 A qualitative assessment of change in cryptosporidiosis risk and climate change to 2050.

| Mechanism | Scenario | Change in Risk | Change in Cost |
|---|---|---|---|
| Increased solids in water treatment plant intake | Wet or dry | Small (+ may decrease) | Low to moderate |
| Increased shellfish contamination | Wet | Small (+ may decrease) | Low |
| Increased contamination of ground/well water | Wet | + | Low to moderate |
| Elevated water tables | Wet | + | Low-high |
| Contaminated recreational bathing water | Wet | Small+ | Low |
| Lowered water tables | Dry | + | Moderate |
| Longer recreational swimming season | Wet or dry | Small+ | Low |
| Use of wastewater on agricultural products | Dry | Small+ | Moderate |
| Increased disaster frequency/severity | Disaster | Small+ | Moderate-high |

Casman, E., Fischhoff, B., Small, M., Dowlatabadi, H., Rose, J., & Morgan, M. G. (2001). Climate change and cryptosporidiosis: A qualitative analysis. *Climatic Change, 50*(1), 219–249.

analysis underestimated the risks borne by people with lower adaptive capacity or risks resulting from massive failures of coping mechanisms. Furthermore, the conclusions of this analysis should not be extended to less industrialized countries or to other diseases.

## Climate Mitigation and Air Pollution

In 1999, during preparation of the IPCC's latest report, Hadi faced a major challenge in messaging about GHG mitigation. The leadership in the IPCC had decided to frame all GHG mitigation as having one or more "co-benefits" (i.e., other non-climate benefits that result from acting to mitigate climate change and its impacts). To be sure, there are many cases where GHG mitigation has co-benefits. For example, fossil fuels differ in their life cycle impacts on health, and if natural gas replaces coal in electricity production, there is both a GHG reduction *and* an air-quality improvement.[9] Moreover, there are many papers showing empirical evidence of specific environmental quality and health benefits in the wake of coal plant shutdowns.[10] However, not all things that can be done to lower GHG emission pathways have co-benefits. This final short section describes a predicted adverse health outcome associated with what Hadi calls "half-baked" climate mitigation policies.

In 2001, the UK acted on the insight that diesel cars produce 10–15% less GHGs compared to gasoline vehicles of similar performance. They implemented two new policies: (a) a carbon excise tax at the time a vehicle is purchased reflecting its lifetime emissions and (b) taxing company cars as a personal benefit. These policies led to a rapid rise in new diesel car registrations from 14% in 2001 to 43% by 2007 and 50% in 2011. It should be noted that this is unambiguously the most effective carbon tax policy known to us. But Hadi was concerned that the fleet would have much higher emissions of $PM_{2.5}$ and $NO_x$ and exact a toll in air quality and health in the tradeoff to reduce GHG emissions.

Eric Mazzi took on the challenge of quantifying the impacts of fleet changes in the UK. He developed a model of the UK fleet composition and estimated emissions and change in air quality in the UK. Eric and Hadi showed that the successful reduction of roughly 600kt of $CO_{2e}$ per year was associated with additional 90–300 premature deaths from poorer air quality per year.[11]

For those living in North America, it has always been surprising that so many European countries have such love for diesel cars. Europe was slow to abandon lead in gasoline. It has been slow to limit pollution from diesel cars. And many cities are now suffering such horrendous air pollution that they are banning diesel cars. This has been an avoidable tragedy that escapes our comprehension.

## Technical Details for Chapter 7

The editorial that Hadi wrote for the journal *Climatic Change* is:

Dowlatabadi, H. (1997). Assessing the health impacts of climate change-an editorial essay. *Climatic Change, 35*, 137–144.

**Abstract:** In many environmental change problems, concern about risk to human health is the key factor turning the torpor of public indifference into a tide for intervention. The climate change debate appears to be following this same pattern. However, two factors make the assessment of human health impacts from climate change very challenging: (1) understanding the role of climate in health, (2) projecting the future patterns of change in climate and other determinants of public health. It is important to interpret the literature on health impacts of climate change in the broader context of public health issues in a changing global environment. This broader interpretation does not always support the conclusion that intervention to halt or reverse climate change is the best option. The role of causes including air pollution and effects is examined.

Here is the book that Liz and Hadi edited, with chapters written by many of the participants in the workshop we ran in Lausanne, Switzerland:

Casman, E. A., & Dowlatabadi, H. (eds.). (2002). The *Contextual Determinants of Malaria*. Resources for the Future Press, 382pp.

This book does not have an Abstract, but Helen Elsey began her brief review in the *International Journal of Epidemiology*, 32(3) May 2003 with the following summary:[12]

As the physicist Niels Bohr put it, "prediction is very difficult ... especially about the future." This book takes the only logical approach and begins by analyzing the history of malaria around the globe and attempting to identify the current determinants of malaria risk. The authors identify determinants far beyond the purely biomedical to include climate, population movement, demography, and economic and social development. The lessons drawn from the past and from current thinking form the basis for the development of a systematic integrated assessment framework to aid in the prediction of malaria risk. The major premise of the book is that while climate and subsequent changes in climate will affect a region's potential for malaria, it is the actions of humans that determine whether the risk of malaria is realized. As a social scientist working in the public health field, this recognition of economic and social factors as determinants in the spread and severity of diseases is a sight for sore eyes.

The paper that Hadi wrote with Richard Tol while Richard was a visitor at Carnegie Mellon is:

Tol, R. S., & Dowlatabadi, H. (2001). Vector-borne diseases, development & climate change. *Integrated Assessment*, *2*(4), 173–181.

**Abstract:** Vector-borne diseases are feared to extend their range in a future where global warming has occurred. There is considerable concern about scourges such as malaria re-invading currently temperate regions and reaching into higher altitudes in Africa. In this paper we examine the various factors thought to determine potential infectivity of malaria, and its actual outbreak in the context of a dynamic integrated assessment model. We quantify: (i) the role of demographics in placing a larger population in harms way; (ii) the role of climate change in increasing the potential geographic range and severity of the risk of infection; and (iii) the role of economic and social development in limiting the occurrence of malaria. We then explore the climate and economic implications of various climate policies in their effectiveness to limit potential infectivity of malaria. In illustration of these issues we present the climate-related and economics-related impacts of unilateral $CO_2$ control by OECD on incidence of malaria in non-OECD nations. The model presented here, although highly stylized in its representation of socio-economic factors, provides strong evidence of the role of socio-economic factors in determination of malaria incidence. The case study offers insights into unintended adverse consequences of well-meaning climate policies.

The paper that summarizes much of the work that we did on modeling cryptosporidiosis is:

Casman, E., Fischhoff, B., Small, M., Dowlatabadi, H., Rose, J., & Morgan, M. G. (2001). Climate change and cryptosporidiosis: A qualitative analysis. *Climatic Change*, *50*(1), 219–249.

**Abstract:** The effects of climate change on drinking-waterborne cryptosporidiosis transmission in the United States are analyzed using an influence diagram representation of epidemic development. Results from a systematic qualitative analysis indicate that climate change will have little effect on cryptosporidiosis incidence if the United States continues to be wealthy and maintains its commitment to public health. The major impact will, instead, be the additional costs of adapting to new climate regimes in order to avoid drinking-waterborne disease risk. These costs, for the most part, will be from improved monitoring and treatment of drinking water. The consequences of disaster scenarios are also considered. These, too, suggest that climate change per se will be a poor predictor of waterborne cryptosporidiosis in countries with high standards of living. Rather, the risk of epidemics will depend on the interplay between population, public health investment, infrastructure maintenance, emergency planning/response capabilities, water-treatment technologies, drinking-water regulations, and climate.

A slightly earlier paper on cryptosporidiosis that laid some of the groundwork for the climate paper is:

Casman, E. A., Fischhoff, B., Palmgren, C., Small, M. J., & Wu, F. (2000). An integrated risk model of a drinking-water–borne cryptosporidiosis outbreak. *Risk Analysis, 20*(4), 495–512.

**Abstract:** A dynamic risk model is developed to track the occurrence and evolution of a drinking-water–borne cryptosporidiosis outbreak. The model characterizes and integrates the various environmental, medical, institutional, and behavioral factors that determine outbreak development and outcome. These include contaminant delivery and detection, water treatment efficiency, the timing of interventions, and the choices that people make when confronted with a known or suspected risk. The model is used to evaluate the efficacy of alternative strategies for improving risk management during an outbreak, and to identify priorities for improvements in the public health system. Modeling results indicate that the greatest opportunity for curtailing a large outbreak is realized by minimizing delays in identifying and correcting a drinking-water problem. If these delays cannot be reduced, then the effectiveness of risk communication in preemptively reaching and persuading target populations to avoid exposure becomes important.

The work on West Nile virus was presented to the annual meeting of the American Public Health Association in November of 2006:

Elmieh, N., Dowlatabadi, H., & Casman, E. (2006). Role of lifestyle characteristics and risk perceptions in preventing West Nile virus disease. *APPH Annual Meeting.*

**Abstract:** There are three measures through which a community can limit cases of West Nile virus (WNV): (1) avoidance: behavioral measures among individuals to reduce exposure to vectors, (2) repulsion: application of chemicals to deter mosquito bites, and (3) extermination: the application of pesticides to kill vectors. In order to better understand how people conceptualize WNV prevention measures, a questionnaire was administered to individuals and public health experts in four areas across Canada with varying risk criteria. The questionnaire was designed to quantitatively estimate perceptions associated with WNV, pesticide spray campaigns using malathion, and potential alternatives (including, no action and use of DEET). The results of the study were used to parameterize PAMSE, a Probabilistic Assessment model of Malathion Spray Exposures, for calculating malathion exposure from WNV control spraying programs. Results show that certain lifestyle characteristics can lead to pesticide exposure levels over the allowable Margin of Exposure (MOE). Such a systematic characterization may lead to more effective interventions than has been possible to date, and should be helpful in targeting future risk communications towards behaviors and attitudes that put people at risk of adverse health impacts from WNV control pesticide applications.

Finally, the paper on air quality and climate policy that demonstrates that policies to reduce climate change do not always have positive "co-benefits" is:

Mazzi, E. A., & Dowlatabadi, H. (2007). Air quality impacts of climate mitigation: UK policy and passenger vehicle choice. *Environmental Science and Technology, 41*(2), 387–392.

**Abstract:** In 2001–2002 the UK began taxing vehicles according to $CO_2$ emission rates. Since then, there has been a significant increase in consumer choice of small cars and diesel engines. We estimate $CO_2$ reductions and air quality impacts resulting from UK consumers switching from petrol to diesel cars from 2001 to 2020. Annual reductions of 0.4 megatons (Mt) of $CO_2$ and 1 million barrels of oil are estimated from switching to diesels. However, diesels emit higher levels of particulate matter estimated to result in 90 deaths annually (range 20–300). We estimate 570, 460, and 0 additional deaths per Mt of $CO_2$ abated, for Euro III, Euro IV, and post-Euro IV emission class vehicles, respectively. $CO_2$ policies are suspected to have contributed substantially to diesel growth, but the magnitude of impact has yet to be quantified rigorously. To the extent that $CO_2$ policies contribute to diesel growth, coordinating $CO_2$ controls with tightening of emission standards would save lives. This research shows that climate policy, while reducing fuel use and $CO_2$, does not always ensure ancillary health benefits. Lessons from the UK can help inform policies designed elsewhere which strive to balance near-term ambient air quality and health with long-term climate mitigation.

### Some notable writings by authors not affiliated with our Centers

Carlson, C. J., et al. (2022). Climate change increases cross-species viral transmission risk. *Nature.*

Kalkstein, L. S. (1993). Direct impacts in cities. *The Lancet, 342*(8884), 1397–1399.

Molyneux, D. H. (1997). Patterns of change in vector-borne diseases. *Annals of Tropical Medicine & Parasitology, 91*(7), 827–839.

Strauss, J., & Thomas, D. (1998). Health, nutrition, and economic development. *Journal of Economic Literature, 36*(2), 766–817.

Walsh, J. F., Molyneux, D. H., & Birley, M. H. (1993). Deforestation: Effects on vector-borne disease. *Parasitology, 106*(S1), S55–S75.

## Notes

1  For example, rather than assuming that people will just keep dying as heat waves become more common in more northerly cities like Chicago, consider how those cities might adapt to heat waves like Atlanta and Houston already have.

2  As explained in earlier chapters, the IPCC is the Intergovernmental Panel on Climate Change, a large joint effort by the UN and the World Meteorological Association that performs periodic assessments of the state of climate science, climate change, and its impacts.

3  The OECD is the Organization for Economic Co-operation and Development, a group of 38 developed countries.

4  For details, see Macdonald, G. (1957). *The Epidemiology and Control of Malaria,* Oxford University Press, 201pp.

5  Casman, E., & Dowlatabadi, H. (eds.). (2002). *The Contextual Determinants of Malaria,* RFF Press, 382pp.

6  ENSO stands for El Niño-Southern Oscillation. This is a periodic oscillation of waters of the Pacific Ocean that exerts influences on many parts of the world's climate.

7  We note that not everyone at the workshop agreed with the conclusions outlined in this section. Specifically, David Bradley asked that his name not be included in the workshop proceedings Conclusion chapter.

8  In his 1798 book, *An Essay on the Principle of Population,* Thomas Malthus predicted that the world's population would quickly become enormous and drive the world into poverty. The Club of Rome produced a study called *The Limits to Growth* in 1972 that argued that the world's societies were quickly approaching resource limits to their ability to grow and prosper.

9  Markandya, A., & Wilkinson, P. (2007). Electricity generation and health. *The Lancet, 370*(9591), 979–990.

10 Yang, M., & Chou, S. Y. (2018). The impact of environmental regulation on fetal health: Evidence from the shutdown of a coal-fired power plant located upwind of New Jersey. *Journal of Environmental Economics and Management, 90,* 269–293.

11 Mazzi, E. A., & Dowlatabadi, H. (2007). Air quality impacts of climate mitigation: UK policy and passenger vehicle choice. Environmental Science & Technology, 41(2), 387–392.

12 The full review can be found at: https://academic.oup.com/ije/article/32/3/473/637121

# 8

# ENERGY EFFICIENCY

*Inês Azevedo*

It was an unusually cold and foggy May morning in Lisbon, Portugal. Since it was spring, students like me at IST were expecting nice, mild mornings that would allow us to leisurely enjoy a decent espresso and a toast with butter in the patio of the university cafeteria. But this morning I had an insanely early morning meeting (by Portuguese standards) – 7 or 8 am, I can't quite recall – with this famous American professor who was visiting IST for an evaluation of IST's Master program on Innovation and Management of Technology. The professor had kindly fit me in to his busy schedule for breakfast. "Obelix[1] would say these Gaulois are crazy," I grunted to myself, "and I would say these Americans are crazy. Who starts working this early?" Little did I know that I was about to meet the person who would have the largest intellectual influence on my professional life and become one of my dearest friends.

Granger Morgan was already waiting for me at the hotel café, and he was already working on his laptop while sipping his coffee. We quickly started talking about climate change and strategies to decarbonize the energy system. We had our first lively research discussion where we disagreed fiercely – one of many, many (did I already say many?) fun and intellectually stimulating discussions yet to come until present day. I kept on arguing that we needed large-scale renewables, fundamental changes in people's behavior and energy efficiency to address the climate problem, and that we needed to stop the use of coal altogether. Granger argued that there was no way we would quickly stop using coal and that some serious thought and funding would need to be invested in carbon capture and sequestration.

Granger is this way. He will make time to meet with prospective students and be happy to engage in discussions with people who are just starting to think about these problems and have virtually no knowledge of the issue – namely when compared to his extensive knowledge. He is particularly careful to meet with people who have different backgrounds and nationalities, and with women, to ensure that they

DOI: 10.4324/9781003330226-8

engaged in research topics related to engineering and public policy, and namely on issues related to climate change, energy, and the environment.

I left the meeting with the certainty that this sort of challenging discussion and broad thinking about different aspects of the energy system were exactly what I needed for my PhD. And so, I applied to the PhD program in Engineering and Public Policy at Carnegie Mellon University. This was the only PhD program I applied to.

Before I get on with the main story, here is a small aside. A few months elapsed between receiving my acceptance to the EPP PhD program and the start of the academic year. At some point during that period, I remember sending an email to Granger with some questions. He didn't respond right away. This was surprising, as he would generally respond very quickly. A few days later, I received an email from Granger. He was at the hospital. He'd been hit by a truck while heading out for lunch after chairing a meeting of the EPA Science Advisory Board in Houston, TX. But he was already working, and his email included a PowerPoint drawing that explained where he was hit and the angle from the truck. I remember later talking about this with another professor in EPP (who will remain nameless), about how impressed I was that he would be back at work so fast and provide all these details to an incoming student. The colleague responded, "Well, this was the only time where my response time to emails was faster than Granger's."

Thinking about pursuing my PhD in the United States didn't come without some doubts about the decision. It was far away from my family. It was different. People cared less about the environment, I thought. There are guns. There is income disparity. I didn't know if I would like it.

When I arrived in Pittsburgh in August 2005, research in the Department of Engineering and Public Policy was heavily dominated by issues related to coal, and given the climate change and environmental focus, it focused also on carbon capture and sequestration. Indeed, many of the dissertations in EPP between 2004 and 2006 were on coal-related issues. This made sense at the time, since coal accounted for a high percentage of both the U.S. national and Pennsylvania's electricity generation. Furthermore, much of the work in the department focused on U.S.-related issues.

There were a few exceptions. Jay Apt was already then working on wind- and solar-related issues, as he describes in Chapter 10. Costa Samaras and Kyle Misterling were already working with Granger Morgan on life cycle emissions from electrified vehicles as compared to conventional gasoline vehicles (more on that in Chapter 14). Importantly, Lester Lave was then also co-chairing with Maxine Savitz a National Academy of Sciences report on the Real Prospects for Energy Efficiency in the United States.

At the beginning, I wanted to work on changing people's behaviors. Granger and I had a protracted argument about just how feasible that would be and whether it would be a viable topic for a PhD thesis. I also wanted to build a decision tool using energy-efficiency supply curves or carbon-abatement supply curves to guide policy decisions but found that the data to build that tool was lacking.

Granger had gotten interested in the potential for solid-state lighting (SSL) based on light-emitting diodes (LEDs) as a potential emerging technology that could reduce energy consumption and thus greenhouse gas emissions. It looked like the technology would become cost-effective in the future, but a lot of uncertainty prevailed on how upfront costs, efficiency, light quality, and public acceptance would evolve in the next few years. Lighting accounted for nearly 20% of overall U.S. electricity consumption and 18% of U.S. residential electricity consumption. A transition to alternative energy-efficient technologies could provide an important decarbonization strategy.

We struck a deal: we would look at the LEDs issue first to get some serious research going, and then I would have the green light (pun intended) to study the energy-efficiency potential for the U.S. residential sector.

## Solid-State Lighting

At the time, a 60W-equivalent LED that could be used in a desk lamp, for example, was costing the modest sum of $150 (in 2005 dollars). It was an ugly piece of technology. It wasn't a sleek design, and the potential for market adoption was questionable. We acquired one of these bulbs, which I still keep and bring to classes when I introduce the concepts of technological learning and consumer adoption.

We were surprised about how little was known at the time about the shape and pace of diffusion of new emerging energy technologies as well as how the costs and performance would likely evolve. In this work, we performed an extensive review of the literature on lighting quality and technologies, and we performed estimates of the cost of ownership for LEDs when compared to other lighting technologies. This included reviewing the estimates from performance of LEDs at the time as well as projections for the future (see Figures 8.1 and 8.2).

At the time, Granger's son, Fritz Morgan, was the CTO at Color Kinetics, a firm that was developing LED solutions. He joined us as co-author, providing important technical expertise. In addition, Kevin Dowling, also at Color Kinetics, provided key advice. Importantly, we argued that given the likely decreases in costs and improvements in performance, the levelized annual costs (i.e., the annualized cost of ownership) for white solid-state lighting would be lower than incandescent *and* lower than compact fluorescent lamps (CFLs) by 2012, even if the lamps were used just 2 hours per day and a high implicit discount rate was used to represent the impatience of consumers to recoup the costs. Projections are generally always wrong (see more on this in Chapter 15 regarding some of the work our group did on energy projections), but in this case we were right. The article we wrote made the cover of *The Proceedings of the IEEE*.

Around this time, Lester Lave and Maxine Savitz were chairing the important National Academies report on the Real Prospects for Energy Efficiency in the United States, and they featured quite a bit of this work on LEDs in their report (see Figure 8.3).

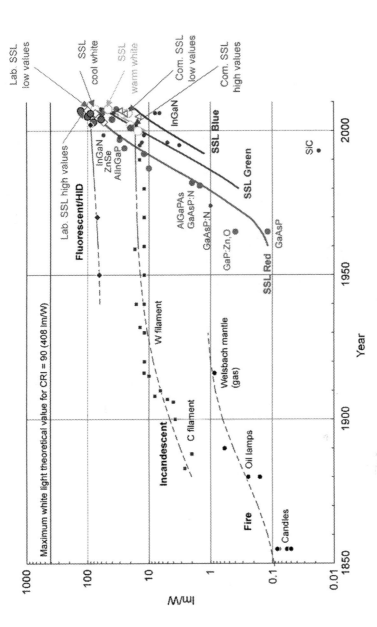

**FIGURE 8.1** Efficiencies of selected lighting technologies between 1850 and 2006, together with future projections. Figure from J. Y. Tsao, Sandia National Laboratory, with added values for SSL technologies from a review I did at the time.

Azevedo, I. L., Morgan, M. G., & Morgan, F. (2009). The transition to solid-state lighting. *Proceedings of the IEEE, 97*(3), 481–510.

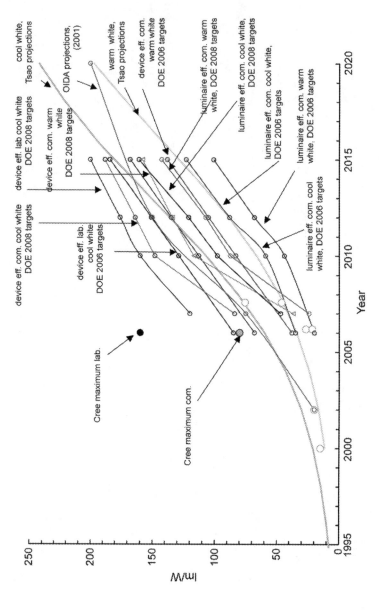

**FIGURE 8.2** Projections of how cold and warm white LEDs would evolve at the time of my research.

Sources: OIDA, DOE, and Tsao. For details, see Azevedo, 2009.

Azevedo, I. L., Morgan, M. G., & Morgan, F. (2009). The transition to solid-state lighting. *Proceedings of the IEEE, 97*(3), 481–510.

### BOX 1.2 *The Benefits of More Efficient Products: An Illustration*

Appendix E of this report provides information on how to calculate the net costs and benefits of energy savings. The figure below illustrates that the overall energy efficiency of providing light using incandescent lamps—starting from the burning of coal to produce electricity and continuing through to the production of visible light—is about 1.3 percent: about two-thirds of the energy in the coal is lost in generating electricity, about 9 percent is lost in transmitting and distributing the electricity, and an incandescent lightbulb's efficiency in transforming electricity to visible light is only 4 percent (Tsao et al., 2009).

In comparison, compact fluorescent lamps (CFLs) are about four times as efficient in transforming electricity to light as is an incandescent lamp (Azevedo et al., 2009; Tsao et al., 2009). Across the residential, commercial, and industrial sectors, a switch from incandescent lighting to CFLs today would save nearly 6 percent of the total amount of electricity generated in the United States today.[1] With further R&D, solid-state lamps (light-emitting diodes, LEDs) are expected to become 10 times as efficient as an incandescent lamp (Azevedo et al., 2009; Tsao et al., 2009).[2]

Across the residential, commercial, and industrial sectors, a switch from incandescent lighting to CFLs today would save approximately 228 terawatt-hours (TWh) of electricity per year relative to today's consumption, or nearly 6 percent of the total amount of electricity generated in the United States.

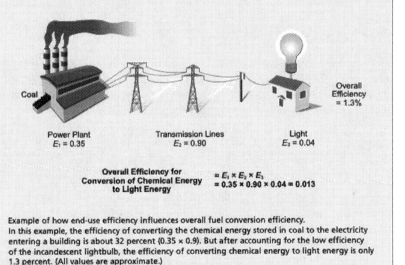

Example of how end-use efficiency influences overall fuel conversion efficiency. In this example, the efficiency of converting the chemical energy stored in coal to the electricity entering a building is about 32 percent (0.35 × 0.9). But after accounting for the low efficiency of the incandescent lightbulb, the efficiency of converting chemical energy to light energy is only 1.3 percent. (All values are approximate.)
Source: Adapted and updated from Hinrichs and Kleinbach, 2006.

**FIGURE 8.3** Box on the benefits of efficiency from the 2010 National Academies Report.

Republished with permission of The National Academies Press, from *Real Prospects for Energy Efficiency in the United States*, America's Energy Future Energy Efficiency Technologies Subcommittee, National Academy of Sciences, National Academy of Engineering, National Research Council, 2009; permission conveyed through Copyright Clearance Center, Inc.

## Energy-Efficiency Supply Curves

I was very keen to understand the potential for energy efficiency and behavioral strategies in the residential sector. The stars aligned in 2007 when economists Karen Palmer and Dallas Burtraw at Resources for the Future (RFF) became interested in understanding the potential for energy efficiency in the United States, and to develop further the portion of the HAIKU model from Resources for the Future that represents energy efficiency. They had a summer internship position to work on these issues, so I went off to Washington D.C. for the summer of 2007. It was a wonderful and very collegial experience as I learned a lot about how economists address issues related on climate and energy policy. The collaboration was productive, and in the following summer I was invited to join again as a summer intern.

I continued to develop the work on energy efficiency supply curves as part of my PhD work, and this resulted in a publication in *Environmental Science & Technology*. We modeled the stock of equipment in the U.S. sector, as well as scenarios for the future composition of that stock. We then modeled the potential for energy and emissions reductions across different types of policies and for different objective functions, as shown in Figure 8.4.

## DC for Buildings

In 2008, a lively and bright young woman named Brinda Thomas joined the PhD program in Engineering and Public Policy to work with Granger and me on demand-side issues. I become her primary advisor. She had an undergrad degree in physics from Stanford University and a deep interest in interdisciplinary research. We delved into an issue that had been debated since the beginnings of electrification.

As we wrote in the introduction of our paper in *Energy Policy* in 2012:

> In 1891, as the "Battle of the Currents" was coming to a close, the board for the Chicago World's Fair received two bids to illuminate the world's first all-electric fair: General Electric proposed a $1.8 million (later reduced to $554,000) direct current (DC) generator and distribution network, while the Westinghouse Electric Company submitted the winning bid of $399,000 for an alternating current (AC) system (all costs in 1891 dollars). The years that followed saw the decline of Thomas Edison's pioneering 110 V DC distribution systems. AC transmission and distribution became standard because the AC transformer made it possible to step voltage up for long distance power transfer and then back down for end use. High voltage AC achieved much greater efficiency for electric power transmission than low voltage DC, since resistive power losses grow as the square of the current, while the amount of power transferred is proportional to the product of voltage and current. In the early 20th century, high voltage DC transmission was not possible due to the lack of a DC transformer.

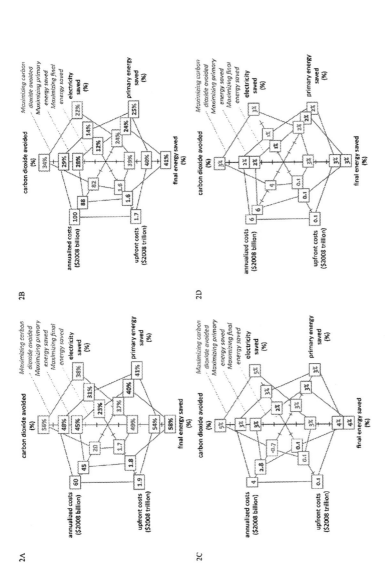

**FIGURE 8.4** Results for the year 2009 from several simulations using my Regional Residential Energy Efficiency Model (RREEM) model, assuming a discount rate of 7%. Each figure shows the results for optimizations that would maximize either carbon dioxide avoided, primary energy saved, or final energy saved, so that the maximum technological potential is obtained in each case. Further details can be found in Azevedo et al. 2013.

Lima Azevedo, I., Morgan, M. G., Palmer, K., & Lave, L. B. (2013). Reducing US residential energy use and CO2 emissions: How much, how soon, and at what cost? *Environmental Science & Technology, 47*(6), 2502–2511.

Together, we were wondering if it would be cost-effective to consider lighting systems for commercial buildings powered by DC circuits and LEDs, namely when the source of power would also be generated in DC. The use of DC in buildings could reduce or eliminate the proliferation of power supplies that convert AC grid power to various DC voltages. Furthermore, if a DC source like solar PV was used, one could avoid the DC-to-AC inverter altogether, improving the efficiency of the system. Figure 8.5 shows the systems that we considered.

Using DOE performance targets for LEDs and solar PV, we concluded that before 2012, a system with LEDs and solar PV using DC circuits would be the system with the lowest costs (see Figure 8.5). However, design, installation, permitting, and regulatory barriers could prevent this solution from materializing.

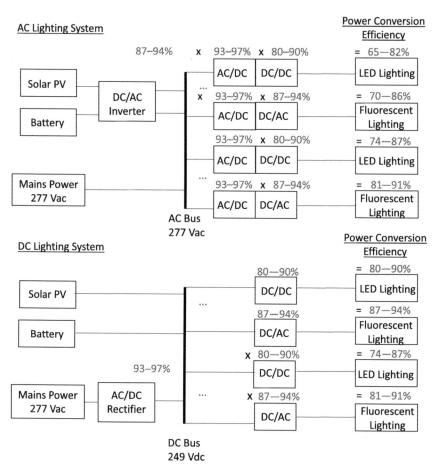

**FIGURE 8.5** A schematic diagram of the commercial office building lighting system of the sort that we analyzed to compare AC with DC circuits for lighting. Figure from Thomas et al., 2012.

Thomas, B. A., Azevedo, I. L., & Morgan, G. (2012). Edison Revisited: Should we use DC circuits for lighting in commercial buildings?. *Energy Policy, 45,* 399–411.

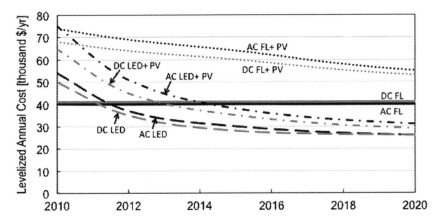

**FIGURE 8.6** Projections of the cost of grid-connected AC versus DC fluorescent and LED lighting systems. Figure from Thomas et al., 2012.

Thomas, B. A., Azevedo, I. L., & Morgan, G. (2012). Edison Revisited: Should we use DC circuits for lighting in commercial buildings?. *Energy Policy, 45*, 399–411.

Brinda went on to work for GE as a lead data analyst when she finished her PhD in 2014 and then moved to Tesla, then Whisker Labs as a staff data scientist, followed by being a data scientist at Stitch Fix. More recently, she joined PG&E as a Principal Data Scientist and continues to also work on sustainable energy related issues in her role as a community advisory committee member for the East Bay Community Energy.

## Consumer Choices for Energy-Efficient Technologies

In 2010, a shy Korean student called Jihoon Min joined EPP to work with me on efficiency-related issues after an undergraduate degree in Korea University and MESc in Environmental Economics at Yale. While extremely shy, Jihoon had an incredible sense of humor, and once you allowed him to shine, the conversation was incredible. For example, after his PhD, he joined the International Institute for Applied System Analysis (IIASA) in Austria as a Research Scholar. In a phone call where we were catching up, he joked on how funny it was that IIASA was actually locked and closed during the weekends and that if it were up to him, he would rather be there and work since at home, a crying infant with poor sleeping habits was waiting for him. Jihoon has incredible modeling skills, and I quickly learned that as we brainstormed ideas and settled on a research goal, he was able to quickly turn it into a lovely piece of analysis. He was very versatile in his modeling skills, being well equipped to handle both social sciences and engineering methods (very much in the vein of many of our EPP PhD students).

Jihoon tackled two distinct issues related to lighting in his PhD work. The first was to try to understand how people would choose different lighting technologies based on their cost, lighting appearance, lifetime, and efficiency. More specifically, we wanted to understand what difference it made to provide information about the lamp's annual operating cost to consumers – a piece of information that had recently been mandated in lighting labels federally. We used a choice-based conjoint field experiment and estimated discrete choice models from these data. We found that providing estimated annual cost information to consumers reduced their implicit discount rate by a factor of five – i.e., it could be a very important mechanism to induce the adoption of energy-efficient lighting technologies.

Jihoon and I then turned to a different question. Could it be the case that the energy savings and $CO_2$ emissions reductions would be greatly reduced when transitioned to more efficient lighting because as one would adopt a more efficient light, less heat was provided in the winter, which would need to be now provided by the heating system? Would this effect be balanced out by the fact that in the summer, less AC would be needed to cool a house? To understand this issue, we modeled the net energy consumption, $CO_{2e}$ emissions, and savings in energy bills for single-family detached houses across the U.S. We found that these heating and cooling effects from more efficient lighting could undermine up to 40% of originally intended primary energy savings, erode anticipated carbon savings completely, or lead to 30% less household monetary savings than intended. However, the size of the effect is dependent on the climate of the region and the types of technologies used for heating and cooling. The effect was generally small, corresponding at most to 1% of either total emissions or energy consumption by a house.

## Technical Details for Chapter 8

The report of the National Academies committee on energy efficiency that Lester Lave co-chaired with Maxine Savitz is:

> National Research Council. (2010). *Real prospects for energy efficiency in the United States.* National Academies Press, 330pp.

Our work on LEDs is reported in:

> Azevedo, I. L., Morgan, M. G., & Morgan, F. M. (2009). The transition to solid-state lighting. *The Proceedings of the IEEE, 97*(3), 481–510.

**Abstract:** Lighting constitutes more than 20% of total U.S. electricity consumption, a similar fraction in the European Union, and an even higher fraction in many developing countries. Because many current lighting technologies are highly inefficient, improved technologies for lighting hold great potential for energy savings and for reducing associated greenhouse gas emissions. Solid-state lighting shows great promise as a source of efficient, affordable, color-balanced white light. Indeed, assuming market discount rates, engineering-economic analysis demonstrates that white solid-state lighting already has a lower levelized annual cost (LAC) than incandescent bulbs. The LAC for white solid-state lighting will be lower than that of the most efficient fluorescent bulbs by the end of this decade. However, a large literature indicates that households do not make their decisions in terms of simple expected economic value. After a review of the technology, we compare the electricity consumption, carbon emissions, and cost-effectiveness of current lighting technologies, accounting for expected performance evolution through 2015. We then simulate the lighting electricity consumption and implicit greenhouse gases emissions for the U.S. residential and commercial sectors through 2015 under different policy scenarios: voluntary solid-state lighting adoption, implementation of lighting standards in new construction, and rebate programs or equivalent subsidies. Finally, we provide a measure of cost-effectiveness for solid-state lighting in the context of other climate change abatement policies.

Here are papers on energy-efficiency supply curves and price elasticity:

Azevedo, I. L., Morgan, M. G., Palmer, K., & Lave, L. B. (2013). Reducing U.S. residential energy use and $CO_2$ emissions: How much, how soon, and at what cost?. *Environmental Science & Technology*, *47*(6), 2502–2511.

**Abstract:** There is growing interest in reducing energy use and emissions of carbon dioxide from the residential sector by deploying cost-effectiveness energy efficiency measures. However, there is still large uncertainty about the magnitude of the reductions that could be achieved by pursuing different energy efficiency measures across the nation. Using detailed estimates of the current inventory and performance of major appliances in U.S. homes, we model the cost, energy, and $CO_2$ emissions reduction if they were replaced with alternatives that consume less energy or emit less $CO_2$. We explore trade-offs between reducing $CO_2$, reducing primary or final energy, or electricity consumption. We explore switching between electricity and direct fuel use, and among fuels. The trade-offs between different energy efficiency policy goals, as well as the environmental metrics used, are important but have been largely unexplored by previous energy modelers and policy-makers. We find that overnight replacement of the full stock of major residential appliances sets an upper bound of just over 710 Å~ $10_6$ tonnes/year of $CO_2$ or a 56% reduction from baseline residential emissions. However, a policy designed instead to minimize primary energy consumption instead of $CO_2$ emissions will achieve a 48% reduction in annual carbon dioxide emissions from the nine largest energy consuming residential end-uses. Thus, we explore the uncertainty regarding the main assumptions and different policy goals in a detailed sensitivity analysis.

Azevedo, I. M. L., Morgan, M. G., & Lave, L. (2011). Residential and regional electricity consumption in the U.S. and EU: How much will higher prices reduce $CO_2$ emissions?. *The Electricity Journal*, *24*(1), 21–29.

**Abstract:** Results of our analysis suggest that, given the price-inelastic behavior in both the U.S. and EU regions, public policies aimed at fostering a transition to a more sustainable energy system in order to address the climate change challenge will require more than an increase in electricity retail price if they are to induce needed conservation efforts and the adoption of more efficient technologies by households.

Some of our other papers on energy efficiency include:

Thomas, B. A., Azevedo, I. L., & Morgan, M. G. (2012). Edison revisited: Should we use DC circuits for lighting in commercial buildings?. *Energy Policy*, *45*, 399–411.

**Abstract:** We examine the economic feasibility of using dedicated DC circuits to operate lighting in commercial buildings. We compare light-emitting diodes (LEDs) and fluorescents that are powered by either a central DC power supply or traditional AC grid electricity, with and without solar photovoltaics (PV) and battery back-up. Using

DOE performance targets for LEDs and solar PV, we find that by 2012 LEDs have the lowest levelized annualized cost (LAC). If a DC voltage standard were developed, so that each LED fixture's driver could be eliminated, LACs could decrease, on average, by 5% compared to AC LEDs with a driver in each fixture. DC circuits in grid-connected PV-powered LED lighting systems can lower the total unsubsidized capital costs by 4–21% and LACs by 2–21% compared to AC grid-connected PV LEDs. Grid-connected PV LEDs may match the LAC of grid-powered fluorescents by 2013. This outcome depends more on manufacturers' ability to produce LEDs that follow DOE's lamp production cost and efficacy targets, than on reducing power electronics costs for DC building circuits and voltage standardization. Further work is needed to better understand potential safety risks with DC distribution and to remove design, installation, permitting, and regulatory barriers.

Min, J., Azevedo, I. L., Michalek, J., & Bruine de Bruin, W. (2014). Labeling energy cost on light bulbs lowers implicit discount rates. *Ecological Economics*, *97*, 42–50.

**Abstract:** Lighting accounts for nearly 20% of overall U.S. electricity consumption and 18% of U.S. residential electricity consumption. A transition to alternative energy-efficient technologies could reduce this energy consumption considerably. To quantify the influence of factors that drive consumer choices for light bulbs, we conducted a choice-based conjoint field experiment with 183 participants. We estimated discrete choice models from the data, and found that politically liberal consumers have a stronger preference for compact fluorescent lighting technology and for low energy consumption. Greater willingness to pay for lower energy consumption and longer life was observed in conditions where estimated operating cost information was provided. Providing estimated annual cost information to consumers reduced their implicit discount rate by a factor of five, lowering barriers to adoption of energy efficient alternatives with higher up-front costs; however, even with cost information provided, consumers continued to use implicit discount rates of around 100%, which is larger than that experienced for other energy technologies.

Min, J., Azevedo, I. L., & Hakkarainen, P. (2015). Assessing regional differences in lighting heat replacement effects in residential buildings across the United States. *Applied Energy*, *141*, 12–18.

**Abstract:** Lighting accounts for 19% of total U.S. electricity consumption and 6% of carbon dioxide equivalent ($CO_{2e}$) emissions. Existing technologies, such as compact fluorescent lamps and light emitting diodes, can substitute low-efficiency technologies such as incandescent lamps, while saving energy and reducing energy bills to consumers. For that reason, lighting efficiency goals have been emphasized in U.S. energy efficiency policies. However, incandescent bulbs release up to 95% of input energy as heat, impacting the overall building energy consumption: replacing them increases demands for heating service that needs to be provided by the heating systems and decreases demands for cooling service that needs to be provided by the cooling systems. This work investigates the net energy consumption, $CO_{2e}$ emissions, and savings in energy bills for single-family detached houses across the U.S. as one adopts more efficient lighting systems. In some regions, these heating and cooling effects from more efficient lighting can undermine up to 40% of originally intended primary energy savings, erode anticipated carbon savings completely, or lead to 30% less household monetary savings than intended. The size of the effect depends on regional factors such as climate, technologies used for heating and cooling, electricity fuel mix, emissions factors, and electricity prices. However, we also find that for moderate lighting efficiency interventions, the overall effect is small in magnitude, corresponding at most to 1% of either total emissions or of energy consumption by a house.

**Some notable writings by authors not affiliated with our Centers:**

Brown, M. A., Southworth, F., & Stovall, T. K. (2005). *Towards a climate-friendly built environment*. Pew Center on Global Climate Change, 78pp.

Gillingham, K., Newell, R. G., & Palmer, K. (2009). Energy efficiency economics and policy. *Annual Review of Resource Economics, 1*(1), 597–620.

National Academy of Sciences, National Academy of Engineering, and National Research Council. (2010). *Real Prospects for Energy Efficiency in the United States*. National Academies Press.

National Research Council. (2013). *Assessment of advanced solid-state lighting*. National Academies Press.

Rosenfeld, A. H. (1999). The art of energy efficiency: Protecting the environment with better technology. *Annual Review of Energy and the Environment, 24*(1), 33–82.

Rubin, E. S., Cooper, R. N., Frosch, R. A., Lee, T. H., Marland, G., Rosenfeld, A. H., & Stine, D. D. (1992). Realistic mitigation options for global warming. *Science, 257*(5067), 148–266.

# Note

1  Obelix is a character in the French comic book series *Asterix*.

# 9

# ENERGY REBOUND

*Granger Morgan and Brinda Thomas*

As previous chapters have noted, reducing global greenhouse gas emissions to limit warming to 2 or 1.5 °C and avoid catastrophic climate change is a monumental task that will require that virtually all possible strategies for emission abatement to be simultaneously deployed. One of those strategies is to increase energy efficiency throughout the economy, but especially in the electricity, transportation, and buildings industries. As countries have worked to promote energy efficiency, some researchers and journalists have argued that such efforts are not as valuable as they appear at first glance. For example, they have suggested that if people buy cars with better gas mileage and save money, there is a good chance they will use the savings to drive more. If they have to spend less money to heat their home, they might use the money they saved to burn a lot of jet fuel to fly halfway around the world to nice vacation spots. These ideas are called the "rebound effect."[1]

These rebound effect arguments echo an older concept called the Jevons Paradox, named after William Stanley Jevons, who in 1865 observed that technology improvements that increased the efficiency of coal consumption led to increased use of coal in industries that adopted them, rather than reducing consumption. If Jevons' paradox applied to energy efficiency investments of all kinds and not just resource use, is it possible that energy consumption could actually increase to a level higher than before the efficiency investments were made?

As an EPP PhD student, Brinda Thomas, working with Inês Azevedo, decided to get to the bottom of these arguments, so we conducted extensive reviews of the literature, commissioned a number of experts to write brief "think pieces," and then organized two international workshops, one in the U.S. and one in Germany to find out what people meant when they talked about rebound, and what was known about how big it might be. We found that when using the word, "rebound" in the context of energy efficiency, different people meant different things. We also

DOI: 10.4324/9781003330226-9

found that in most cases the amount of rebound was pretty modest so that in general energy efficiency can achieve real and substantial savings and reductions in the emission of climate changing greenhouse gases.

Through these workshops, we realized that when people talk about rebound, they are often talking about four rather different ideas. Without first sorting and classifying studies in terms of these different ideas, it is not easy to compare study results or determine when and whether rebound effects are large or small. The four different kinds of rebound that people talk about, often without being very clear about which one they have in mind, are:

1.  **Direct rebound effect:** This is the rebound effect that is most commonly discussed and easiest to understand. This is what happens if someone buys a more efficient hybrid electric car or electric car instead of a gasoline one and then drives it more because of the lower fuel costs. This type of effect could happen with any type of energy service – for example, leaving energy-efficient LED lights on longer because they cost less to use than the energy-guzzling incandescent bulbs, or doing more loads of laundry in a more efficient washer and dryer because of their lower energy costs.

2.  **Indirect rebound effect:** This type of rebound effect has to do with spending energy savings from an energy-efficient purchase, whether a more efficient car, home appliance, or light bulb. Any good or service in the economy requires energy for its production or releases greenhouse gas emissions through its life cycle. The additional income that is freed up by saving energy costs can be used for other energy- or carbon-intensive consumption. For example, the income gained by installing an efficient furnace and insulating one's house could be spent on additional air travel, leading to an overall increase in GHG emissions.

    The direct and indirect rebound effects can be related to two familiar economic effects:

    **Substitution effect:** Efficiency gains in a particular energy service lead to a shift into more consumption of that service and out of other goods and services.

**FIGURE 9.1** Two views of the workshop we ran in Washington, DC in order to sort out the very different definitions and ways of thinking that people use when they talk about energy rebound. We ran a second similar workshop in Germany.

Images taken by Granger Morgan.

**Income effect:** Efficiency gains in a particular energy service make available additional income from energy-cost savings that can be used for greater consumption overall, in both energy and other goods and services.

After accounting for substitution and income effects, the direct rebound effect focuses on net changes in energy-service consumption, while the indirect rebound effect focuses on the net changes in other non-energy goods or service consumption.

3. **Changes in consumption patterns:** Substitution and income effects may lead to overall changes in consumption patterns. For example, energy-efficiency gains lead to changes in behavior (such as buying more frozen food when energy-efficient freezers are available). Measurements of the direct and indirect rebound effects would also include these changes in consumption patterns.

4. **Economy-wide rebound effect:** Energy-efficiency investments lead to changes in prices of goods and services, which lead to structural changes in the economy, resulting in a new equilibrium in the consumption of energy and other goods and services.

In addition to finding that the word "rebound" is used to describe different things by different people, we also found that in most cases the amount of rebound was pretty modest, so, in general, energy efficiency can achieve real and substantial savings and reductions in the emission of climate-changing greenhouse gases.

How people respond to energy-efficiency technologies and policies can play a big role in whether and how energy demand is affected, and whether there is significant rebound. A few of the things that shape people's energy and other consumption include:

- How people perceive and respond to prices (they may not even notice small changes).
- Whether, and how much, people engage in conspicuous consumption (buying and doing things to "keep up with the Joneses").
- People's attitudes and values toward things like the environment (buying "green," "consuming less," etc.).
- How well people understand and make proper use of energy-efficient devices (complicated "smart thermostats").
- Differences in people's lifestyles (travel, food, clothes, etc.).
- People's social and personal norms and habits.
- Whether, as economists often assume, people always want more. For many services, there's a ceiling to how much someone consumes, and demand saturates.

The quantitative range of estimates of the energy rebound effect varied greatly in the literature but were generally less than 100%, meaning that energy-efficiency policies did lead to energy reductions, although not as much as planners may originally anticipate without considering the rebound effect (Figure 9.2).

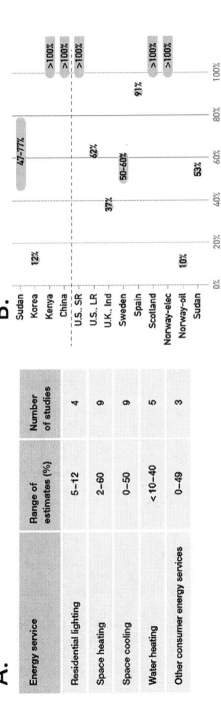

**FIGURE 9.2** **A. (left)** Range of estimates of direct rebound effects. **B. (right)** Economy-wide rebound estimates for different countries. Images are from IRGC, 2013 (where the literature on which these results are based is also cited).

International Risk Governance Council (IRGC). The Rebound Effect: Implications of Consumer Behaviour for Robust Energy Policies: A review of the literature on the rebound effect in energy efficiency and report from expert workshops, 35pp. Available online at: https://irgc.org/wp-content/uploads/2018/09/IRGC_ReboundEffect-FINAL.pdf

Based on the outcomes of the two workshops we ran and the several "think pieces" that participants had prepared, in a report published by the International Risk Governance Council (IRGC), Inês and we summarized a number of policy implications for coping with rebound effects:

1.  Economists often propose reducing the negative side effects (a.k.a., externalities in economic jargon) of energy use by increasing energy prices by introducing energy taxes, carbon prices, etc. Other strategies include "energy budgets" or "caps" placed on energy consumption. If all negative side effects could be incorporated in the energy price, the only rebound effects would be those that improve welfare.
2.  Rebound effects are neglected or insufficiently included in many energy scenarios and models. This omission raises the risk of underestimating future energy demand. Including rebound effects is difficult because knowledge in specific contexts is still uncertain. Some scenarios and models take into account rebound effects implicitly but use a different wording. Here, clear definitions and a common wording are needed (Naegler and Vögele "think pieces").
3.  Among middle- and upper-income consumers in the U.S. who display lower price elasticities, direct rebound effects appear to be modest, falling in the range of 3% (in the short run) to 22% (in the long run) for transportation (Small and van Dender, 2005) and 29–37% for residential electricity demand in the U.S. (Reiss and White, 2005).
4.  There is very little evidence of rebound effects exceeding 100% (backfire) for household energy-efficiency investments in developed countries.
5.  Rebound effects can be large in the developing world and among low-income groups, and could be large in the production sector of the economy; there has been too little study of these groups ("think pieces" by Chi, Polimeni and Thomas). Especially in developing countries and low-income groups, it is crucial to gain a better understanding of the extent to which rebound effects lead to enhanced individual well-being and desirable socioeconomic or macroeconomic co-benefits ("think piece" by Golde).
6.  It is important that policymakers understand that their policy strategies to increase the energy efficiency of goods and services may not be as effective as simple direct analysis suggests. At the same time, care should be taken that energy-efficiency policies are not called into question in general. Energy-efficiency policies could be improved by explicitly taking into account rebound effects.
7.  In situations in which empirical analysis suggests that rebound effects are greater than a few percent, these effects should be considered in the design of policy programs.
8.  The UK systematically takes into account direct rebound effects (i.e., people increase the temperature of their homes due to financial savings from installed energy-efficiency measures in building designs). The U.S. Environmental Protection Agency (EPA) also assumes a 10% rebound effect in vehicle miles

traveled in assessing the regulatory impacts of fuel economy standards (2009). Elsewhere, rebound effects are generally neglected in policymaking. As more countries begin to pursue serious energy-efficiency policies, the consideration of rebound effects will become increasingly important.

9.   For local or regional energy-efficiency policies, such as utility-efficiency programs and state-level energy-efficiency resource standards, energy-savings estimates from engineering estimates should be reduced by the estimate of the direct rebound effect. Energy-demand changes from indirect and economy-wide rebound effects are not yet attributable at less than national scales.

10.  For national energy-efficiency policies, such as appliance and vehicle standards and rebates and tax credits for energy efficiency, engineering estimates of energy savings should be reduced by estimates of the direct and indirect rebound effect. However, it is also important to account for the improvement in household well-being or increase in firms' profits that is possible from re-spending energy-cost savings for greater consumption or production.

11.  Intervention strategies, such as the introduction of feedback mechanisms on energy consumption (smart metering) or contracting models for heating, etc., promise to foster energy-conservation behavior. However, in order for them to lead to significant changes in consumption, multiple intervention strategies must be applied.

## Technical Details for Chapter 9

We ran the two workshops on the rebound effect in collaboration with the Swiss-based International Risk Governance Council. The report we produced with them is:

IRGC (2013). The Rebound Effect: Implications of Consumer Behaviour for Robust Energy Policies: A review of the literature on the rebound effect in energy efficiency and report from expert workshops), 35pp.
Available online at: https://irgc.org/wp-content/uploads/2018/09/IRGC_ReboundEff ect-FINAL.pdf

Here are the two papers we wrote in order to better define energy rebound and suggest improved strategies for assessing its impact:

Thomas, B. A., & Azevedo, I. L. (2013). Estimating direct and indirect rebound effects for U.S. households with input–output analysis Part 1: Theoretical framework. *Ecological Economics, 86,* 199–210.

**Abstract:** This is the first part of a two-part paper providing an analytical model of the indirect rebound effect, given a direct rebound estimate, that integrates consumer demand theory with the embodied energy of household spending from environmentally extended input–output analysis. The second part applies the model developed in part one to simulate the direct and indirect rebound for the average U.S. household in terms of primary energy, $CO_{2e}$, $NO_x$, and $SO_2$ emissions and for energy efficiency investments in electricity, natural gas, or gasoline services. Part one provides a critical review of the largely independent economic and industrial ecology literatures on the indirect rebound.

By studying the two-goods case and the *n*-goods case, we demonstrate that the indirect rebound is bounded by the consumer budget constraint, and inversely related to the direct rebound. We also compare the common proportional spending and income elasticity spending assumptions with our model of cross-price elasticities including both substitution and income effects for the indirect rebound. By assuming zero incremental capital costs and the same embodied energy as conventional technologies for efficient appliances, we model an upper bound of the indirect rebound. Future work should also consider the increase in consumer welfare possible through the rebound effect.

Thomas, B. A., & Azevedo, I. L. (2013). Estimating direct and indirect rebound effects for U.S. households with input–output analysis. Part 2: Simulation. *Ecological Economics*, *86*, 188–198.

**Abstract:** This is the second part of a two-part paper that integrates economic and industrial ecology methods to estimate the indirect rebound effect from residential energy efficiency investments. We apply the model developed in part one to simulate the indirect rebound, given an estimate of the direct rebound, using a 2002 environmentally-extended input–output model and the 2004 Consumer Expenditure Survey (in 2002$) for the U.S. We find an indirect rebound of 5–15% in primary energy and $CO_2$ emissions, assuming a 10% direct rebound, depending on the fuel saved with efficiency and household income. The indirect rebound can be as high as 30–40% in $NO_x$ or $SO_2$ emissions for efficiency in natural gas services. The substitution effect modeled in part one is small in most cases, and we discuss appropriate applications for proportional or income elasticity spending assumptions. Large indirect rebound effects occur as the U.S. electric grid becomes less-carbon intensive, in households with large transportation demands, or as energy prices increase. Even in extreme cases, there is limited evidence for backfire, or a rebound effect greater than 100%. Enacting pollution taxes or auctioned permits that internalize the externalities of energy use would ensure that rebound effects unambiguously increase consumers' welfare.

This is the review paper that summarizes what we and others had learned about energy rebound:

Azevedo, I. M. (2014). Consumer end-use energy efficiency and rebound effects. *Annual Review of Environment and Resources*, *39*, 393–418.

**Abstract:** Energy efficiency policies are pursued as a way to provide affordable and sustainable energy services. Efficiency measures that reduce energy service costs will free up resources that can be spent in the form of increased consumption—either of that same good or service or of other goods and services that require energy (and that have associated emissions). This is called the rebound effect. There is still significant ambiguity about how the rebound effect should be defined, how we can measure it, and how we can characterize its uncertainty. Occasionally the debate regarding its importance reemerges, in part because the existing studies are not easily comparable. The scope, region, end-uses, time period of analysis, and drivers for efficiency improvements all differ widely from study to study. As a result, listing one single number for rebound effects would be misleading. Rebound effects are likely to depend on the specific attributes of the policies that trigger the efficiency improvement, but such factors are often ignored. Implications for welfare changes resulting from rebound have also been largely ignored in the literature until recently.

**Some notable writings by authors not affiliated with our Centers:**

Alcott, B. (2005). Jevons' paradox. *Ecological Economics*, *54*(1), 9–21.

Brown, M. A. (2001). Market failures and barriers as a basis for clean energy policies. *Energy Policy*, *29*(14), 1197–1207.

Blumstein, C., Krieg, B., Schipper, L., & York, C. (1980). Overcoming social and institutional barriers to energy conservation. *Energy*, *5*(4), 355–371.

Jaffe, A. B., Stavins, R. N. (1994). The energy efficiency gap: What does it mean? *Energy Policy*, *22*, 804–810.

Jevons, W. S. (1865). *The Coal Question. An Inquiry Concerning the Progress of the Nation, and the Probable Exhaustion of Our Coal-Mines*. Macmillan.

Reiss, P. C. and White, M. W. (2005). *Review of Economic Studies*, *72*(3), 853–883.

Small, K. A., & Van Dender, K. (2005). The effect of improved fuel economy on vehicle miles traveled: Estimating the rebound effect using US state data, 1966–2001.

Sorrell, S. (2009). Jevons' Paradox revisited: The evidence for backfire from improved energy efficiency. *Energy Policy*, *37*(4), 1456–1469.

# Note

1 Portions of the text in this chapter are modified from the IRGC report: The Rebound Effect: Implications of Consumer Behaviour for Robust Energy Policies: A review of the literature on the rebound effect in energy efficiency and report from expert workshops which is available online at: https://irgc.org/wp-content/uploads/2018/09/IRGC_ReboundEffect-FINAL.pdf; and from, Azevedo, I. M. (2014). Consumer end-use energy efficiency and rebound effects. *Annual Review of Environment and Resources*, *39*, 393–418.

# 10

# ENERGY FROM THE WIND AND THE SUN

*Jay Apt*

In the fall of 2003, Katerina Dobesova, a Fulbright Scholar from the Czech Republic, arrived at Carnegie Mellon looking for a problem on which to do her MS degree research. In those days, wind turbines generated 3/10ths of a percent of U.S. electricity from 6 gigawatts of installed capacity (out of a total of roughly 1000 GW of U.S. generator capacity of all types). That was a tiny fraction of what hydro-electric power generated (7.2% that year), and it was even below the generation from geothermal power. Coal generated more than half the nation's power in 2003, emitting oxides of sulfur and nitrogen, mercury, and carbon dioxide.

In 1999, Texas governor George W. Bush had signed a bill that restructured Texas' electric power system and included a renewable portfolio standard (RPS). Industrial customers in Texas had fought to block this standard because they were worried about the effect wind power would have on their electricity bills.[1] The RPS requirements began at the start of 2002, and I thought that it would be interesting to examine the costs of the RPS as a way to meet a goal of lowering both conventional and greenhouse gas pollution. Since Katerina had good under-graduate economics training, we decided to work together on the problem, and we pulled in Lester Lave to keep us straight on the economics.

Over the years, we have found that enthusiastic young graduate students can talk people into giving them data that would be hard or impossible for senior faculty to get. The Texans who ran the renewables market were no exceptions: Katerina was able to learn the costs of each transmission project that served the wind farms and the cost of administering the RPS, as well as public data on the costs associated with the production tax credit and with curtailing a wind farm's output when it was too much for the transmission system.

It turned out that the 7/10ths of a percent of Texas generation provided by wind in 2002 cost about 6 cents per kilowatt-hour (kWh) more than electricity being generated by plants that burned coal and oil when we included in the additional

DOI: 10.4324/9781003330226-10

costs for the wind turbines, new transmission lines, tax subsidies, and the cost of administering the RPS. Because there were markets that put prices on the allowed emissions of $SO_2$, $NO_x$, and mercury, we were able to estimate the cost savings from the reduced pollution from the wind generation and found it was about 1 cent per kWh. That meant that wind was reducing emissions of $CO_2$ at a cost of roughly $55 per metric ton of carbon dioxide. This was one of the first calculations based on observed data of the cost of reducing greenhouse gas pollution from the electric power industry. Our cost estimate was comparable to the very preliminary estimates of the cost of such reduction from adding $CO_2$ capture and sequestration to coal or natural gas plants (see Chapter 11).

If we adjust for inflation, the installed cost of a wind project today is very close to the same as it was in 2002. However, the performance of new wind turbines has improved a lot, so today the average cost of power from wind (as reflected in power purchase agreements) is about 85% of what it was in 2002, again adjusting for inflation.[2] Meanwhile, there are only a handful of coal facilities that can capture $CO_2$, and the cost of the power they produce is much higher than the estimates that were being made a couple of decades ago.

Since we did the early work in Texas, the installed capacity of wind generation has grown dramatically. Today, wind generation facilities in the U.S. are 20 times larger in total gigawatts installed than they were in 2002 and now supply almost 7% of our electricity. As Katerina's work suggested, wind remains a quite reasonable economic choice to avoid pollution, and Katerina went on to earn her PhD in economics.

In the course of looking at the details of the early installations of wind power, we got to know a few companies that were willing to share power output data from their wind turbines. In dealing with policy decisions that are based on underlying technical realities, it is very useful to look at the actual data. Up to that point, only a very few analyses of the output of individual turbines, or the total output of all the turbines in a wind farm, had appeared. Most of those were not in the peer-reviewed literature, which made the values a little suspect. Almost all data in the literature were from devices that measured wind speed (anemometers) rather than from the turbines themselves. With the help of friends I had made in wind companies, in 2007 I published data taken every second for ten days for a single wind power turbine, and for 6- and 10-turbine wind farms. Figure 10.1 shows the data for a single turbine.

My training is in physics. When a physicist looks at a time series like those in Figures 10.1 or 10.2, the question that comes up is: what are the relative contributions of fast changes in the power as opposed to slow ones? If you are listening to a symphony orchestra, the double bass string instrument may be contributing more to the sound than the high-pitched piccolo, or the piccolo may be louder. Today, with modern computers, figuring that out is pretty easy. There is software on any device that plays music (such as your phone) that decomposes the music into low- and high-frequency components and displays the result so that you can boost the bass or treble. A plot with the different frequencies that make up the

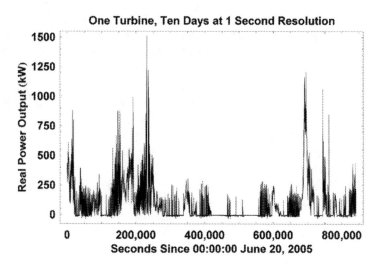

**FIGURE 10.1** Real power output (kW) sampled once every second for a single 1.5 MW turbine at one wind farm for 10 days. The occasional small negative values are due to turbine electrical loads. I had hoped that by combining the output from many turbines this variability would smooth out, but when I combined the output from over a hundred turbines, large and deep fluctuations in the overall power output remained (Figure 10.2).

Apt, J. (2007). The spectrum of power from wind turbines. *Journal of Power Sources*, *169*(2), 369–374.

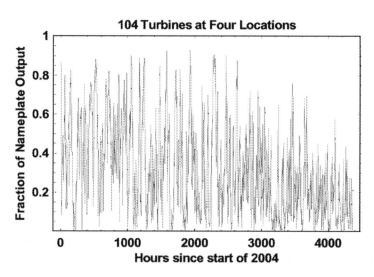

**FIGURE 10.2** Real power output as a percent of installed (nameplate) capacity sampled once an hour for the sum of 104 turbines at four wind farms from January 1 to June 30, 2004.

Apt, J. (2007). The spectrum of power from wind turbines. *Journal of Power Sources*, *169*(2), 369–374.

music on the horizontal axis and strength or amplitude of those frequencies on the vertical axis would show quite different characteristics for symphonic music than for heavy-metal rock music. A plot like this is called a "power spectrum."

Specialists call the mathematical tool that works out how loud the signal is in each frequency band a Fourier transform, named after Joseph Fourier who worked out the details in 1822. Modern software running on the chips in a laptop can do the trick in an instant. So, what happens when we look at the low and high frequencies in wind power? The result is in Figure 10.3.

The first observation we can make from Figure 10.3 is that the amplitude of fluctuations from wind generators is very much smaller at high frequencies than at low frequencies. This characteristic has important consequences for the generators that must be used to compensate for wind's variability. Suppose the plot were flat – the same amplitude at all frequencies (we call that white noise). In this case, wind power would have equally large fluctuations at periods of, say, a minute as it does at periods of several days. This would mean that large numbers of very fast generators or energy storage devices would be required to smooth things out. Nature has kindly arranged things so that the largest variability is at low frequency (the power

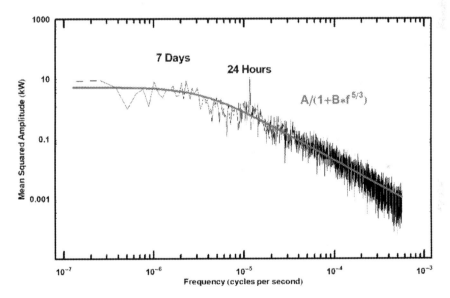

**FIGURE 10.3** Plot of amplitude vs. frequency for 1 year of data sampled at 15-minute time resolution for a 63 MW wind plant in Texas. Both scales are logarithmic; the mean squared size of the variability at 7 days is 10,000 times stronger than the variability at 30 minutes. The data plotted span frequencies corresponding to 3 months on the left to 30 minutes on the right. The peak at a frequency corresponding to 24 hours is because the wind at most locations in the U.S. blows more strongly almost every night than during the day.

Drawn by Jay Apt.

variations at time scales of a minute are many times smaller than those at time scales of a day). Thus, slow-ramping plants (coal or combined-cycle natural gas, for example) can compensate for most of wind's variability.

Second, at frequencies below ~5 x $10^{-6}$ Hz (corresponding to ~2.5 days), the power spectrum of output from wind plants becomes quite flat. The basic physics is that the turbulence that dominates the higher frequencies is replaced at low frequencies with very large-scale phenomena such as the ~5-day planetary wave – things like the large weather systems that move across the country.

Third, there is a remarkably linear region of the power spectrum (exponential in frequency), covering several orders of magnitude in frequency, where the amplitude of the variations in output power of the wind plant decreases with increasing frequency $f$, as $f^{-5/3}$ (smooth line in Figure 10.3). This characteristic is observed in data taken from wind plants in many regions of the world.

By chance, I attended a seminar by Mahesh Bandi, a recently graduated PhD from the University of Pittsburgh, our neighboring university. That happy chance led to an explanation of why the characteristics of wind as shown in Figure 10.3 look as they do. Mahesh is now a tenured professor and a brilliant physicist trained in fluid dynamics. We've traded data and explanations and collaborated on several papers. In a 2017 paper, Mahesh gave the first physical explanation of the frequency spectra that I'd first published a decade before. He explained that the relative weakness of high-frequency fluctuations in wind turbine output power happens because of the large-scale influence of atmospheric turbulence. Power fluctuations in an individual turbine contain signatures of the largest scales of atmospheric turbulence. Mahesh has constructed a solid physical foundation for the data we observe from wind turbines.

In 2007 and 2008, a remarkable group of students began their PhD studies at the department of Engineering and Public Policy at Carnegie Mellon University. Warren Katzenstein had graduated from the excellent engineering program at Harvey Mudd, where he was elected president of the student body. He spent several years using computers to model the flow of air (computational fluid dynamics) at SAIC before applying to grad school. Steve Rose, after getting his MS at Georgia Tech, had been one of the members of the small team that designed the control system for General Electric's large new 1.5-megawatt wind turbine. He walked to Pittsburgh from GE's facility in South Carolina, hiking a portion of the Appalachian Trail. Emily Fertig, a geoscience major from Williams College, had edited an oil and gas journal in Moscow after working on organic farms in Europe. Brandon Mauch, with mechanical engineering degrees from Kansas and Wisconsin, had been a security engineer with the State Department in Serbia and Senegal after teaching math and physics in Tanzania. Working together with me, this crew of outstanding students discovered crucial details of how wind power can be integrated into the power grid.

One of the important uncertainties surrounding the deployment of wind in those early years was the degree to which the fluctuations in wind turbine output might be smoothed by adding together the output of a number of different wind

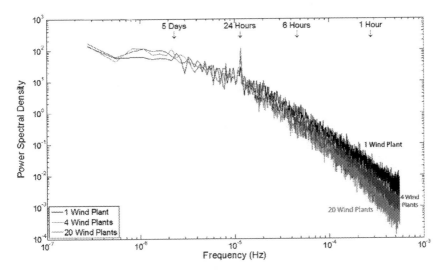

**FIGURE 10.4** The size of wind turbine generator output power fluctuations (on the vertical axis) vs. the frequency of the fluctuations (horizontal axis). The output of a single wind plant is in the top line. When 4 plants' outputs are summed together (middle line), there is some reduction at high frequencies (fast fluctuations), and there is more reduction when 20 plants are summed together (bottom line). But the large variability on time scales of a few hours to a few days isn't reduced much by geographic smoothing.

Katzenstein, W., Fertig, E., & Apt, J. (2010). The variability of interconnected wind plants. *Energy Policy, 38*(8), 4400–4410.

farms that were located far apart. In 2009, Warren, Emily, and I worked with the largest wind plant operator in the U.S. to get data from 20 wind farms in Texas measured every 15 minutes. These plants contained roughly 600 individual wind turbines spread out over 500 km (300 miles).

We found two things that changed the way people thought about geographic smoothing. First, most of the smoothing is of the very fast fluctuations in wind power (those at high frequencies). The large, deep, and slow fluctuations at time scales of a few hours to a few days are not reduced very much at all by combining the outputs of widely separated wind turbines. That means that the strongest variability in wind power can't be smoothed that way (Figure 10.4). Second, connecting more and more wind farms together quickly reaches a point of diminishing returns. At any time scale, the benefits from interconnecting dispersed wind turbines quickly disappear after about a half-dozen wind plants have been added together (Figure 10.5).

We expanded our team with a freshly graduated PhD, Paulina Jaramillo; and with funding from the Doris Duke Charitable Trust, the Richard King Mellon Foundation, the U.S. Department of Energy, and our NSF Center for Climate and Energy Decision Making, we began a project to take an objective look at the technical, regulatory, and policy requirements and implications of high levels of

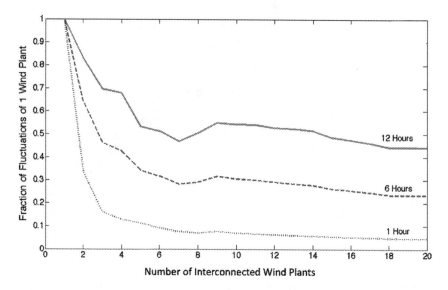

**FIGURE 10.5** The point of diminishing returns for reducing the variability of wind power at periods of 1, 6, and 12 hours is quickly reached (after about 6 plants have been connected together). Ninety-five percent of the fluctuations at a period of one hour are reduced by geographic smoothing. But only half of the much more important deep fluctuations at a period of 12 hours are reduced, even when 20 plants are connected.

Katzenstein, W., Fertig, E., & Apt, J. (2010). The variability of interconnected wind plants. *Energy Policy*, *38*(8), 4400–4410.

renewable electricity in the United States. A team with technical and policy experts at Carnegie Mellon University, the University of Vermont, Vermont Law School, and the Washington, DC law firm of Van Ness Feldman was formed to survey previous research on the integration of variable renewable electricity and undertake a series of engineering and economic analyses that could provide the basis for recommending the regulatory and policy measures needed to accommodate a large penetration of renewable generation. Paulina became the executive director and I the director of the RenewElec (renewable electricity) project.

We began by writing a series of 15 white papers on regulatory, legislative, and technical aspects of renewable power generation, including a preliminary review of previous integration studies. These were provided to participants in an integration and policy workshop held at Carnegie Mellon University in late 2010. Then, a family of models was developed to assess the interactions between renewable generators and the power grid that the workshop identified as most critical. A first series of policy materials was then prepared. These included comments made to the Federal Energy Regulatory Commission's Notice of Proposed Rule Making on Variable Energy Resources. A second integration and policy workshop was held in the fall of 2011 to ensure that operational experience from systems with moderate levels of variable generation was incorporated and to focus new research on issues that emerged as important. A second round of more quantitative materials describing this research was prepared and presented to the National Association of Regulatory Utility Commissioners, the Department of Energy, and the Federal Energy Regulatory Commission as well as to staff of the U.S. Congress.

Our team looked at geographic smoothing across most of the western U.S.[3] What we found was similar to the results in Texas. Interconnecting just a few wind plants reduces the average changes in the amount of power produced at any time scale, but interconnection across huge regions provides little further reduction in variability.

However, there is one benefit to interconnecting large regions: it doubles the "firm" (i.e., reliable) power output as compared to that from a single region of the country (we defined "firm" output to be at least some generation [92%] of all hours or more). In later work, Mark Handschy of Enduring Energy LLC, Steve Rose, and I looked at geographic smoothing's effect on reducing the number of times that very low power is produced by the sum of many wind plants.[4] We found that the occurrence of low-power events is reduced exponentially with the number of wind plants whose wind regimes are not correlated with each other. So, while geographic smoothing doesn't do much to reduce the variability of wind power, it does some-what decrease the number of hours when no power at all is produced by a set of wind plants dispersed over a large region.

Getting power output data for wind plants was difficult, but it was easy compared to getting similar data for solar plants. When we started our work, the few firms that operated solar facilities considered the data extremely sensitive. Colleen Lueken joined our PhD program after two years as an officer in the U.S. Air Force and three years as an analyst at the Science and Technology Policy Institute looking at Department of Energy programs. Together, we finally managed to get a full year of data for the power generated by a 4.5 MW solar photovoltaic (PV) array in Arizona measured every minute, the same sort of data from a 20 MW PV plant in Colorado, and output measured every 5 minutes for a year from a 75 MW solar thermal generation facility in Nevada. What we found was that solar PV plants have substantially more power output variability at high frequencies (short time scales) than do wind plants.

In contrast to plants that use large PV arrays, solar thermal plants operate by concentrating sunlight onto black tubes containing a synthetic oil. The oil gets very hot and is then run through a heat exchanger to produce steam that then powers a generator. Because the synthetic oil retains heat for a while, it tends to smooth out the moments when a cloud passes over the sun. For this reason, we found that solar thermal plants had smaller high-frequency variability than either PV or wind.[5]

Working with Kelly Klima (a former EPP PhD student now at RAND) and Clyde Loutan from the California Independent Systems Operator (CAISO), we partnered with SoCore Energy to look at the geographic smoothing of solar PV generation from 15 utility-scale plants in California, Nevada, and Arizona and of 19 PV installations on commercial buildings in California.[6] We found that utility-scale and commercial rooftop plants exhibited similar geographic smoothing, with 10 combined plants reducing the amplitude of fluctuations at 1 hour to about

20% of those seen for a single plant. We also found that combining a few PV sites together reduces fluctuations, but the point of quickly diminishing returns is reached after roughly five sites have been combined; for all the locations and plant sizes considered, PV does not exhibit as much geographic smoothing as can be achieved by combining wind plants.

In two papers, we examined the implications of the variability in wind and solar power for the costs of integrating those types of generators into the grid. Electricity markets are not just for energy (kilowatt-hours) but also for services that help to integrate variable power. These services include fast-starting generators and generators that can ramp their output up or down to follow changes in demand for power or to follow changes in the output of other generators. We used the prices for those services to determine the costs of integrating wind and solar power. For wind, Warren Katzenstein and I estimated that the 20 interconnected wind plants we examined in Texas had a variability cost of roughly 0.4 cents per kWh, or about 5–10% of the total cost of wind power.[7] Wind plants in locations where the wind blows strongly are less costly to integrate than plants in less windy sites. Colleen Lueken, Gilbert Cohen from Eliasol Energy, and I found about the same costs for integrating wind in several sites, and we found that solar thermal power also cost about half a cent per kWh to integrate. However, we found that the faster variability of solar photovoltaic power cost 0.8 to 1.1 cents per kWh to smooth out.

Understanding the uncertainty inherent in forecasts of wind power generator output can allow operators to schedule enough other generators to ensure power is always available without expensive over-scheduling. Brandon Mauch and I worked with Pedro Carvalho, an electrical engineer at the Technical University of Lisbon, Portugal, and EPP statistics expert Mitch Small to characterize the uncertainties in forecasts of wind generator output. We found that when forecasts indicated that there would be low wind power, the forecasts over-predicted the actual wind power produced. Conversely, when the forecast was for high wind power, the forecast tended to under-predict the actual power produced. Partly as a result of our work, most wind power forecast companies now have recognized this systematic error and correct for it. Due to Mitch and Brandon's statistical skill, we were able to determine that the confidence in a forecast depends on the forecasted wind power. If the forecast says it is likely to be calm, it turns out that grid operators can have high confidence in that forecast. The same is true if the forecast calls for strong winds and very high wind power output. But between those two extremes, the uncertainty in the forecasted wind turbine output power reaches a maximum (Figure 10.6).

Some of the scatter in Figure 10.6 is due to the inherent variability of the power output of wind turbines. Mahesh Bandi and I were able to get data on wind speed and wind turbine output taken five times every second for three weeks.[8] The curve that relates wind speed to wind turbine output power is often shown as a single line, rising smoothly from zero output power (when the wind is blowing too weakly to spin the blades) to the maximum power the turbine can produce. The curve is, per requirements of the standards bodies, plotted in terms of the mean wind speed, designated $\bar{v}$. What we found was that the output power of real

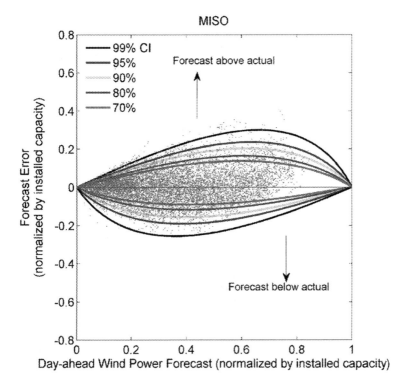

**FIGURE 10.6** Wind power forecast errors for the Mid-Continent Independent Systems Operator (MISO) territory as a function of the wind power forecast. Dots represent each data point. Confidence intervals (CI) for the forecast errors between 70% and 99% are shown.

Mauch, B., Apt, J., Carvalho, P., & Small, M. J. (2013). An effective method for modeling wind power forecast uncertainty. *Energy Systems*, 4(4), 393–417.

wind turbines fluctuates around that smooth line, as the mechanical and electrical systems in the turbine react with varying responses to the inherent variability in wind (Figure 10.7). The smooth curve that is often depicted in textbooks is the average output power ($\bar{P}$) versus the average wind speed $\bar{v}$ (large black dots in Figure 10.7). In reality, the output power responds to the fast fluctuations in wind by varying a lot at a given time-varying wind speed $v(t)$ (small dots in Figure 10.7). This adds to the uncertainty in forecasting wind power, but grid operators quickly learned how to incorporate that uncertainty.

Grid planners have always needed to account for uncertainty in predicting the demand for electric power. Adding variable renewable generation adds to that uncertainty, and the grid must have enough reserve generation standing by to cover the summed uncertainties of load and wind or solar. Brandon, Pedro, Paulina, and I were able to quantify the additional reserve generators that need to be online in both Texas (the ERCOT territory) and the Midwest (the MISO territory) to account for the

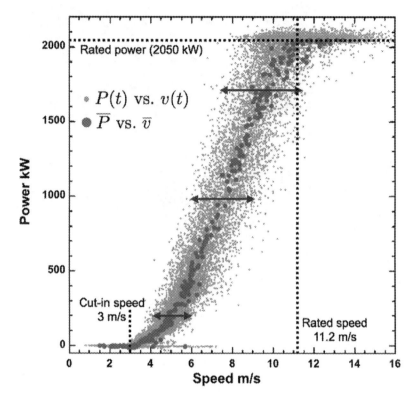

**FIGURE 10.7** Instantaneous power $P(t)$ versus instantaneous wind speed $v(t)$ (large dots) and time-averaged power $\overline{P}$ versus time-averaged wind speed $\overline{v}$ (small dots) for one wind turbine. Considerable scatter in $P(t)$ versus $v(t)$ occurs about the time-averaged power curve. The scatter increases with mean speed $\overline{v}$, as qualitatively shown with arrows at $\overline{v}$ = 5, 7, and 9 m/s.

Bandi, M. M., & Apt, J. (2016). Variability of the wind turbine power curve. *Applied Sciences*, 6(9), 262.

additional uncertainties introduced by wind power.[9] We found that the wind fore-cast uncertainty is smaller in MISO than in ERCOT because MISO's much larger geographic footprint allows some forecast errors to cancel each other. We found that for each MW of additional wind power capacity for ERCOT, about 0.25 MW of dispatchable capacity is needed to compensate for wind uncertainty based on day-ahead forecasts. For MISO (with its more accurate forecasts), the requirement is about 0.1 MW of dispatchable capacity for each MW of additional wind capacity.

Wind and solar energy do not give rise to any direct emissions of carbon dioxide. However, there is a great deal of variation in the amount of carbon dioxide being emitted by today's power systems in different parts of the country. For example, today the power systems in the Southwest and in California emit a great deal less carbon dioxide than the power system in the Northeast. While the amount of sunlight available is higher in the Southwest than in the Northeast, the amount of carbon dioxide emissions that a new solar plant will offset if it's installed in the Northeast is greater than the amount that the same plant would offset in the Southwest. Working with then PhD student Kyle Siler-Evans, Inês Azevedo performed an analysis that examined how much reduction in carbon dioxide emissions could be achieved for wind and solar plants located across different parts of the country. The results of that work are reproduced in Figure 10.8.

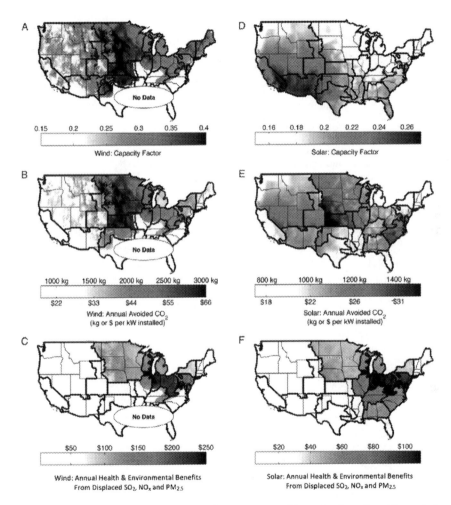

**FIGURE 10.8** Estimates of the health and environmental benefits from installing wind turbines (A–C) and solar panels (D–F) to replace power that is generated by the existing regional power system.

Source: Siler-Evans et al., 2013.

Siler-Evans, K., Azevedo, I. L., Morgan, M. G., & Apt, J. (2013). Regional variations in the health, environmental, and climate benefits of wind and solar generation. Proceedings of the National Academy of Sciences, *110*(29), 11768–11773.

## Technical Details for Chapter 10

This is the paper that kicked off our work looking at wind and solar:

Dobesova, K., Apt, J., & Lave, L. B. (2005). Are renewables portfolio standards cost-effective emission abatement policy?. *Environmental Science & Technology, 39*(22): 8578–8583.

**Abstract:** Renewables portfolio standards (RPS) could be an important policy instrument for 3P and 4P control. We examine the costs of renewable power, accounting for the federal production tax credit, the market value of a renewable credit, and the value of producing electricity without emissions of $SO_2$, $NO_x$, mercury, and $CO_2$. We focus on Texas, which has a large RPS and is the largest U.S. electricity producer and one of the largest emitters of pollutants and $CO_2$. We estimate the private and social costs of wind generation in an RPS compared with the current cost of fossil generation, accounting for the pollution and $CO_2$ emissions. We find that society paid about 5.7 ¢/kWh more for wind power, counting the additional generation, transmission, intermittency, and other costs. The higher cost includes credits amounting to 1.1 ¢/kWh in reduced $SO_2$, $NO_x$, and Hg emissions. These pollution reductions and lower $CO_2$ emissions could be attained at about the same cost using pulverized coal (PC) or natural gas combined cycle (NGCC) plants with carbon capture and sequestration (CCS); the reductions could be obtained more cheaply with an integrated coal gasification combined cycle (IGCC) plant with CCS.

This paper reports on our first look at the output of wind turbines in the frequency domain:

Apt, J. (2007). The spectrum of power from wind turbines. *Journal of Power Sources, 169*(2), 369–374.

**Abstract:** The power spectral density of the output of wind turbines provides information on the character of fluctuations in turbine output. Here both 1-second and 1-hour samples are used to estimate the power spectrum of several wind farms. The measured output power is found to follow a Kolmogorov spectrum over more than four orders of magnitude, from 30 s to 2.6 days. This result is in sharp contrast to the only previous study covering long time periods, published 50 years ago. The spectrum defines the character of fill-in power that must be provided to compensate for wind's fluctuations when wind is deployed at large scale. Installing enough linear ramp rate generation (such as a gas generator) to fill in fast fluctuations with amplitudes of 1% of the maximum fluctuation would oversize the fill-in generation capacity by a factor of two for slower fluctuations, greatly increasing capital costs. A wind system that incorporates batteries, fuel cells, supercapacitors, or other fast-ramp-rate energy storage systems would match fluctuations much better, and can provide an economic route for deployment of energy storage systems when renewable portfolio standards require large amounts of intermittent renewable generating sources.

This paper describes how we applied the tools of frequency domain analysis to the problem of reducing wind power variability by geographic smoothing:

Katzenstein, W., Fertig, E., & Apt, J. (2010). The variability of interconnected wind plants. *Energy Policy, 38*(8), 4400–4410.

**Abstract:** We present the first frequency-dependent analyses of the geographic smoothing of wind power's variability, analyzing the interconnected measured output of 20 wind plants in Texas. Reductions in variability occur at frequencies corresponding to times shorter than ~24 hours and are quantified by measuring the departure from a Kolmogorov spectrum. At a frequency of $2.8 \times 10^{-4}$ Hz (corresponding to 1 hour), an 87% reduction of the variability of a single wind plant is obtained by interconnecting 4 wind plants. Interconnecting the remaining 16 wind plants produces only an additional 8% reduction. We use step-change analyses and correlation coefficients to compare our results with previous studies, finding that wind power ramps up faster than it ramps down for each of the step change intervals analyzed and that correlation between the power output of wind plants 200 km away is half that of co-located wind plants. To examine variability at very low frequencies, we estimate yearly wind energy production in the Great Plains region of the United States from automated wind observations at airports covering 36 years. The estimated wind power has significant inter-annual variability and the severity of wind drought years is estimated to be about half that observed nationally for hydroelectric power.

Our paper analyzing the uncertainties in forecasts of wind plant output power is here:

Mauch, B., Apt, J., Carvalho, P. M., & Small, M. J. (2013). An effective method for modeling wind power forecast uncertainty. *Energy Systems*, *4*(4), 393–417.

**Abstract:** Wind forecasts are an important tool for electric system operators. Proper use of wind power forecasts to make operating decisions must account for the uncertainty associated with the forecast. Data from different regions in the USA with forecasts made by different vendors show the forecast error distribution is strongly dependent on the forecast level of wind power. At low wind forecast power, the forecasts tend to under-predict the actual wind power produced, whereas when the forecast is for high power, the forecast tends to over-predict the actual wind power. Most of the work in this field neglects the influence of wind forecast levels on wind forecast uncertainty and analyzes wind forecast errors as a whole. The few papers that account for this dependence, group wind forecasts by the value of the forecast bin wind forecast data and fit parametric distributions to actual wind power in each bin of data. In the latter case, different parameters and possibly different distributions are estimated for each data bin. We present a method to model wind power forecast uncertainty as a single closed-form solution using a logit transformation of historical wind power forecast and actual wind power data. Once transformed, the data become close to jointly normally distributed. We show the process of calculating confidence intervals of wind power forecast errors using the jointly normally distributed logit transformed data. This method has the advantage of fitting the entire dataset with five parameters while also providing the ability to make calculations conditioned on the value of the wind power forecast.

This is the paper that examined how large a reduction in carbon dioxide emissions could be achieved when a solar or wind plant is located to offset generation by existing power plants in different parts of the United States:

Siler-Evans, K., Azevedo, I. L., Morgan, M. G., & Apt, J. (2013). Regional variations in the health, environmental, and climate benefits of wind and solar generation. *Proceedings of the National Academy of Sciences*, *110*(29), 11768–11773.

**Abstract:** When wind or solar energy displace conventional generation, the reduction in emissions varies dramatically across the United States. Although the Southwest has the greatest solar resource, a solar panel in New Jersey displaces significantly more sulfur dioxide, nitrogen oxides, and particulate matter than a panel in Arizona, resulting in 15 times more health and environmental benefits. A wind turbine in West Virginia displaces twice as much carbon dioxide as the same turbine in California. Depending on location, we estimate that the combined health, environmental, and climate benefits from wind or solar range from $10/MWh to $100/MWh, and the sites with the highest energy output do not yield the greatest social benefits in many cases. We estimate that the social benefits from existing wind farms are roughly 60% higher than the cost of the Production Tax Credit, an important federal subsidy for wind energy. However, that same investment could achieve greater health, environmental, and climate benefits if it were differentiated by region.

**Some notable writings by authors not affiliated with our Centers:**

Ahmadi, M. H., & Yang, Z. (2021). On wind turbine power fluctuations induced by large-scale motions. *Applied Energy*, *293*, 116945.

Bandi, M. M. (2017). Spectrum of wind power fluctuations. *Physical Review Letters*, *118*(2), 028301.

Liu, H., Jin, Y., Tobin, N., & Chamorro, L. P. (2017). Towards uncovering the structure of power fluctuations of wind farms. *Physical Review E*, *96*(6), 063117.

Sørensen, P., Hansen, A. D., & Rosas, P. A. C. (2002). Wind models for simulation of power fluctuations from wind farms. *Journal of Wind Engineering and Industrial Aerodynamics*, *90*(12–15), 1381–1402.

Stevens, R. J., & Meneveau, C. (2017). Flow structure and turbulence in wind farms. *Annual Review of Fluid Mechanics*, *49*, 311–339.

Xie, L., Carvalho, P. M., Ferreira, L. A., Liu, J., Krogh, B. H., Popli, N., & Ilić, M. D. (2010). Wind integration in power systems: Operational challenges and possible solutions. *Proceedings of the IEEE*, *99*(1), 214–232.

# Notes

1 Wiser, R. H., & Langniss, O. (2001). *The renewables portfolio standard in Texas: An early assessment* (No. LBNL-49107). Lawrence Berkeley National Lab. (LBNL), Berkeley, CA (United States). Available at: www.osti.gov/biblio/790029

2 2017 Wind Technologies Market Report, LBNL. Available at: www.energy.gov/eere/wind/downloads/2017-wind-technologies-market-report

3 Fertig, E., Apt, J., Jaramillo, P., & Katzenstein, W. (2012). The effect of long-distance interconnection on wind power variability. *Environmental Research Letters*, *7*(3), 034017.

4 Handschy, M. A., Rose, S., & Apt, J. (2017). Is it always windy somewhere? Occurrence of low-wind-power events over large areas. *Renewable Energy*, *101*, 1124–1130.

5 Lueken, C., Cohen, G. E., & Apt, J. (2012). Costs of solar and wind power variability for reducing $CO_2$ emissions. *Environmental Science & Technology*, *46*(17), 9761–9767.

6 Klima, K., Apt, J., Bandi, M., Happy, P., Loutan, C., & Young, R. (2018). Geographic smoothing of solar photovoltaic electric power production in the Western USA. *Journal of Renewable and Sustainable Energy*, *10*(5), 053504.

7 Katzenstein, W., & Apt, J. (2012). The cost of wind power variability. *Energy Policy*, *51*, 233–243.

8 Bandi, M. M., & Apt, J. (2016). Variability of the wind turbine power curve. *Applied Sciences*, *6*(9), 262.

9 Mauch, B., Apt, J., Carvalho, P. M., & Jaramillo, P. (2013). What day-ahead reserves are needed in electric grids with high levels of wind power?. *Environmental Research Letters*, *8*(3), 034013.

# 11

# CAPTURING AND DISPOSING OF CARBON DIOXIDE FROM POWER PLANTS

*Granger Morgan and Dalia Patino-Echeverri*

## Capturing Carbon Dioxide from Power Plants

### Granger Morgan

The U.S. has lots of coal, oil, and natural gas. The U.S. Energy Information Agency (EIA) reports that in 2017, the U.S. generated roughly 30% of its electricity from coal, 32% from natural gas, and less than 1% from oil.[1] That same year, 20% of U.S. electricity was generated by nuclear power (see Chapter 12). While wind and solar are growing in importance, in 2017 EIA reports that wind generated only 6.3% of all electricity and solar generated about 1.3%. As Chapter 10 explained, wind and solar have the additional complication that they come and go over time – that is, they are variable and intermittent.

Most of us who have studied the problem of meeting the nation's electricity needs while also eliminating emissions of carbon dioxide think that we are going to have to continue to use some coal, gas, and oil – but we will need to do that in a way that avoids putting the $CO_2$ into the atmosphere. That means we need to capture and dispose of the carbon dioxide before it is released. Once we can do that, we can also use electricity to help decarbonize cars and light trucks (see Chapter 14).

As Figure 11.1 indicates, there are basically three ways to capture and dispose of $CO_2$ from power plants. In a set of activities related to, but separate from, our NSF-supported work in CEDM, Ed Rubin and his students and research assistants have spent years building detailed computer models of all these systems.

These models are collectively known as Integrated Environmental Control Model (IECM). This family of models has been made publicly available and has been used by thousands of public and private sector analysts all around the world. You can learn all about IECM and gain access to the models at www.cmu.edu/epp/iecm/. Much of the work on building and refining the IECM models has been supported by the U.S. Department of Energy.

DOI: 10.4324/9781003330226-11

**1.** Burn coal or gas in the normal way and then scrub the carbon dioxide out of the "flue gas" (which is mainly made up of nitrogen since that is what makes up most of air).

**2.** Separate oxygen from the air and burn the coal or gas in oxygen. That way the flue gas is mostly carbon dioxide so it's easy to capture.

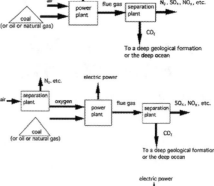

**3.** Extract the hydrogen from coal and run the power plant on hydrogen. When hydrogen burns it combines with oxygen to make H₂O, which of course is just water.

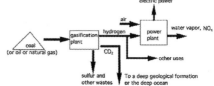

**FIGURE 11.1** Three ways to continue to use fossil fuel while not releasing carbon dioxide to the atmosphere.

Drawn by Granger Morgan.

IECM contains a large number of separate models of all the different parts that go into making up a power plant. Many of these are built up from chemical-engineering-style process models built in a modeling language called ASPEN. As illustrated in Figure 11.2, IECM has an interface that allows a user to connect a bunch of these separate pieces together to make a full power plant system. The latest best estimates of the value for all the key quantities in each module (things like efficiency, energy use, and cost) are included. However, users can change those values if they want to explore the implications of using different ones.

Much like the ICAM integrated assessment model described in Chapter 3, IECM can run stochastic simulations, which means that rather than producing a single number for something like the cost of a system, it can produce a probability distribution that gives both a best estimate of what the number is as well as an indication of how well the modelers believe that the number can be known in light of all the uncertainties in our present knowledge.

Over the years, Ed has become one of the world's leaders on assessing technology to capture carbon dioxide from power plants. For example, in 2005, when the International Panel on Climate Change or IPCC[2] produced a special report titled *Carbon Dioxide Capture and Storage* (CCS), Ed was the expert they asked to write the technical summary. References to this and several other things that Ed has written on CCS are listed at the back of this chapter.

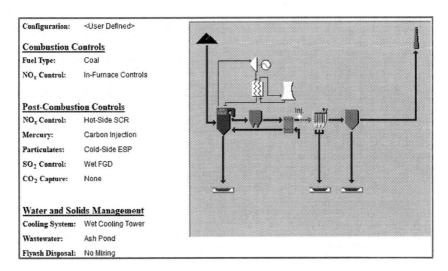

| Configuration: | <User Defined> |
| --- | --- |
| **Combustion Controls** | |
| Fuel Type: | Coal |
| $NO_x$ Control: | In-Furnace Controls |
| **Post-Combustion Controls** | |
| $NO_x$ Control: | Hot-Side SCR |
| Mercury: | Carbon Injection |
| Particulates: | Cold-Side ESP |
| $SO_2$ Control: | Wet FGD |
| $CO_2$ Capture: | None |
| **Water and Solids Management** | |
| Cooling System: | Wet Cooling Tower |
| Wastewater: | Ash Pond |
| Flyash Disposal: | No Mixing |

**FIGURE 11.2** Example of the user interface for the Integrated Environmental Control Model (IECM) system. Users can select a variety of components and link them together to simulate a wide variety of different kinds of power plants and then run the resulting model to estimate things like the performance, air and other emissions, and cost of the system.

Rubin, Ed, Getting Started, Integrated Environmental Control Model, www.cmu.edu/ epp/iecm/videos/01_GettingStarted.mp4

Once $CO_2$ has been captured, something has to be done with it. We're talking about an enormous amount of stuff (mass). A large (~2000 MW) coal-fired power plant, running at full capacity, burns roughly the amount of coal carried by 230 100T rail cars every day (that's a coal train just a bit less than two miles long). Coal is not all carbon. It also has ash and water. While the fraction of coal that is carbon varies for different types of coal, roughly 60% of the coal in that train is carbon (0.6 × 100T/car × 230 cars/day = 13,800 T of carbon/day). When it burns the coal, the plant turns that carbon into an invisible gas ($CO_2$) by adding two oxygen atoms to each atom of carbon. Each oxygen atom weighs one and a third times as much as a carbon atom, so each day the plant produces about 38,000 T of $CO_2$. CCS systems can't economically capture every last ton, but they can typically capture 80–90%. Thirty-four thousand tons of carbon dioxide every day is a lot of stuff (mass) to get rid of, and that is just from one plant. There are hundreds of such plants across the U.S. and all around the world. Society uses some $CO_2$ to do things like add the bubbles in soft drinks, or act as a solvent in some high-tech industrial processes, but the market for those uses is tiny compared to the amount of $CO_2$ that power plants produce. So far, the best strategy that experts have been able to come up with to get rid of all the captured $CO_2$ is to compress it into a "supercritical fluid" and pump it down deep underground. The U.S. is fortunate in that we have lots of underground geological formations (many filled with water that is too salty to be useful) into which the $CO_2$ could be pumped.

There is an interesting story about the S in CCS. Most people say it stands for "storage." The use of that word is not an accident. When people first started working on this topic, they thought the word storage sounded benign and would make the technology more acceptable to people. However, the word storage implies that something is being set aside for later use. If CCS starts being used on a large scale, nobody is ever going to try to take much of the $CO_2$ back out of the ground and use it. Indeed, once it has been there for a few years, most of it will have become trapped and couldn't be gotten out even if we wanted it. So, the right word is "disposal" rather than "storage." To stick with the S in CCS, some of us use the word "sequestration" to imply that it has been put away in a secure place.

Not every power plant is located close to a place that has the right underground geology (porous rocks like sandstone) to dispose of lots of captured $CO_2$. For example, our PhD student Olga Popova found that the amount of underground capacity in Pennsylvania is much smaller than in Ohio, so power plants capturing $CO_2$ in Pennsylvania would probably have to pipe it out to Ohio for disposal. One of Ed's PhD students named Sean McCoy did his PhD building cost and performance models of such pipelines.

Ed and his co-workers have been very interested in understanding how commercial-scale environmental control technologies get developed. An example of their work on learning is included in the additional readings at the end of this chapter. Economists often argue that as more and more of something gets built, the people doing the manufacturing will learn and the costs will go down. When Ed and his colleagues looked at how costs changed over time for technologies to control the sulfur and nitrogen air pollution from power plants, that is not what they found. Their results are shown on the right side of Figure 11.3. Notice that for the first several plants, the cost went *up* and not down. This is because early designs turned out not to work very well and so various technical refinements had to be made. Finally, once engineers produced a design that worked, and more and more systems were installed, "learning" gradually brought the costs down.

The key lesson here for CCS is that if we are going to have cost-effective systems available to add to power plants when we decide we need to use them, we need to start now to refine the processes. There have been a few pilot projects built both in the U.S. and elsewhere around the world, but, because in most of the world there are still no stringent limits (if any) on putting $CO_2$ into the atmosphere, progress on developing capture technology has been much too slow.

As we did more work in this area, we began to understand that the technical issues of capturing $CO_2$ at a power plant, moving it around in pipelines, and injecting it underground were only a part of the problem. The other important topic involves all the legal and regulatory issues associated with doing underground disposal. Building on Ed's work and work we'd done in our NSF-supported Centers, we were fortunate to be able to persuade the Doris Duke Charitable Foundation to

**A.** Commonly expected
shape for learning curves.

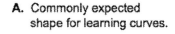

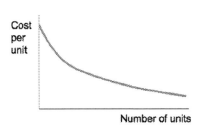

**B.** Actual learning curves for
SO$_2$ scrubbers (above)
and NO$_x$ control (below).

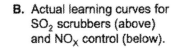

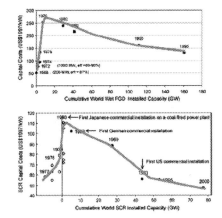

**FIGURE 11.3** While economists typically expect the cost of something to start high and then get lower and lower as producers gain more experience (curve **A** on the left), in the case of both systems to control SO$_2$ (**B**, above) and NO$_x$ (**B**, below) from power plants, costs actually *rose* before finally starting down because the first few designs did not work as well as engineers had expected. It is reasonable to expect a similar experience with CO$_2$ capture technology. This means that it is important to start building commercial-scale capture systems today so that they will be ready, reliable, and cost-effective when the country decides we need them. (Figure B modified from Rubin et al., *International Journal of Greenhouse Gas Control*.)

Rubin, E. S., Yeh, S., Antes, M., Berkenpas, M., & Davison, J. (2007). Use of experience curves to estimate the future cost of power plants with CO2 capture. *International Journal of Greenhouse Gas Control, 1*(2), 188–197.

provide a grant to support what we call the CCSReg project, and to persuade Sean McCoy to coordinate the effort.

To study and develop recommendations on all the regulatory and legal issues, we put together an interdisciplinary team of engineers, lawyers, and social scientists. Members of the team are listed in Table 11.1. We were especially fortunate to have major involvement from two very experienced lawyers. Bob Nordhaus had a long history of work in the Federal Government which included serving as Chief Council for the U.S. Department of Energy. Mike Dworkin had chaired the Vermont Public Utilities Commission and ran the energy program at Vermont Law, which was ranked as the leading environmental law school in the country.

In much of the world, the deep underground region is controlled by the national government. This is not true in the United States. In the U.S., the legal situation varies from state to state. People who own the land at the surface sometimes also

**TABLE 11.1** : Members of the interdisciplinary team of engineers, lawyers, and social scientists who made up the CCSReg project.

| From Carnegie Mellon: | From the law firm of Van Ness Feldman: | From the Humphrey School of Public Affairs at the University of Minnesota: |
|---|---|---|
| Granger Morgan, PI | Bob Nordhaus | Elizabeth Wilson |
| Sean McCoy, Proj. Mgr. | Emily Pitlick | Melisa Pollak |
| Jay Apt | Ben Yamagata | And several MS students |
| Sue Day | | |
| Paul Fischbeck | From Vermont Law School: | |
| Edward Rubin | Michael Dworkin | |
| And PhD students: | Don Kreis | |
| Lee Gresham | And over a dozen law students | |
| Olga Popova | | |

Drawn by Granger Morgan.

have legal control of the ground under them, but not always. For example, often there are others who own the mineral rights.

While the U.S. Environmental Protection Agency (EPA) has the authority under the Underground Injection Control (UIC) program to permit and regulate the underground disposal of $CO_2$, the Agency does not have the authority to consider issues of who owns space underground. One of the things our project developed was a proposal for a regulatory framework for underground disposal that balances the interests of private property owners with the public benefit of disposing of $CO_2$. Our approach was designed to also reduce the possibility of interference with other productive uses of the underground (subsurface), like mining, that are also in the public interest. We also developed recommendations for dealing with $CO_2$ pipelines.

Done properly, in well-engineered projects, underground disposal of $CO_2$ should be permanent. However, since nothing is ever perfectly predictable, another issue we addressed was "long-term stewardship and liability." This included a proposal to set up a fund to deal with any site remediation, should that ever become necessary. To handle this, we proposed the creation of a Federal Geological Sequestration Board.

Because underground disposal of large quantities of $CO_2$ is still rather new, we proposed a regulatory approach that could evolve as we learned more. We recommended that every eight years, a committee of the National Academies of Science, Engineering, and Medicine review all the evidence gathered from projects all around the world and make recommendations for any needed changes in the regulations.

Throughout the course of the four years of the CCSReg project, we produced and distributed policy briefs on each of the topics we were addressing. We also ran briefing sessions on Capitol Hill and at the U.S. EPA, and we conducted various workshops to tell people about our thinking, get their advice, and revise our ideas.

At the end of the project, we published a book that provides a detailed introduction to all the different parts of the topic, lays out the arguments we had developed, and makes a range of recommendations, which included:

- Create a federal opt-in program for $CO_2$ pipelines.
- Establish an integrated federal approach to geological sequestration.
- Learn and adapt with experience.
- Establish a federal framework for access to pore space.
- Establish a federal strategy for long-term stewardship and liability.
- Establish a $CO_2$ accounting framework for all phases of CCS activities.
- Stop procrastinating and make CCS a reality.

Each of these proposals is summarized in a separate chapter and then summarized in a final concluding chapter.

Because the project team included two lawyers who'd had a lot of government experience, we also prepared and included (as an appendix) a 50-page draft of a bill to show how the Congress could implement the ideas we had developed. Of course we did not expect the Congress to pick it up and pass it, but we know that bills are written by cutting and pasting things together and we wanted to help make that process easier for any future legislators. We distributed our work to all House and Senate offices on Capitol Hill.

In the years since we completed the CCSReg project, interest in reducing the country's emissions of $CO_2$ has waxed and waned. Most members of Congress and their constituents have not seen the regulation of underground disposal of $CO_2$ as a pressing national issue, so the Congress has not acted. In the meantime, several states have set up different strategies to deal with some of the issues we addressed. The unfortunate consequence of this is that if geological disposal of $CO_2$ is ever something the nation decides to do on a large scale, the people and organizations who do this will have to deal with a wide variety of different laws and regulations in different parts of the nation. Of course, the geology underground does not pay attention to state boundaries, so developing such projects could turn out to be much more complicated than it needs to be.

## Flexible Mandates to Incentivize Investment in CCS

### Dalia Patino-Echeverri

The slow progress that has been achieved in developing and adopting CCS, compounded with the urgent need to limit greenhouse gas emissions from fossil-fired power plants, creates a dilemma for policymakers. On the one hand, it could seem to be a good idea to mandate that all new coal power plants install CCS. In this way, new coal-fired power plants would emit a minuscule amount of $CO_2$ while reliably providing dispatchable electricity. On the other hand, the technology's immaturity and high cost could result in the unintended consequence of delaying the replacement of old plants with low emission alternatives. A policy mandate intended to reduce $CO_2$ and other air emissions from the combustion of fossil fuels could cause old power plants to continue operating at a dirtier level of performance than would their replacement in the absence of such regulation. At the same time, not mandating CCS for new plants could be equally disastrous. New coal or natural gas plants would have a long operating life, effectively locking in their technical design and assuring the emission of large quantities of $CO_2$ for decades. Also, not mandating CCS would delay the learning-by-doing process necessary to reduce its costs and achieve a required level of deployment that could retain it as an abatement option, should it be needed as one of the wedges of the planet's energy system.

So, what to do then? What is the right policy to incentivize the deployment of a necessary but immature technology?

Motivated by this question, I teamed with Dallas Burtraw and Karen Palmer of Resources for the Future (RFF) to set up a model to test the design of flexible-compliance regulations for coal-fired power plants. Of the three attributes of regulations: stringency, certainty, and flexibility, the last one is the most straightforward and most easily adjusted by regulators. Changing the stringency of a mandate would require establishing new environmental goals, which is not easily done. For the same reason, changing the level of certainty of regulation is also not easy to do. Instead, flexible-compliance rules could attain the same or better environmental outcomes at a lower cost than traditional, inflexible mandates.

We simulated investors' behavior using a stochastic optimization model that minimized the cost of complying with air emissions regulations during the lifetime of their power plant under different policies. The model considered fuel price uncertainty and uncertainty about other policies that might be enacted in the future, such as imposing taxes on $CO_2$ emissions. We compared the outcomes of a conventional strict standard that required new plants' emissions to be below a threshold (in lbs of $CO_2$ per MWh) that could only be achieved with CCS installation, with two types of flexible mechanisms. The first flexible mechanism provided the opportunity to pay an emissions surcharge for plants that failed to meet the maximum $CO_2$ emission rate standard. The second flexible mechanism allowed revenue from the surcharge to be held in an escrow account and used to fund later retrofit investment in CCS technology.

We demonstrated that introducing a new inflexible emissions rate standard could delay investment in new power plants, thereby increasing cumulative $CO_2$ emissions over a 30-year horizon while at the same time delaying CCS cost reductions. We also found that introducing flexibility with an opportunity to pay a surcharge for emissions above the emissions standard would lead to earlier investments than under the inflexible standard, with lower aggregate emissions and greater profits to investors. When funds from the payment of the surcharge held in escrow were made available to pay for part of the capital costs of CCS retrofits, the investment would occur most quickly. Under this policy, aggregate emissions were the lowest, and profits to investors the highest. This study was summarized in an RFF discussion paper[3] and published in the *Journal of Regulatory Economics*.

After we published this work, in September of 2013 the U.S. EPA proposed a $CO_2$ emissions standard of 1000–1100 lbs/MWh for coal and small natural gas plants depending on their size. The standard was designed with some "flexibility" allowing coal-fired power plants to comply by achieving the standard over a 7-year operating period. This flexibility made it possible for a new coal-fired power plant to operate without CCS. However, it implicitly required that the construction of CCS began soon after the plant commenced operations so that this equipment could be operational within the necessary time frame to limit the 7-year average $CO_2$ emissions rate below the standard. The intended flexibility of the rule, therefore, was nil.

This proposal that de-facto mandated CCS installation in addition to other policies constraining coal-fired power plants' air emissions and water use created disincentives for installing new coal power plants or keeping the existing ones. Also, the significant reduction in natural gas prices, and the expectation that these prices would stay low for some decades, made the gas-for-coal replacement much more attractive than CCS in a retrofit or new plant installation. This would have been great news if natural-gas-fired electricity were a solution commensurate with the urgent need to reduce greenhouse emissions from the electric sector to near zero. But far from that, the combustion of natural gas in a natural gas combined cycle power plant still emits between one-half and one-third of the $CO_2$ emitted by a coal plant. Unless great care is taken, the processes of extracting, processing, and transporting natural gas can result in high methane emissions. Methane is a potent greenhouse gas – in the short-term causing roughly 30 times more warming than an equivalent amount of $CO_2$. If even modest amounts of methane leak from the natural gas system, that can eliminate most or all of the benefits from replacing coal power with natural gas power.

Given that a combination of regulatory and market factors point investors toward natural gas, and that lock-in of this sub-optimal technology is counterproductive in the race to decarbonization, there is a need for a better policy; one that reaps at least identical emissions reductions as that from a coal-to-gas switch while still incentivizing a technology that truly addresses climate change. In a paper titled "Feasibility of flexible technology standards for existing coal-fired power plants and their implications for new technology development" that I published in the *UCLA Law Review*, I presented the conditions for the existence of such policy.

I considered the fact that natural gas price uncertainties and the high likelihood of stringent limits on carbon dioxide emissions make it difficult for owners of coal-fired power plants to make compliance decisions. This difficulty implies that investors would happily pay for the option to delay their investment decision and continue operating their extant coal plant until some uncertainty is resolved.

Is there a win-win policy that allows investors to pay to wait before deciding on a compliance investment, that at the same time incentivizes a better technology, and that, at least in expectation, results in the same or better environmental outcomes (as measured by cumulative greenhouse gas emissions) than a traditional policy? The answer is yes. Suppose there is a reasonably high probability that a superior technology will become commercially attractive in the next few years. In that case, investors will be willing to pay a fee to keep operating their existing coal plants for a few years as they wait for the new technology to become available. At the end of such a waiting period, in the best case, the technology will be ready for adoption, and its low operating cost will more than offset the money paid in the fees. In the worst-case scenario, the technology will not be ready, and they will need to replace their plant with a conventional natural gas plant. From the regulator's perspective, in the best case, the near-zero-emissions new technology will more than offset the extra emissions incurred during the years when the coal plant was allowed to operate under business-as-usual conditions. In the worst case, the new technology will not materialize. Still, at least, the payments for each ton of $CO_2$ emitted by the conventional plant during the first few years may finance other carbon mitigation alternatives.

While the fees paid to operate the coal-plant emitting $CO_2$ above a threshold (equal to the emissions of a gas plant) look similar to a tax, they are different in that they are merely a mechanism that investors can use to delay investment for a short period. One way or the other, at the end of that period investors must reduce emissions, regardless of whether new technologies are in place. The commitment to invest in new technology, if it materializes, will help to advance the development of such technology and increase its chance of success.

## Technical Details for Chapter 11

The technical summary that Ed Rubin wrote for the IPCC Special Report on CCS can be downloaded at:

www.ipcc.ch/pdf/special-reports/srccs/srccs_technicalsummary.pdf Accessed July 22, 2018.

A more recent summary of cost and performance of CCS is:

Rubin, E. S., Davison, J. E., & Herzog, H. J. (2015). The cost of $CO_2$ capture and storage. *International Journal of Greenhouse Gas Control, 40,* 378–400.

**Abstract:** The objective of this paper is to assess the current costs of $CO_2$ capture and storage (CCS) for new fossil fuel power plants and to compare those results to the costs reported a decade ago in the IPCC Special Report on Carbon Dioxide Capture and Storage (SRCCS). Toward that end, we employed a similar methodology based on review and analysis of recent cost studies for the major CCS options identified in the SRCCS, namely, post-combustion $CO_2$ capture at supercritical pulverized coal (SCPC) and natural gas combined cycle (NGCC) power plants, plus pre-combustion capture at coal-based integrated gasification combined cycle (IGCC) power plants. We also report current costs for SCPC plants employing oxy-combustion for $CO_2$ capture—an option that was still in the early stages of development at the time of the SRCCS. To compare current CCS cost estimates to those in the SRCCS, we adjust all costs to constant 2013 US dollars using cost indices for power plant capital costs, fuel costs and other O&M

costs. On this basis, we report changes in capital cost, levelized cost of electricity, and mitigation costs for each power plant system with and without CCS. We also discuss the outlook for future CCS costs.

Here is a paper by Olga Popova and colleagues that shows that the geology of Pennsylvania has rather limited capacity to sequester $CO_2$. Other work shows that the capacity in Ohio is much larger so that power plants that capture $CO_2$ in Pennsylvania would probably need to send their $CO_2$ in a pipeline to dispose of it in Ohio:

Popova, O. H., Small, M. J., McCoy, S. T., Thomas, A. C., Rose, S., Karimi, B., ... & Goodman, A. (2014). Spatial stochastic modeling of sedimentary formations to assess $CO_2$ storage potential. *Environmental Science & Technology*, 48(11), 6247–6255.

**Abstract:** Carbon capture and sequestration (CCS) is a technology that provides a near-term solution to reduce anthropogenic $CO_2$ emissions to the atmosphere and reduce our impact on the climate system. Assessments of carbon sequestration resources that have been made for North America using existing methodologies likely underestimate uncertainty and variability in the reservoir parameters. This paper describes a geostatistical model developed to estimate the $CO_2$ storage resource in sedimentary formations. The proposed stochastic model accounts for the spatial distribution of reservoir properties and is implemented in a case study of the Oriskany Formation of the Appalachian sedimentary basin. Results indicate that the $CO_2$ storage resource for the Pennsylvania part of the Oriskany Formation has substantial spatial variation due to heterogeneity of formation properties and basin geology leading to significant uncertainty in the storage assessment. The Oriskany Formation sequestration resource estimate in Pennsylvania calculated with the effective efficiency factor, $E = 5\%$, ranges from 0.15 to 1.01 gigatonnes (Gt) with a mean value of 0.52 Gt of $CO_2$ ($E = 5\%$). The methodology is generalizable to other sedimentary formations in which site-specific trend analyses and statistical models are developed to estimate the $CO_2$ sequestration storage capacity and its uncertainty. More precise $CO_2$ storage resource estimates will provide better recommendations for government and industry leaders and inform their decisions on which greenhouse gas mitigation measures are best fit for their regions.

This paper describes Sean McCoy's work on modeling pipelines to move $CO_2$ from power plants to places where it can be sequestered deep underground:

McCoy, S. T., & Rubin, E. S. (2008). An engineering-economic model of pipeline transport of $CO_2$ with application to carbon capture and storage. *International Journal of Greenhouse Gas Control*, 2(2), 219–229.

**Abstract**: Carbon dioxide capture and storage (CCS) involves the capture of $CO_2$ at a large industrial facility, such as a power plant, and its transport to a geological (or other) storage site where $CO_2$ is sequestered. Previous work has identified pipeline transport of liquid $CO_2$ as the most economical method of transport for large volumes of $CO_2$. However, there is little published work on the economics of $CO_2$ pipeline transport. The objective of this paper is to estimate total cost and the cost per tonne of transporting varying amounts of $CO_2$ over a range of distances for different regions of the continental United States. An engineering-economic model of pipeline $CO_2$ transport is developed for this purpose. The model incorporates a probabilistic analysis capability that can be used to quantify the sensitivity of transport cost to variability and uncertainty in the model input parameters. The results of a case study show a pipeline cost of US$ 1.16 per tonne of $CO_2$ transported for a 100 km pipeline constructed in the Midwest handling 5 million tonnes of $CO_2$ per year (the approximate output of an 800 MW coal-fired power plant with carbon capture). For the same set of assumptions, the cost of transport is US$ 0.39 per tonne lower in the Central US and US$ 0.20 per tonne higher in the Northeast US. Costs are sensitive to the design capacity of the pipeline and the pipeline length. For example, decreasing the design capacity of the Midwest US pipeline to 2 million tonnes per year increases the cost to US$ 2.23 per tonne of $CO_2$ for a 100 km pipeline, and US$ 4.06 per tonne $CO_2$ for a 200 km pipeline. An illustrative probabilistic analysis assigns uncertainty distributions to the pipeline capacity factor, pipeline inlet pressure, capital recovery factor, annual O&M cost, and escalation factors for capital cost components. The result indicates a 90% probability that the cost per tonne of $CO_2$ is between US$ 1.03 and US$ 2.63 per tonne of $CO_2$ transported in the Midwest US. In this case, the transport cost is shown to be most sensitive to the pipeline capacity factor and the capital recovery factor. The analytical model elaborated in this paper can be used to estimate pipeline costs for a broad range of potential CCS projects. It can also be used in conjunction with models producing more detailed estimates for specific projects, which requires substantially more information on site-specific factors affecting pipeline routing.

The paper that examines "learning curves" for the development of sulfur and nitrogen controls from power plants is:

Rubin, E. S., Yeh, S., Antes, M., Berkenpas, M., & Davison, J. (2007). Use of experience curves to estimate the future cost of power plants with $CO_2$ capture. *International Journal of Greenhouse Gas Control, 1*(2), 188–197.

**Abstract:** Given the dominance of power plant emissions of greenhouse gases, and the growing worldwide interest in $CO_2$ capture and storage (CCS) as a potential climate change mitigation option, the expected future cost of power plants with $CO_2$ capture is of significant interest. Reductions in the cost of technologies as a result of learning-by-doing, R&D investments and other factors have been observed over many decades. This study uses historical experience curves as the basis for estimating future cost trends for four types of electric power plants equipped with $CO_2$ capture systems: pulverized coal (PC) and natural gas combined cycle (NGCC) plants with post-combustion $CO_2$ capture; coal-based integrated gasification combined cycle (IGCC) plants with pre-combustion capture; and coal-fired oxyfuel combustion for new PC plants. We first assess the rates of cost reductions achieved by other energy and environmental process technologies in the past. Then, by analogy with leading capture plant designs, we estimate future cost reductions that might be achieved by power plants employing $CO_2$ capture. Effects of uncertainties in key parameters on projected cost reductions also are evaluated via sensitivity analysis.

The review of learning rates for new technology is:

Rubin, E. S., Azevedo, I. M., Jaramillo, P., & Yeh, S. (2015). A review of learning rates for electricity supply technologies. *Energy Policy, 86*, 198–218.

**Abstract:** A variety of mathematical models have been proposed to characterize and quantify the dependency of electricity supply technology costs on various drivers of technological change. The most prevalent model form, called a learning curve, or experience curve, is a log-linear equation relating the unit cost of a technology to its cumulative installed capacity or electricity generated. This one-factor model is also the most common method used to represent endogenous technical change in large-scale energy-economic models that inform energy planning and policy analysis. A characteristic parameter is the "learning rate," defined as the fractional reduction in cost for each doubling of cumulative production or capacity. In this paper, a literature review of the learning rates reported for 11 power generation technologies employing an array of fossil fuels, nuclear, and renewable energy sources is presented. The review also includes multi-factor models proposed for some energy technologies, especially two-factor models relating cost to cumulative expenditures for research and development (R&D) as well as the cumulative installed capacity or electricity production of a technology. For all technologies studied, we found substantial variability (as much as an order of magnitude) in reported learning rates across different studies. Such variability is not readily explained by systematic differences in the time intervals, geographic regions, choice of independent variable, or other parameters of each study. This uncertainty in learning rates, together with other limitations of current learning curve formulations, suggests the need for much more careful and systematic examination of the influence of how different factors and assumptions affect policy-relevant outcomes related to the future choice and cost of electricity supply and other energy technologies.

The website that provides the various policy briefs produced by the CCSReg Project can be found at:

www.CCSReg.org. Accessed July 23, 2018 July.

The book that summarizes the results of the CCSReg Project is:

M. Granger Morgan, Sean T. McCoy and 15 co-authors. (2012) *Carbon capture and sequestration: Removing the legal and regulatory barriers.* Routledge, 274pp.

The table of contents of this book reads:

This is the paper Dalia published with colleagues at RFF on the topic of flexible mandates for emissions control:

Patino-Echeverri, D., Burtraw, D., & Palmer, K. (2013). Flexible mandates for investment in new technology. *Journal of Regulatory Economics, 44*(2), 121–155.

**Abstract:** Environmental regulators often seek to promote forefront technology for new investments; however, technology mandates are suspected of raising cost and delaying investment. We examine investment choices under an inflexible (traditional) emissions rate performance standard for new sources. We compare the inflexible standard with a flexible one that imposes an alternative compliance payment (surcharge) for emissions in excess of the standard. A third policy allows the surcharge revenue to fund later retrofits. Analytical results indicate that increasing flexibility leads to earlier introduction of new technology, lower aggregate emissions and higher profits. We test this using multi-stage stochastic optimization for introduction of carbon capture and storage, with uncertain future natural gas and emissions allowance prices. Under perfect foresight, the analytical predictions hold. With uncertainty these predictions hold most often, but we find exceptions. In some cases investments are delayed to enable the decision maker to discover additional information.

And this is Dalia's law review article that further explores the same topic:

Patino-Echeverri, D. (2013). Feasibility of flexible technology standards for existing coal-fired power plants and their implications for new technology development. *UCLA L. Rev.*, *61*, 1896.

**Abstract:** This article explores the feasibility of adding flexibility to mandates for existing power plants in order to foster technology innovation and reduce compliance costs and emissions. Under new and proposed EPA rules a significant portion of the coal-fired electricity generating capacity will require multi-billion dollar investments to retrofit and comply with emission standards on $SO_2$, $NO_X$, PM, mercury, toxic metals, acid gases, coal combustion residuals, and cooling water rules. A number of plant owners may find it preferable to replace these plants with new units that run with today's low-cost natural gas. Massive retrofit or replacement of the current coal-fired power generation fleet with today's solutions will harm the conditions for research and development of path-breaking fossil-fired power generation technologies. This would not be a serious problem if the current retrofit and replacement technologies were in fact a solution to the many environmental externalities posed by the coal fired power plants that are now candidates for expensive retrofitting or retirement. But the technologies available today are far from being a solution commensurate with the climate and environmental risks that fossil-fired generation poses in the United States and the world. This article finds that a policy with a flexible compliance payment that allows investors to delay the decision of retrofitting or replacing and hence, maintains incentives for innovation in retrofit a new plant technologies, can perform an inflexible mandate by reducing compliance costs and improving environmental performance.

**Some notable writings by authors not affiliated with our Centers:**

Benson, S. M., & Orr, F. M. (2008). Carbon dioxide capture and storage. *MRS Bulletin*, *33*(4), 303–305.

Bergerson, J. F., & Keith, D. (2010). The truth about dirty oil: Is CCS the answer? *Environmental Science & Technology*, *44*, 6010–6015.

Herzog, H. J. (2011). Scaling up carbon dioxide capture and storage: From megatons to gigatons. *Energy Economics*, *33*(4), 597–604.

House, K. Z., Baclig, A. C., Ranjan, M., van Nierop, E. A., Wilcox, J., & Herzog, H. J. (2011). Economic and energetic analysis of capturing $CO_2$ from ambient air. *Proceedings of the National Academy of Sciences*, *108*(51), 20428–20433.

Metz, B., Davidson, O., De Coninck, H. C., Loos, M., & Meyer, L. (2005). *IPCC special report on carbon dioxide capture and storage*. Cambridge University Press.

Psarras, P., He, J., Pilorgé, H., McQueen, N., Jensen-Fellows, A., Kian, K., & Wilcox, J. (2020). Cost analysis of carbon capture and sequestration from U.S. natural gas-fired power plants. *Environmental Science & Technology, 54*(10), 6272–6280.

Wilcox, J. (2012). *Carbon capture*. Springer Science & Business Media.

## Notes

1 See www.eia.gov/tools/faqs/faq.php?id=427&t=3. Accessed July 20, 2018.
2 The IPCC is a joint undertaking of the World Meteorological Organization and the Environment program of the United Nations. It assembles teams of leading experts drawn from all around the world to perform assessments of climate-related issues. All their reports are available for free at www.ipcc.ch
3 RFF DP 12-14 *Flexible Mandates for Investment in New Technology,* March 2012, available online at https://media.rff.org/documents/RFF-DP-12-14.pdf

# 12

# CAN NUCLEAR POWER HELP SOLVE THE CLIMATE PROBLEM?

*Granger Morgan and Ahmed Abdulla*

Nuclear power faces many economic, political, and institutional challenges. However, it is indisputably capable of producing large amounts of low-carbon electricity. Moreover, after decades of operational experience with light water reactors, several countries have managed to substantially boost the reliability of these plants, enabling them to produce electricity virtually day and night. From the beginning of the atomic age to the 1980s, the U.S. occupied a leadership role in the development and deployment of civilian nuclear power reactors: domestically, this buildout has meant that for the last three decades, roughly 20% of all electricity generated in the U.S. came from nuclear power plants, a fraction that is shrinking as plants approach retirement age and are forced to compete both in a deregulated electricity market and with other low-cost generation options.

In the first decade of the 21st century, there was optimistic talk in the U.S. about a "nuclear renaissance." Among utilities and established vendors, talk of this renaissance revolved around a push to build new, large, light water reactors to replace the nation's aging plants. A combination of old and new problems soon quelled this enthusiasm: there was no evidence that the enormous cost overruns and delays that plagued nuclear power were resolved; the utilities that would have to finance these plants were operating in a deregulated electricity market, not under the old rate-based paradigm; there was still no federal solution for the waste challenge; and the decades-long lull in reactor construction had left the nation with depleted nuclear supply chains and human capital – there was too little infrastructure and too few skilled engineers and workers to follow through. The number of proposed new plants slowly dwindled. Eventually, only two progressed on the sites of existing nuclear power plants: Virgil C. Summer in South Carolina and Vogtle in Georgia. The owners of both were regulated utilities. In retrospect, it was unsurprising when an industry that had failed to systematically address its history of cost overruns and

DOI: 10.4324/9781003330226-12

delays faced many more billions of dollars of cost overruns, with some eventually abandoned altogether.

Several hypotheses have been offered to explain nuclear power's troubles. A popular hypothesis is that power companies in the U.S. managed the construction of each nuclear plant as a separate project, and each of those integrated incremental refinements in design over the previous one – both because of growing experience and due to morphing regulation. The industry had argued that design standardization would achieve the level of cost control required to make nuclear power economically competitive. This argument is speculative, and it was always unclear the extent to which it would apply in a country as vast as the U.S., where designs were often tweaked to suit site-specific considerations. One enterprise that adopted this philosophy was the French nuclear industry, which managed to standardize the designs of its reactors. This did not prevent significant cost escalation in their program. Despite the French experience, many proponents of nuclear power still argue that design standardization remains an appealing cost-control strategy. In 1993, we noted that if the U.S. manufactured airplanes the way it was building nuclear power plants, none of us would be able to afford to fly across the country, let alone across continents. If we wanted new nuclear power to become a viable option for reducing emissions from the power sector, we had to investigate ways of reducing costs – or at least making them predictable – using an alternative deployment paradigm.

After completing a BS in chemical engineering at Princeton, Ahmed Abdulla joined us in 2010 to do a PhD in EPP. He decided to investigate the possibility that small reactors built in a factory using standardized designs might offer a way for new nuclear power plants to contribute to decarbonizing the energy system over the course of the next few decades.

Ahmed went to work learning about various small modular reactor (SMR) technologies that were being developed across the world. Once he'd become very knowledgeable, he adopted the methods of "expert elicitation" that we had developed for our work with climate scientists (see Chapter 4). He created a detailed set of technical descriptions of two light water SMR designs that were then in a more advanced stage of development, including their major subcomponents, and used them to conduct 16 extended face-to-face interviews with nuclear experts. Most of these experts were actively involved in commercial SMR development. In order not to compromise proprietary company information, he was careful to only show them material or ask them questions that were based on publicly available blueprints, though most of the experts were able to draw upon their experience and design knowledge in answering his questions.

Because the SMRs Ahmed studied were much smaller than existing large nuclear reactors (SMRs are smaller than 300 $MW_e$; Ahmed looked at reactors ranging from 45 $MW_e$ to 225 $MW_e$, compared to the 1000 $MW_e$ to 1600 $MW_e$ light water reactors that are currently on the market), it is not surprising that the estimates he got for the overall cost of an SMR plant were much lower than the cost of large nuclear power plants. However, when he compared the specific capital cost of these plants – and the cost of electricity they were likely to produce – it was comparable to, but generally higher than, the cost of electricity the experts anticipated for power produced by a newly constructed regular large nuclear plant (Figure 12.1).

In Ahmed's judgment, generating these cost estimates was one of two valuable contributions that the expert elicitation made to the literature. Up until he published his elicitation, only point estimates were available for SMR costs, most of which were anchored on large reactors that are very different. Also valuable was our new understanding of the extent to which the nuclear engineers designing SMRs disagreed fervently about their eventual cost. Many of them remained quite unconvinced of the speculative advantages of "going small" that the industry and its proponents claimed.

Still, even if most experts believed that the cost per unit output of an SMR would be greater, the lower initial cost to buy an SMR expanded the pool of potential customers who could now purchase nuclear power: generating companies

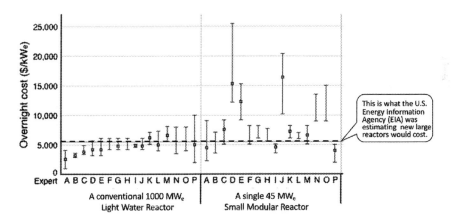

**FIGURE 12.1** An example of the sorts of cost estimates for conventional and small nuclear reactors that Ahmed Abdulla got from the expert elicitations he did with 16 leading nuclear experts. The costs assume that each type of plant has already been built several times and that this is how much a buyer would have to pay to take delivery on a new one. MW$_e$ stands for megawatts electric. The letters A through P along the bottom of the plot show the 16 different experts. The solid dots show each expert's best estimate of what the plant would cost to build. The vertical lines show the range of each expert's uncertainty. Notice that the experts did not agree on what a new, large plant would cost (expert M's estimate of the cost of a new plant is more than twice expert A's estimate). Most experts thought that the cost of an SMR, per unit of power output, would be higher than that from a large plant, although a few, like expert P, thought that their cost could be similar to or lower than a large nuclear plant's. Five of the 16 experts thought the cost of an SMR would be *much* higher than from a conventional nuclear plant.

Abdulla, A., Azevedo, I. L., & Morgan, M. G. (2013). Expert assessments of the cost of light water small modular reactors. Proceedings of the National Academy of Sciences, *110*(24), 9686–9691.

would only need to invest a fraction of the cost of building a large conventional nuclear plant. As the need for power grew in their service territory, and once they had paid off the first reactor, they could add more SMRs at a cost that did not involve "betting the future of the company."

In 2013, we reported our findings from this work in a paper titled "Expert Assessments of the Cost of Light Water Small Modular Reactors" in the *Proceedings of the National Academy of Sciences*. We also gave briefings on the work to several key groups including the annual gathering of reactor operators run by the Electric Power Research Institute (EPRI).

Dismayed that a chance to revivify nuclear power was, in his judgment, being squandered by an industry that had not learned from past mistakes, Ahmed turned to what he considered the second most important reason to develop SMRs: to prevent developing nations from building long-lived fossil fuel infrastructure – power plants, pipelines, and oil fields – and then getting "locked into" a carbon-intensive energy system. Lowering the cost of nuclear development by deploying smaller, standardized systems could catalyze the development of a market for SMRs across the world, preempting the need to develop more polluting infrastructure.

However, many developing nations are unlikely to have the capacity to create and maintain the infrastructure and institutions that are required to safely manage nuclear power. Lowering barriers to nuclear development with SMRs – and doing so prudently – struck us as a serious problem that deserved attention. Along with colleagues at the University of Maryland and Princeton, we each obtained grants from the MacArthur Foundation to assess different aspects of that risk and devise ways by which it might be managed. We were especially concerned about the possibility that spent fuel from an SMR might be diverted for military or other purposes during refueling. Proliferation is a risk inherent in all nuclear development, but the prospect of deploying thousands of small reactors, including in locations that might be ill-prepared to cope with accidents or attempted sabotage, was unnerving.

As part of that work, we organized an invitational workshop later that year at the Paul Scherrer Institute, the Swiss Federal Nuclear Lab. This invitational workshop was attended by 40 leading SMR experts from all over the world, including China, France, Korea, Russia, and the U.S. (Figure 12.2). The workshop adopted

**FIGURE 12.2** Views of the workshop we held at the Paul Scherrer Institute in Switzerland with a group of 40 leading nuclear experts from all around the world.

Photos taken by Granger Morgan.

a very unusual format that made it different from others that we had previously organized: instead of focusing on presentations or roundtable discussions, we developed a detailed, 38-page workbook that included technical descriptions of six different SMR designs. We began each session at the workshop with a few short introductory talks and some discussion. Then we asked the experts to use the workbook we had developed to record their specific quantitative and qualitative judgments in response to a carefully developed set of questions. We ran the workshop with help from Shikha Prasad, who had just joined us as a Postdoc after completing her PhD in nuclear engineering at the University of Michigan.

There were interesting paradigms that would enable nuclear power to be deployed safely and securely in more locations across the world, at least in theory. Our colleague John Steinbruner, who founded and directed the Center for International and Security Studies at the University of Maryland, had suggested that the best way to minimize the proliferation risk from SMRs was to seal the core – in other words, to never engage in on-site refueling or any maintenance that involves accessing the core or its fuel. Instead, he argued, the only safe deployment option was to build SMRs that could be fueled in centralized factories operating under International Atomic Energy Agency (IAEA) safeguards. These fully fueled reactors would be shipped to site and then returned safely to centralized fuel handling facilities for refueling or decommissioning at the end of their core life. Throughout the process, the supplier – or vendor – would retain ownership of the reactor, effectively transforming customers from owners to power purchasers. We called this the "build-own-operate-return" (or BOOR) model, and we were under no illusions about how many technical and institutional changes would be required to bring it about.

Still, we found the idea attractive, and we discussed it extensively at our workshop in Switzerland. The expert reactor designers at our workshop threw cold water on the model. The collective response came very close to a curt dismissal: basically, "in your dreams." Several argued that there is no way that an SMR with spent nuclear fuel can be safely removed from a site and shipped back to a secure facility. Some argued that transporting SMRs fully fueled is a no-go, whether the fuel is fresh or spent. Both the workshop and our elicitation emphasized that many of the attractive models of SMR deployment that are offered by nuclear advocates do not stand up to rigorous scrutiny. Scratch the surface, and efforts to simplify nuclear development often generate new risks to which neither the industry nor advocates had paid sufficient attention.

Throughout our studies of SMR proliferation risk, we and colleagues from Maryland and Princeton met regularly to brainstorm and share results. In 2014, our three groups organized a special session at the annual meeting of the American Association for the Advancement of Science in Chicago titled "Opportunities and Challenges for Nuclear Small Modular Reactors," where we presented our findings.

Ahmed received his PhD in 2014 and stayed on at CMU as a Postdoc. In 2015, he moved to the University of California, San Diego (UC San Diego). After a few years at UC San Diego, he moved back to EPP as Assistant Research Professor. Then, in 2020, he joined Carleton University in Ottawa, Canada, as Assistant Professor in Mechanical and Aerospace Engineering. Throughout the years, the two of us have continued to work together on a variety of nuclear and other energy system-related research questions.

In 2015, we published two papers summarizing all we had learned from our work on nuclear power. In the first, titled "Nonproliferation improvements and challenges presented by small modular reactors," which we published in *Progress in Nuclear Engineering*, we described the six SMR designs that we considered at the workshop, discussed the benefits and risks of each design, and synthesized the quantitative judgments that our experts had made in their workbooks. The second paper, titled "Nuclear power for the developing world," was published in the journal *Issues in Science & Technology*. In that paper, we laid out many of the international regulatory and other issues that we believe will need to be addressed if SMRs are going to be safely adopted across the world. We argued that "the international community should urgently act to create a global control and accounting system for all civilian nuclear materials," and that in selecting reactors, "preference must be given to SMR designs that minimize the need for on-site fuel handling and storage—in general, the fewer times the fuel is handled, the better." In 2015, Ahmed also wrote a report for the International Risk Governance Council (with which we had collaborated for our workshop in Switzerland) titled *Preserving the Nuclear Option: Overcoming the institutional challenges facing small modular reactors.*

We began to investigate other strategies that might be adopted to implement the BOOR model. If it was not safe or technically feasible to pick up an SMR full of spent fuel and transport it back to a secure, internationally supervised facility, what about moving it along with the ground on which it is sited? One way to do this would be to put the SMR on an ocean-going barge that could be anchored offshore in water deep enough to mitigate the risk from earthquakes and tsunamis. We were fortunate to recruit a PhD student, Michael Ford, to explore the technical and economic issues that would be involved in doing that. Mike was an unusual PhD student. He had just retired at the rank of Captain after a full career in the U.S. Navy. His award-winning Navy career included multiple sea commands and extensive Pentagon leadership experience, including two years spent as Chief of Staff for the Deputy Chief of Naval Operations. Mike knew a lot about how to build and operate ships. He was also a mature professional who set out to explore the issue of floating SMRs (fSMRs). Mike's analysis concluded that a double-hulled fSMR platform could be built for less than the cost of preparing a land-based site, but only if it was built to *commercial* shipbuilding standards. If the barges had to comply with the same shipbuilding standards as military vessels in the U.S. Navy's surface fleet, significant cost escalation occurs and the fSMR platform would likely cost one and a half to two times as much as a land-based site. Nonetheless, a floating SMR would have several big advantages. It would not require the host country to develop extensive – and expensive – regulatory and other nuclear management institutions and their associated infrastructure. It might not require as large an "emergency planning zone" as a land-based SMR. And, because the reactor would never be refueled in the host country but instead replaced with a new fSMR once its fuel is spent, the old barge could be towed back to a secure facility for refueling, dramatically reducing the risk of material diversion and proliferation. While no U.S. vendor is developing fSMRs today, Russia, China, and France have built or proposed such systems.

In 2017, we published the results of Mike's analysis of fSMRs in a paper titled "Evaluating the Cost, Safety and Proliferation Risks of Small Floating Nuclear Reactors" in the journal *Risk Analysis*.

Up to this point, our efforts to envision a role for nuclear power in reducing U.S. power sector emissions were not yielding promising results. Perhaps the problems facing nuclear power could be solved through radical innovation that makes some of its "speculative" benefits a reality? We decided to investigate the possibility that a new generation of advanced reactor designs might contribute to reducing the $CO_2$ emissions from the U.S. energy system by mid-century. Today's light water reactors use the same basic design that ushered in the atomic age. For 50 years, the industry has assumed that we would transition to advanced reactor designs when the current fleet is retired. Yet, despite much research and considerable investment, no such design is ready for deployment today.

The entity responsible for advancing the technological readiness of these new reactor designs in the U.S. is the Department of Energy through its Office of Nuclear Energy, which is also known as NE. We decided to start by conducting a retrospective analysis of NE's performance in its charge to develop advanced reactor technologies. To do this, we filed a Freedom of Information Act (FOIA) request for DOE budget justification documents going back to the Department's founding. This request netted more than 400,000 pages of federal and departmental budget data, which we augmented with records derived through library and online searches. These data painted a picture of how public sector expenditures on nuclear power have been spent, down to the level of fundamental research and development in the DOE's National Laboratories.

We found that NE's budget had allocated only a marginal amount to its advanced reactor development mission: on average, 15% of annual spending. Broadly, the rest of NE's money has been spent on three other activities. The first is light water reactor sustainability, which refers to innovations that optimize existing reactor operations. In any other domain, industry would be expected to be responsible for such activities – not government. The second encompasses cross-cutting activities, from modeling and simulation infrastructure to work on sensors, controls, and the rest of the nuclear fuel cycle. The biggest share of NE's budget by far has been spent on overhead, which includes program direction and facility management.

The data were phenomenally detailed, but we wanted to answer other questions, too: how can limited government support for emergent energy technologies be allocated judiciously, and how can DOE better enable nuclear innovation? Answering these questions ultimately requires expert judgment. To answer these questions, we conducted semi-structured interviews with 30 senior experts across the nuclear enterprise, on the condition of anonymity, and asked for their help in filling data gaps, identifying bottlenecks to innovation, and crafting an advanced fission agenda for the U.S. The picture they painted was dismaying, with a frequently evoked image being that of an enterprise "on the brink of death," transformed into a jobs program for highly qualified scientists and engineers.

To yield results, nuclear innovation requires political will, greater appropriations, and highly adept project management; experts bemoaned a clear absence of each. Without dramatic changes on these fronts, we concluded that it is highly unlikely that the advanced reactor R&D effort in the U.S. will generate a deployable technology that will help us drastically reduce power sector emissions in the timeframe necessary to avert the worst consequences of climate change.

In 2017, we reported the results from our budget analysis in a paper titled "A Retrospective Analysis of Funding and Focus in U.S. Advanced Fission Innovation" in the journal *Environmental Research Letters*. The results of our interviews with experts can be found in a paper titled "Expert Assessments of the State of U.S. Advanced Fission Innovation" in the journal *Energy Policy*.

At this point, we decided the time had come to put all the pieces together to see if we could identify some way by which newly built nuclear power plants could make a significant contribution to decarbonizing the U.S. energy system over the course of the next two or three decades. We'd already concluded that because of the high cost, no U.S. power company was going to build any new large conventional nuclear power plant before mid-century. Our work on advanced reactors had left us persuaded that no advanced nuclear plants would become commercially available before mid-century at the earliest. But we thought there was still some possibility that we could identify a large enough market for factory-manufactured light water SMRs to contribute to decarbonization – note that we weren't talking about just two or three plants, but the many dozens it would take to make a dent in emissions of carbon dioxide from the energy system. With support from the Sloan Foundation, we set out to investigate whether we could make a defensible case for the development of such light water SMRs in the next few decades.

There has been some effort since the 1990s on the part of industry, advocates, and the national laboratories to build nice, simple narratives of how nuclear power could satisfy novel energy services or bundles of services. These have included supporting renewable energy sources, desalinating water, and providing process heat. After all, wind and solar power are intermittent, and we still need power when the wind is not blowing or the sun is not shining. There had also been much talk for years about the need to turn to desalination as freshwater grows scarcer due to changes in our climate – this talk is often anchored in a broader discussion of the "energy water nexus." Putting those pieces together, new PhD student Mike Rath performed an analysis to see if it might be possible to develop a market for SMRs that could desalinate water when the wind is blowing or the sun is shining but produce electricity for the grid when renewable energy production falls. His results, published in the journal *Progress in Nuclear Energy*, concluded that this could be done but at a cost that was somewhat higher than a system that uses a natural gas turbine and captures and sequesters its carbon dioxide emissions (see Chapter 11).

Of course, since there are other complications associated with nuclear power, he also concluded that the commercial risks would be higher for the system that used an SMR. There might be an international market for such systems, especially in arid parts of the world, but in the face of competition from Chinese, Korean, and Russian reactor exports, there is not much chance that U.S. SMR manufacturers could capture much of that market.

The other service that nuclear power could provide while displacing greenhouse gas emissions is industrial process heat. After all, generating electricity is not the only thing that makes carbon dioxide; lots of coal, oil, and gas is burned to make heat to run industrial processes. Electrifying some of these can be prohibitively expensive, whereas different nuclear technologies could provide low-carbon process heat at different temperatures. To investigate the size of a potential light water SMR market in this space, we developed a database of all U.S. facilities that rely on the combustion of fossil fuels for the provision of process heat. We determined the process characteristics at these facilities and evaluated the cost and performance of deploying light water SMRs across different classes of facilities.

We eliminated three classes of facilities where light water SMR deployment is infeasible: petroleum refineries; plants that require process heat at temperatures higher than 350 degrees Celsius; and plants in states that have all but closed the door to further nuclear development, such as Hawaii, California, and the Pacific Northwest. We estimated that there exists a fairly large market for light water SMR deployment in industrial facilities: approximately 1000 SMRs – up to 4000, depending on assumptions – could be accommodated by U.S. industry. They would mostly be sited in states with extensive industrial capacity, including Texas, Louisiana, and the Midwest. The 12 highest-emitting facilities account for 10% of the sector's greenhouse gas emissions. And yet, the problems that often confront nuclear power are perhaps more prominent in an industrial setting: nuclear heat costs more than fossil alternatives; few corporations could finance nuclear construction, especially since industry lacks the ability of some electric power companies to pass the risk of cost growth on to their customers. Several siting and regulatory challenges also arise due to the confluence of hazardous materials. Using this market to justify factory SMR production appears to be a particularly implausible strategy, given industry demands for cheap heat, predictable performance, and low commercial risk.

Finally, in recent years various folks have argued that the way to get SMRs commercialized is to have the Department of Defense (DoD) do it. It turns out that DoD has already done a pretty good job of dealing with their energy needs for bases in the U.S. and around the world. Further, if DoD did develop their own SMR, they would probably want it to use more enriched uranium fuel than would be appropriate for the civilian market. Finally, the DoD has its own important missions, and we think it is bad public policy to torque them to meet civilian needs. In 2018, we made all these arguments in a paper we published in *Issues in Science and Technology* titled "Nuclear Power Needs Leadership, but Not from the Military."

When we put all of these and other arguments together, we did not like the answer we ended up with: unless there are some rapid and fundamental changes in current public policy, new nuclear power is not going to be able to contribute much of anything before 2050 to decarbonizing the U.S. energy system.

After 2050, advanced designs may begin to show up, and that could be helpful. But if we are going to avoid catastrophic climate change, we need to start making dramatic reductions in emissions now. In our 2018 paper titled "U.S. Nuclear Power: The vanishing low-carbon wedge," which we published in the *Proceedings of the National Academy of Science*, we summed up this grim picture as follows: "We believe that achieving deep decarbonization of the energy system will require a portfolio of every available technology and strategy we can muster. It should be a source of profound concern for all who care about climate change that, for entirely predictable and resolvable reasons, without immediate and profound changes we appear set to lose one of the most promising candidates for providing a wedge of reliable, low-carbon energy over the next few decades and perhaps even the rest of the century."

While it does not appear that the U.S. is going to play a significant role in the development and export of nuclear power over the next few years, several other countries, including China, Russia, and Korea, are developing and exporting reactors. In the final part of his PhD thesis, Mike Ford explores the issue of where one might be able to safely export reactors without producing significant risks. He gathered extensive data on the energy systems, economic performance, and institutional quality of the world's countries and then used Data Envelopment Analysis to assess how relatively risky it would be to export reactors to each. In a paper published in 2021 in the journal *Risk Analysis*, he and Ahmed reported that "85% of potential low-carbon electricity demand growth is in nations that are in the bottom two quartiles of performance." They offered ideas for how to deploy nuclear power in each of the clusters of nations they had identified if the goal is to mitigate risk – these ideas revolve around changing the deployment paradigm from the traditional "owner-operator" model to models that approximate BOOR. They also demonstrated how some nations have seen dramatic increases in low-carbon energy needs and economic output while the quality of their institutions has been deteriorating: Egypt and Saudi Arabia are two prominent examples. These countries might be considered our "best bet" for building a new, global "nuclear wedge," but they're our worst bet if the goal is mitigating risk.

The demise of the U.S. nuclear industry has not only seriously complicated the task of decarbonizing our energy system, but some national security experts have also begun to argue that it has resulted in a serious decline in U.S. influence in the international control of nuclear materials and counterproliferation. In September of 2018, with EPP PhD student Jessica Lovering, we organized and ran a workshop in Washington, D.C. with 21 nuclear and foreign policy experts. We used a workbook format like the one we had adopted for our SMR workshop in Switzerland. For the Washington workshop, we developed an extensive set of background readings and then crafted six detailed strategies that might be used to strengthen U.S. influence in these matters. After a series of framing presentations and a discussion and evaluation of our operating assumptions, we described and participants discussed each of the six strategies, which they then evaluated individually in their workbooks. On the second day, participants developed four additional strategies. As the abstract reproduced at the end of this chapter notes, "While not all experts agreed that U.S. influence has already declined, most indicated that it likely would decline in the future if present domestic and international trends continue. Although none of the proposed strategies that we advanced or that the experts suggested are likely to be effective in the short term, several warrant ongoing refinements and, if they can be implemented, might have beneficial impacts in coming decades."

For the second part of her PhD thesis, Jessica performed a detailed assessment of the possibility that microreactors might find a market in places like remote arctic communities where importing diesel fuel is extraordinarily expensive. Her results indicate that microreactors can be cheaper and more reliable than 100% renewable systems, and they can also be cost-competitive with diesel where fuel costs are greater than $1/liter and the microreactor capital cost is less than $15,000/kW. A $15,000/kW sticker price might seem high, but it also means that a 1 MW$_e$ microreactor would cost $15 million – a remarkably low sum. It remains quite unclear – and in our judgment seems quite unlikely – that a nuclear reactor could be manufactured and deployed at that low price. The levelized cost of electricity (LCOE) from that microreactor is most sensitive to the initial capital cost. Whether microreactors ever move beyond niche markets will depend on initial costs that are driven by early design choices, as well as the learning effects accrued through factory fabrication and laissez-faire institutional arrangements that do not overcomplicate their deployment at scale.

## Technical Details for Chapter 12

Here is the paper and abstract for the expert elicitation study by Ahmed Abdulla on the likely cost of SMRs.

Abdulla, A., Azevedo, I. L., & Morgan, M. G. (2013). Expert assessments of the cost of light water small modular reactors. *Proceedings of the National Academy of Sciences*, *110*(24), 9686–9691.

**Abstract:** Analysts and decision makers frequently want estimates of the cost of technologies that have yet to be developed or deployed. Small modular reactors (SMRs), which could become part of a portfolio of carbon-free energy sources, are one such technology. Existing estimates of likely SMR costs rely on problematic top-down approaches or bottom-up assessments that are proprietary. When done properly, expert elicitations can complement these approaches. We developed detailed technical descriptions of two SMR designs and then conduced elicitation interviews in which we obtained probabilistic judgments from 16 experts who are involved in, or have access to, engineering-economic assessments of SMR projects. Here, we report estimates of the overnight cost and construction duration for five reactor-deployment scenarios that involve a large reactor and two light water SMRs. Consistent with the uncertainty introduced by past cost overruns and construction delays, median estimates of the cost of new large plants vary by more than a factor of 2.5. Expert judgments about likely SMR costs display an even wider range. Median estimates for a 45 megawatts-electric (MW$_e$) SMR range from

$4,000 to $16,300/kW$_e$ and from $3,200 to $7,100/kW$_e$ for a 225-MW$_e$ SMR. Sources of disagreement are highlighted, exposing the thought processes of experts involved with SMR design. There was consensus that SMRs could be built and brought online about 2 y faster than large reactors. Experts identify more affordable unit cost, factory fabrication, and shorter construction schedules as factors that may make light water SMRs economically viable.

This paper summarizes insights from the workshop we ran in Switzerland:

Prasad, S., Abdulla, A., Morgan, M. G., & Azevedo, I. L. (2015). Nonproliferation improvements and challenges presented by small modular reactors. *Progress in Nuclear Energy, 80,* 102–109.

**Abstract:** Small modular reactors (SMRs) may provide an energy option that will not emit greenhouse gases. From a commercial point-of-view, SMRs will be suitable to serve smaller energy markets with less developed infrastructure, to replace existing old nuclear and coal power plants, and to provide process heat in various industrial applications. In this paper, we examine how SMRs might challenge and improve the existing nonproliferation regime. To motivate our discussion, we first present the opinions gathered from an international group of nuclear experts at an SMR workshop. Next, various aspects of SMR designs such as fissile material inventory, core-life, refueling, burnup, digital instrumentation and controls, underground designs, sealed designs, enrichment, breeders, excess reactivity, fuel element size, coolant opacity, and sea-based nuclear plants are discussed in the context of proliferation concerns. In doing this, we have used publicly available design information about a number of SMR designs (B&W, mPower, SVBR-100, KLT-40S, Toshiba 4S, and General Atomics EM2). Finally, a number of recommendations are offered to help alleviate proliferation concerns that may arise due to SMR design features.

A description of our work on floating small modular reactors can be found in:

Ford, M. J., Abdulla, A., & Morgan, M. G. (2017). Evaluating the cost, safety, and proliferation risks of small floating nuclear reactors. *Risk Analysis, 37*(11), 2191–2211.

**Abstract:** It is hard to see how our energy system can be decarbonized if the world abandons nuclear power, but equally hard to introduce the technology in non-nuclear energy states. This is especially true in countries with limited technical, institutional, and regulatory capabilities, where safety and proliferation concerns are acute. Given the need to achieve serious emissions mitigation by mid-century, and the multi-decadal effort required to develop robust nuclear governance institutions, we must look to other models that might facilitate nuclear plant deployment while mitigating the technology's risks. One such deployment paradigm is the build-own-operate-return model. Because returning small land-based reactors containing spent fuel is infeasible, we evaluate the cost, safety, and proliferation risks of a system in which small modular reactors are manufactured in a factory, and then deployed to a customer nation on a floating platform. This floating SMR would be owned and operated by a single entity and returned unopened to the developed state for refueling. We developed a decision model that allows for a comparison of floating and land-based alternatives considering key International Atomic Energy Agency plant-siting criteria. Abandoning on-site refueling is beneficial, and floating reactors built in a central facility can potentially reduce the risk of cost overruns and the consequences of accidents. However, if the floating platform must be built to military-grade specifications then the cost would be much higher than a land-based system. The analysis tool presented is flexible, and can assist planners in determining the scope of risks and uncertainty associated with different deployment options.

This is the paper in which we argued that it is unlikely that new nuclear power plants will be able to contribute much to carbonizing the U.S. energy system over the next several decades:

Morgan, M. G., Abdulla, A., Ford, M. J., & Rath, M. (2018). US nuclear power: The vanishing low-carbon wedge. *Proceedings of the National Academy of Sciences, 115*(28), 7184–7189.

**Abstract:** Nuclear power holds the potential to make a significant contribution to decarbonizing the US energy system. Whether it could do so in its current form is a critical question: Existing large light water reactors in the United States are under economic pressure from low natural gas prices, and some have already closed. Moreover, because of their great cost and complexity, it appears most unlikely that any new large plants will be built over the next several decades. While advanced reactor designs are sometimes held up as a potential solution to nuclear power's challenges, our assessment of the advanced fission enterprise suggests that no US design will be commercialized before midcentury. That leaves factory-manufactured, light water small modular reactors (SMRs) as the only option that might be deployed at significant scale in the climate-critical period of the next several decades. We have systematically investigated how a domestic market could develop to support that industry over the next several decades and, in the absence of a dramatic change in the policy environment, have been unable to make a convincing case. Achieving deep decarbonization of the energy system will require a portfolio of every available technology and strategy we can muster. It should be a source of profound concern for all who care about climate change that, for entirely predictable and resolvable reasons, the United States appears set to virtually lose nuclear power, and thus a wedge of reliable and low-carbon energy, over the next few decades.

This is the paper that looked at the possibility of using small modular reactors in combination with wind to desalinate water and make electricity:

Rath, M., & Morgan, M. G. (2020). Assessment of a hybrid system that uses small modular reactors (SMRs) to back up intermittent renewables and desalinate water. *Progress in Nuclear Energy, 122,* 103269.

**Abstract:** Because water is easier to store in substantial quantity than electricity, this paper examines the possibility that a U.S. domestic market for factory-manufactured small modular nuclear reactors (SMRs) might be developed to use the constant output of an SMR to perform water desalination when wind or solar power are producing high output and generate electricity for the grid when wind or solar power output is low. In the first part of the paper, we compare powering desalination systems with electricity from SMRs and from natural gas plants that are equipped with a system that performs carbon capture and geological sequestration (NG CCS). We show that mass-produced SMRs could have costs that are comparable to, but probably somewhat higher than those of systems based on NG CCS. We find that the cost of $CO_2$ emissions would have to rise to roughly 200 \$/ton for the SMR solution to be clearly dominant. In the second part of the paper, we examine the uneven water supply situation across the U.S, focusing on the southwestern and western regions, and conclude that over the next several decades serious shortages are likely to develop only in a few local markets, such as West Texas and the Monterey Peninsula. Even if factory mass production of SMR's were more successful in reducing costs than experts have estimated, and costs could be reduced to that of NG CCS systems, the commercial risks and siting difficulties likely to accompany SMRs would probably preclude their wide adoption in the U.S. over the next few decades. Globally there are regions where a significant market for desalination supported by nuclear power might develop. However, aggressive changes in U.S. regulatory and export policy will be needed if U.S. SMR manufacturers are to play a role in those or similar markets in the face of aggressive Chinese, South Korean, and Russian programs of reactor exports.

This paper describes the work on assessing how well countries around the prepared are prepared to accept nuclear power:

Ford, M. J., & Abdulla, A. (2021). New methods for evaluating energy infrastructure development risks. *Risk Analysis.*

**Abstract:** Many energy technologies that can provide reliable, low-carbon electricity generation are confined to nations that have access to robust technical and economic capabilities, either on their own or through geopolitical alliances. Equally important, these nations maintain a degree of institutional capacity that could lower the risks associated with deploying emergent energy technologies such as advanced nuclear or carbon capture and storage. The complexity, expense, and scrutiny that come with building these facilities make them infeasible choices for most nations. This paradigm is slowly changing, as the pressing need for low-carbon electricity generation and ongoing efforts to develop modular nuclear and carbon capture technologies have opened the door for potentially wider markets, including in nations without substantial institutional capacity. Here, using advanced nuclear technologies as our testbed, we develop new methods to evaluate national readiness for deploying complex energy infrastructure. Specifically, we use Data Envelopment Analysis—a method that eliminates the need for expert judgment—to benchmark performance across nations. We find that approximately 80% of new nuclear deployment occurs in nations that are in the top two quartiles of institutional and economic performance. However, 85% of potential low-carbon electricity demand growth is in nations that are in the bottom two quartiles of performance. We offer iconic paradigms for deploying nuclear power in each of these clusters of nations if the goal is to mitigate risk. Our research helps redouble efforts by industry, regulators, and international development agencies to focus on areas where readiness is low and risk correspondingly higher.

This is the paper that describes the workshop we ran in Washington, D.C. on whether the U.S. is losing influence on international nuclear matters and what might be done about it:

Lovering, J. R., Abdulla, A., & Morgan, G. (2020). Expert assessments of strategies to enhance global nuclear security. *Energy Policy*, *139*, 111306.

**Abstract:** Historically, the U.S. has sought to use commercial trade in nuclear technologies to influence international nuclear security standards and promote nonproliferation. Concern has grown that, with a stagnating domestic nuclear industry and declining export industry, the U.S. will lose a significant tool of foreign policy and leverage in maintaining strong international standards. While the issue has been discussed extensively in the policy community and used as a powerful rhetorical tool to motivate tangentially related policies such as subsidizing existing U.S. nuclear plants, no one has systematically assessed the issue, structured the problem and proposed and evaluated potential solutions. Here we briefly analyze the current international state of play, and then outline a set of specific strategies the U.S. might adopt on its own, or promote internationally, to retain its influence. Building on the literature, nuclear security and nuclear power experts assisted us in framing the issues and then, in a participatory workshop, helped us to assess and refine possible strategies. While not all experts agreed that U.S. influence has already declined, most indicated that it likely would decline in the future if present domestic and international trends continue. Although none of the proposed strategies that we advanced or that the experts suggested are likely to be effective in the short term, several warrant ongoing refinements and, if they can be implemented, might have beneficial impacts in coming decades.

Here are several more of our nuclear-related publications:

- Abdulla, A. (2015). *Preserving the nuclear option: Overcoming the institutional challenges facing small modular reactors.* A report of the International Risk Governance Council (IRGC), 41pp. Available online at: www.irgc.org/publications/reports-on-special-issues/
- Ford, M. J., Abdulla, A., Morgan, M. G., & Victor, D. G. (2017). Expert assessments of the state of US advanced fission innovation. *Energy Policy, 108,* 194–200.
- Abdulla, A., Ford, M. J., Morgan, M. G., & Victor, D. G. (2017). A retrospective analysis of funding and focus in U.S. advanced fission innovation. *Environmental Research Letters, 12*(8), 084016.
- Ford, M. J., Abdulla, A., & Morgan, M. G. (Summer 2018). Nuclear power needs leadership, but not from the Military. *Issues in Science & Technology,* 67–72.

**Some notable writings by authors not affiliated with our Centers:**

Ahearne, J. F., Carr Jr, A.V., Feiveson, H. A., Ingersoll, D., Klein, A. C., Maloney, S., Oelrich, I., Squassoni, S., Wolfson, R., & From, F.A.S. (2012). *The future of nuclear power in the United States.* Federation of American Scientists.

Ahmad, A. (2021). Increase in frequency of nuclear power outages due to changing climate. *Nature Energy, 6,* 755–762.

Froese, S., Kunz, N. C., & Ramana, M.V. (2020). Too small to be viable? The potential market for small modular reactors in mining and remote communities in Canada. *Energy Policy, 144,* 111587.

Institute for Nuclear Power Operations is an organization that works to improve the security and reliability of the U.S. nuclear power fleet. Details available online at: www. inpo.info/AboutUs.htm

International Energy Agency (IEA). (2019). *Nuclear Power in a Clean Energy System,* 101pp. Available online at: https://iea.blob.core.windows.net/assets/ad5a93ce-3a7f-461d-a441-8a05b7601887/Nuclear_Power_in_a_Clean_Energy_System.pdf

Krall, L. M., Macfarlane, A. M., & Ewing, R. C. (2022). Nuclear waste from small modular reactors. *Proceedings of the Natural Academy of Sciences, 119*(23), e2111833119.

Nuttall, W. J. (2022). *Nuclear renaissance: Technologies and policies for the future of nuclear power.* CRC Press.

Portugal-Pereira, J., Ferreira, P., Cunha, J., Szklo, A., Schaeffer, R., & Araújo, M. (2018). Better late than never, but never late is better: Risk assessment of nuclear power construction projects. *Energy Policy, 120,* 158–166.

Stewart, W. R., & Shirvan, K. (2021). Capital cost estimation for advanced nuclear power plants. *Renewable and Sustainable Energy Reviews,* 111880.

# 13

# MAKING ELECTRIC POWER MORE RESILIENT

*Jay Apt and Granger Morgan*

As the source of roughly a quarter of all U.S. emissions of carbon dioxide, the electricity system must be a central part of any effective effort to decarbonize the energy system. The elements of such an effort include improving the efficiency with which we use electricity (Chapters 8 and 9), making much greater use of non-emitting wind and solar energy (Chapter 10), and finding ways to continue to use reliably dispatchable fossil fuel generation, especially natural gas, without adding more carbon dioxide to the atmosphere. Not only must we decarbonize the existing system, but because electricity is the best way to decarbonize many other parts of the systems (such as light vehicles and much of industry), in parallel with decarbonizing we also need to figure out how to increase the amount of electricity we generate. As if that were not enough, while we are doing this we also need to work to make the electricity system more reliable and resilient.

Almost all the activities of modern society depend upon electricity. We use it to light, heat, and cool our homes and places of work. It powers communication and our stores and factories as well as most water and sewer systems. Brief interruptions are annoying. They are mainly caused by local events such as thunderstorms, cars or trucks crashing into power poles, or squirrels getting into the wrong place. Users that are critical usually have backup power.

Larger interruptions of long duration are much more disruptive and costly. Such LLD-outages can be caused by a variety of natural[1] and accidental and pernicious human events. Figure 13.1A, which Granger prepared for use in a briefing on a National Academy report on the resilience of the power system, displays the range of causes. Figure 13.1B, which Jay originally assembled and the Academies then updated, shows that LLD-outages are more frequent than one might think.

DOI: 10.4324/9781003330226-13

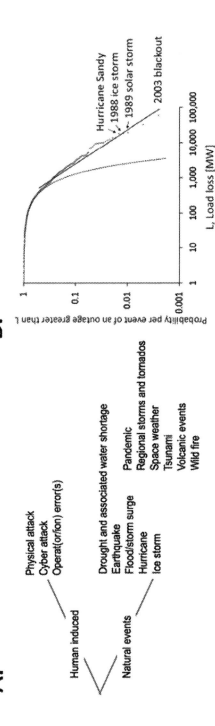

**FIGURE 13.1** **A.** Large power outages of long duration (LLD–outages) can be caused by many natural events as well as both intentional and unintentional human actions. **B.** Such outages are more common than one might think as shown in this plot of the relative frequency of outages in the U.S. bulk power system over the period from 1984 to 2015. The figure, which is reproduced from a report from the U.S. National Academies, includes 1002 events with load loss (loss in electricity demand) greater than 1 MW. The dashed line fits an exponential distribution to the more frequent events with load loss below 500 MW. Note that large outage events do not fit this line and are much more common than one might expect from an extrapolation of the frequency of smaller events. Fitting a curve to the large outage events suggests that at least every century North America will experience a blackout that affects about 25% of all customers.

Source: Data are from EIA (2000–2015), NERC (2000–2009), and NRC (2012).

Drawn by Granger Morgan and Jay Apt.

Republished with permission of The National Academies Press, from Enhancing the Resilience of the Nation's Electricity System, National Academies of Sciences, Engineering, and Medicine, 2017; permission conveyed through Copyright Clearance Center, Inc.

## The Characteristics of Large Blackouts

### Jay Apt

When PhD student Paul Hines, working with Sarosh Talukdar and me, examined all the large blackouts in North America from 1984–2006, we found that the frequency of blackouts remained about the same during that period, whether or not we made a correction for the increasing number of customers. As we had expected, we found that large blackouts are substantially more frequent in the summer and winter than in the spring and fall and that they occur more frequently during mid-afternoon hours. However, we were a bit surprised to find that the amount of time it took to restore power and the size of the blackout were not correlated (Figure 13.2).

Soon after the August 14, 2003 Northeast blackout, five of us published a paper with an early version of Figure 13.1 B above and argued that "Inevitably, there will be future cascading failures." It was a good thing that most of the authors were tenured professors (we were two power engineers, two physicists, and an economist) since that statement seriously annoyed the grid pundits who insisted that the system could be made invulnerable if only we were to spend more money on the grid.[2] But as we wrote, "We cannot be sure of detecting and eliminating major flaws from [proposed] solutions, nor of even reliably distinguishing good solutions from bad ones." Rather, we argued, it is better to figure out how to fulfill the critical missions of the system during a power blackout than to attempt the impossible task of eliminating blackouts. Nearly 20 years later, the February 2021 Texas blackout once again showed that a safety net is more practical than trying to design

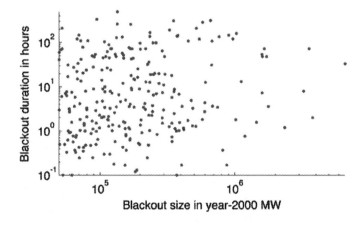

**FIGURE 13.2** Blackout duration vs. blackout size in MW for events larger than 300 MW (after scaling). There is no systematic relationship between how big a blackout is and how long it lasts – that is, the two variables are almost perfectly uncorrelated.

Hines, P., Apt, J., & Talukdar, S. (2009). Large blackouts in North America: Historical trends and policy implications. *Energy Policy, 37*(12), 5249–5259.

an invulnerable grid. Of course, cost-effective measures such as winterization of power plants and fuel supplies should be taken. But we noted that, for example, solar-powered traffic lights in city cores, backup power for water pumps, and heated shelters for people are all much more practical and affordable than attempting to make the grid invulnerable.

In a paper and a book chapter after the 2001 attacks on the World Trade Center and Pentagon, we considered the issues involving a terrorist attack on the power system. We noted that "Competitive markets will force the adoption of the lowest-cost solutions to providing electricity under the stipulated rules…Because security is a classic public good, our expectation is that it will not be an attractive investment. Thus, it is up to government to answer questions concerning how much the nation is willing to pay for additional security, what organizations will be charged with ensuring it, and who should pay for it." As of this writing, there has been considerable progress and a number of grid exercises to restore the system after an attack has occurred. However, we stand by our conclusion that the best posture under the irreducible uncertainties of both natural and human-caused grid failure is to ensure that the essential services normally provided by the grid are available even when grid power goes out.

Natural gas fuels an ever-increasing fraction of electricity produced in North America and is an important fuel for distributed backup-power generation. In 2017, PhD student Gerad Freeman joined us with several years' experience in the natural gas industry. We began investigating the reliability of the gas pipeline system. Together with Michael Dworkin, former chair of the Vermont Public Service Board, we determined that the gas pipeline network has much looser requirements to report outages than does the electric power system. We recommended that all pipeline incidents of sufficient size to trigger a mandatory power plant outage report should be reported and that Congress should follow the example of its requirement for the electric grid by designating a central entity to oversee the reliability of the natural gas delivery system.

Gerad, John Moura from the North American Electric Reliability Corporation (NERC), and I published a paper in 2020 that analyzed 6 years of gas power plant failures, finding that failures in the pipeline network account for no more than 5% of the MWh lost to natural gas fuel shortages. While the pipelines are quite reliable, power plants that did not purchase firm fuel contracts were often out-prioritized for gas. In fairly large regions of the Mid-Atlantic and Midwest, there was always available gas and if they had purchased firm contracts, power plants could have continued operating. The volume of gas needed by power plants to fuel the lost MWh in those regions was only a small fraction of the total volume delivered to potentially non-essential commercial and industrial pipeline customers in those regions, and modest prices there at the times when power plants failed indicate gas was available. The largest region that has a long-standing shortage of natural gas is New England.

Working with Tom Coleman of NERC and EPP graduate and Pennsylvania State University professor Seth Blumsack, Gerad and I found that nearly all of the

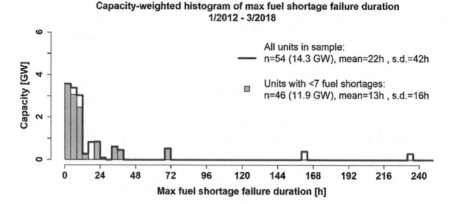

**FIGURE 13.3** Capacity-weighted histogram of the maximum natural gas fuel shortage failure duration for all natural gas-fired electric generation units in New England reporting fuel shortage failures between January 1, 2012 and March 31, 2018.

Freeman, G. M., Apt, J., Blumsack, S., & Coleman, T. (2021). Could on-site fuel storage economically reduce power plant-gas grid dependence in pipeline constrained areas like New England?. *The Electricity Journal*, *34*(5), 106956.

New England power plant outages due to unavailability of natural gas last only a few hours (Figure 13.3) and that dual fuel oil-gas generators can profitably make up for the gas unavailability if compensated with a "reliability adder" of $3–7/MWh during their normal operations. Onsite compressed natural gas is also feasible, at slightly higher cost. We estimate that the capital expenses associated with the fuel storage options would be less expensive than installing battery backup for resource adequacy at current battery prices.

## Disruptions Caused by Extreme Heat and Cold

### Jay Apt

In the decade from 2011–2021, there were four cold-weather events that affected the U.S. bulk electric system generation. From February 1–5, 2011, 14.7 GW of generation in Texas was unavailable due to cold weather, resulting in 5.4 GW of load being shed over 7 hours. January 6–8, 2014 saw the loss of 9.8 GW of generation in PJM with 300 MW of lost load for 3 hours. On January 15–19, 2018, the South Central U.S. saw 15.6 GW of unavailable generation but avoided shedding load. The February 8–20, 2021 cold weather in Texas and the southern Great Plains caused 65.6 GW of generation to be lost, resulting in 23 GW of unserved load for up to 70 hours.

In the fall of 2014, PhD student Sinnott Murphy came to EPP with considerable experience with large data sets. We worked with the North American Electric Reliability Corporation (NERC) and the PJM Interconnection to gain access to information each organization has been gathering on the failures of 85% of all electric power generators in North America. The data were very difficult to process, and it took three years before publication of the first results.[3] The nationwide data covered only four years, but even in that limited period, there were correlated generator outages in nearly every region. This is important because grid operators compute the number of reserve generators they keep online based on the (incorrect) assumption that generators fail independently of each other.

In a subsequent paper, Sinnott, CMU professor Fallaw Sowell, and I used 23 years of data for most of the generators in the country's largest electric power market, the PJM Interconnection, to show that these correlated generator failures tend to happen at extremely low and high outside temperatures (Figure 13.4). Interestingly, there is no dependence of correlated failures on the demand for electric power.

In the fall of 2017, an exceptional PhD student with a deep background in energy markets, Luke Lavin, joined EPP. Luke, Sinnott, and Brian Sergi (all former EPP PhD students) and I published a method whereby PJM and other grid operators can compute the required reserve generation taking into account these temperature-dependent correlated failures.[4] We found in PJM that capacity requirements vary by month, with more than 95% of loss-of-load risk accruing in July. We identified modest resource adequacy risks from potential future climate scenarios, modeled as temperature increases of 1 and 2°C relative to our study period. Holding loads fixed, these scenarios increase capacity requirements by approximately 0.5% and 1.5%, respectively.

After the February 2021 Texas freeze and subsequent blackouts, Luke and I published an op-ed in *The Washington Post* pointing out that until grid operators take these temperature-dependent correlated failures into account, what happened in Texas will keep on happening.[5]

## Assessing the Cost of LLD-Outages

### Granger Morgan

Short power outages are annoying, but most of us can manage without power for a few hours. Critical services like hospitals or police and emergency services have backup power. Cell phone towers, many newer buildings, and others are also required (or choose) to have backup. Similarly, there is battery backup in many life-critical medical devices that people have in their homes.

Utilities work hard to minimize local short outages caused by events in their distribution systems. To do this, they make investments in technologies such as distribution system automation. In order to justify these investments to their regulators, they have conducted many studies to assess the costs to customers. Most of the results of these studies are proprietary, but analysts supported by Lawrence Berkeley National Labs have obtained access to these data and used them to build a system called the Interruption Cost Estimate (ICE) Calculator[6] that anyone can use to

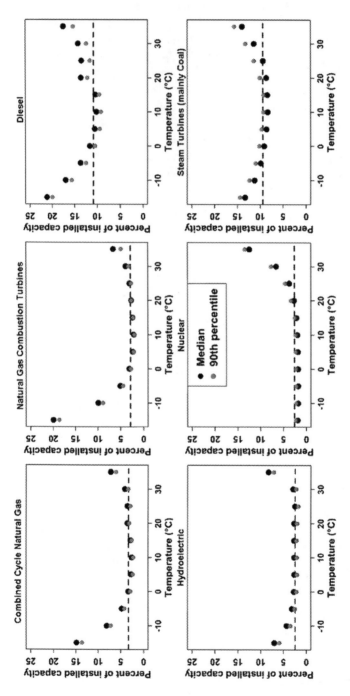

**FIGURE 13.4** Levels of unavailable generator capacity as a function of temperature as modeled from 1995–2018 PJM data from 1845 generators of six types. Black dots were calculated using median load from temperature neighborhood; red dots were calculated using 90th percentile load from temperature neighborhood. The dashed lines indicate the average failure level, without considering temperature, for each generator type.

Murphy, S., Sowell, F., & Apt, J. (2019). A time-dependent model of generator failures and recoveries captures correlated events and quantifies temperature dependence. *Applied Energy, 253*, 113513.

get an estimate of the cost of short disruptions to different types of customers in different parts of the country at different times of the year.

Almost all of the data on which the ICE Calculator has been built is for outages that last only a few hours. As a result, it cannot provide any insight about the costs (often called the value of lost load, or VLL) for long duration outages. This is especially true for outages that black out a large area, which means customers can't just temporally go to a nearby place that still has power.

Working with me and EPP faculty member Alex Davis, EPP PhD student Sunhee Baik designed a series of studies to obtain estimates of VLL for residential customers experiencing large outages of long duration. While many people will answer almost any question you put in a survey, the answers that people give to questions about large outages of long duration are not likely to be meaningful unless they are given the opportunity to think systematically about what such an outage would involve. In a first study, which was conducted using face-to-face interviews (each of which took roughly an hour) with 45 residents of Allegheny County, PA, Sunhee explored what people would be willing to pay to avoid going without electricity for an entire 24 hours on a summer or cold winter weekend. She also asked about willingness to pay for limited backup service. To pose these questions, she provided detailed information about what would and would not work as well as consequences such as losing food in refrigerators and freezers. She also developed and had people play a "card-stacking game" in which they were asked to build up the electricity demand for their house at different times of the day using cards whose height was proportional to the amount of current needed by a range of household appliances (see Figure 13.5).

While the face-to-face interviews worked very well, they were too time-consuming to allow us to run studies with a large number of participants that would allow us to get better statistical power. To address this problem, Sunhee built a web-based system that supported people doing similar things online. To help

Example of cards people used to construct the "load profile" for their home

Typical make-up load at 4 times during a day

After creating the profile, Sunhee would then apply a 20 Amp limit and ask people to think about what they could run if they only had this partial back-up

**FIGURE 13.5** Example of the cards Sunhee used in her card-stacking game. The height of each card showed the amount of current each appliance or device used. She could then apply a 20A limit for backup service and ask respondents to think about what they would still want to operate under those conditions.

Photos taken by Granger Morgan.

people think about their needs for electricity and the consequences of not having power, the system included a number of short video briefings. Using the web-based system, she conducted a series of studies to examine people's willingness to pay for LLD-outages of 10 days' duration during cold winter weather.

Sunhee collected a sample of 483 residential customers across the Northeast U.S. She found that people were willing to pay US$1.7–2.3 per kWh to sustain their private demands during such an outage – roughly ten times more than the typical cost of electricity in the Northeast, which runs roughly 15–20¢ per kWh. She also found that people were willing to pay an additional US$19–29 per day to support critical social services and low-income residents in their communities. She found that the answers people gave did not depend on whether they had previously experienced a long-duration outage (e.g., during one of the big hurricanes that hit the East Coast) or whether the cause of the outage was natural or human-caused (e.g., a large solar storm or a terrorist attack).

You can find details on these studies in the three papers listed at the end of the chapter.

## Strategies to Increase Resilience to LLD-Outages

### Granger Morgan

Considering modern technologies for distribution system automation and distributed generation, EPP PhD students Doug King, Anu Narayanan, and Angelena Bohman have conducted studies of the technologies and costs of using distributed generation and microgrids to keep the power on when the main power grid is not operating. Mike Rath has examined the tradeoffs between cost and environmental impacts of microgrids.

In 2007, Doug found that even under existing electricity rate structures, there were a number of situations in which switching some kinds of large customers to their own distributed generation could be cost-effective.

Anu asked the question, "How might distribution automation combined with distributed generation be used to reduce potential vulnerability to LLD-outages?" She examined the system shown in Figure 13.6 and concluded that the costs for this system ranged from $9 to $22 per year per household, for annual outage probabilities of 0.01, 0.001, and 0.0001 for the different scenarios and DG configurations she studied. She noted that even her upper-bound estimate comprised a very small fraction of median annual household income, making the proposed strategy potentially worthwhile.

In conclusion, Anu and I argued that our cost analysis suggested that:

> … at least a few regions might find it reasonable to invest in a system of the type we have outlined to secure critical social services in the event of a large, long duration outage that occurs in a temperate season of the year. Clearly, no electric utility will make these investments on its own. However, if a

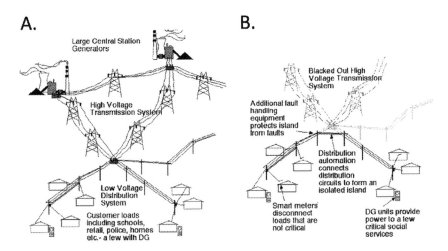

**A.**

Large Central Station
Generators

High Voltage
Transmission System

Low Voltage
Distribution
System

Customer loads
including schools,
retail, police, homes
etc.- a few with DG

**B.**

Blacked Out High
Voltage Transmission
System

Additional fault
handling
equipment
protects island
from faults

Distribution
automation
connects
distribution
circuits to form an
isolated island

Smart meters
disconnnect
loads that are
not critical

DG units provide
power to a few
critical social
services

FIGURE 13.6 Simplified diagrams of **A,** the electric power transmission and distribution system under normal operation, and **B,** the islanded distribution system analyzed by Anu Narayanan during a large outage of long duration in which DG units serve local critical social services. Smart meters have disconnected loads that are not critical. Feeders have been reconfigured to form an isolated "island" using distribution automation and added low-power fault-handling equipment.

Narayanan, A., & Morgan, M. G. (2012). Sustaining critical social services during extended regional power blackouts. *Risk Analysis: An International Journal, 32*(7), 1183–1193.

> public utility commission (PUC) concluded that installing such capabilities constituted a prudent investment, then in regulated distribution companies nondepreciated capital costs and operation costs could be recovered through the rate base with the approval of the regulator. Alternatively, local, county, or state government might choose to fund the project with tax revenue, contracting with the local distribution utility and other parties to implement the changes.

The paper briefly elaborated this issue of alternative ways to cover the cost of investments to improve power-system resilience and inspired some additional discussion of these issues in two National Academy reports and in Angelena Bohman's subsequent work.

Building on some of the ideas in Anu's work, Angelena Bohman worked with Ahmed Abdulla and me to construct a hypothetical community in the upper Connecticut River Valley, a region that has experienced both hurricanes and ice storms in the past. Her focus was on comparing the costs of individual strategies (local backup generators at each house) and a variety of collective strategies that could be used to enhance resilience in the face of a possible LLD-outage. Figure 13.7 shows the alternatives she considered.

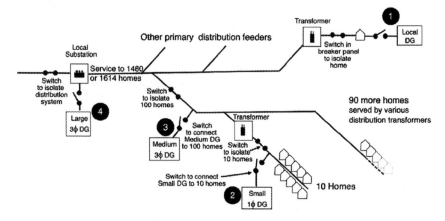

**FIGURE 13.7** Diagram of the hypothetical distribution feeder and associated backup strategies that Angelena Bohman considered in her work. Four plausible DG deployment strategies are shown that involve alternative ways to supply a limited amount of power to households during an LLD-outage. They are: (1) individual generators at each house; (2) a single-phase generator that serves 10 homes operated as an islanded microgrid; (3) a three-phase generator that serves 100 homes as an islanded microgrid; and (4) a three-phase generator at the substation that can supply the entire feeder. It was assumed that sectionalizing switches and smart meters can be used to create microgrids or to isolate portions of the feeder that are not intact because of storm damage.

Bohman, A. D., Abdulla, A., & Morgan, M. G. (2022). Individual and collective strategies to limit the impacts of large power outages of long duration. *Risk Analysis*, *42*(3), 544–560.

The analysis assessed the cost of serving the basic emergency load during outages that last 5, 10, and 20 days in either winter or summer. We find that the cost of collective solutions could be as much as 10 to 40 times less than the cost of individual solutions (less than $2 per month per home). Of course, implementing collective solutions requires that a variety of arrangements be made with the town government, the power company, regulators, and possibly the state legislature. In addition to its engineering economic analysis, the paper discusses a number of these issues.

The proponents of microgrids often argue that they have environmental benefits. When they are used to implement combined heat and power systems, like some of the microgrid systems discussed by Doug King, the efficiency with which fossil fuel is converted to useful output is increased and in that sense, there are benefits. However, if the objective is to completely decarbonize the energy system, then microgrids that use natural gas are not environmentally attractive. Working with Ahmed Abdulla, Jerry Cohon, and me, EPP PhD student Mike Rath developed a multi-objective model that explored the tradeoff between cost and environmental impacts for microgrids. This analysis examines several potential microgrid applications, all located in Southern California, and found that as the relative importance assigned to cost versus environmental was changed, only a small set of alternative designs emerged. It did not appear that microgrids based solely on solar photovoltaics and storage would ever be attractive in the applications that we've studied.

## Technical Details for Chapter 13

Before investigators in our Centers began to consider the potential that weather and climate held to disrupt the power system, we wrote several papers that examined the potential for disruptions of the power system that could be caused by terrorism. Two examples of this work are:

Farrell, A. E., Lave, L. B., & Morgan, G. (2002). Bolstering the security of the electric power system. *Issues in Science & Technology, 18*(3), 49–56.

Lave, L. B., Apt, J., & Morgan, G. (2007). 13. Worst-case electricity scenarios: The benefits and costs of prevention. In *The Economic Costs and Consequences of Terrorism*, Edward Elgar Publishing.

Here are two papers we wrote that examined the characteristics of blackouts and whether prevention is practical. The first of these contains the earliest version of the exceedance curve of blackout size shown in Figure 13.1B above:

Talukdar, S. N., Apt, J., Ilic, M., Lave, L. B., & Morgan, M. G. (2003). Cascading failures: Survival versus prevention. *The Electricity Journal, 16*(9), 25–31.

**Abstract:** While measures can be taken to reduce the number of large-scale power losses due to failures of the generation and high-voltage transmission grid, such failures cannot be eliminated. The survival of essential missions is a more tractable problem than the prevention of all large cascading failures, and its solutions are verifiable. Thus, serious attention should be directed towards assuring the continuation of essential missions even after the grid has failed.

Hines, P., Apt, J., & Talukdar, S. (2009). Large blackouts in North America: Historical trends and policy implications. *Energy Policy, 37*(12): 5249–5259.

**Abstract:** Using data from the North American Electric Reliability Council (NERC) for 1984–2006, we find several trends. We find that the frequency of large blackouts in the United States has not decreased over time, that there is a statistically significant increase in blackout frequency during peak hours of the day and during late summer and mid-winter months (although non-storm-related risk is nearly constant through the year) and that there is strong statistical support for the previously observed power-law statistical relationship between blackout size and frequency. We do not find that blackout sizes and blackout durations are significantly correlated. These trends hold even after controlling for increasing demand and population and after eliminating small events, for which the data may be skewed by spotty reporting. Trends in blackout occurrences, such as those observed in the North American data, have important implications for those who make investment and policy decisions in the electricity industry. We provide a number of examples that illustrate how these trends can inform benefit-cost analysis calculations. Also, following procedures used in natural disaster planning we use the observed statistical trends to calculate the size of the 100-year blackout, which for North America is 186,000 MW.

Here are five papers that examine how the reliability of the power system is influenced by weather and climate:

Murphy, S., Sowell, F., & Apt, J. (2019). A time-dependent model of generator failures and recoveries captures correlated events and quantifies temperature dependence. *Applied Energy*, *253*, 113513.

**Abstract:** Most current approaches to resource adequacy modeling assume that each generator in a power system fails and recovers independently of other generators with invariant transition probabilities. This assumption has been shown to be wrong. Here we present a new statistical model that allows generator failure models to incorporate correlated failures and recoveries. In the model, transition probabilities are a function of exogenous variables; as an example we use temperature and system load. Model parameters are estimated using 23 years of data for 1845 generators in the USA's largest electricity market. We show that temperature dependencies are statistically significant in all generator types, but are most pronounced for diesel and natural gas generators at low temperatures and nuclear generators at high temperatures. Our approach yields significant improvements in predictive performance compared to current practice, suggesting that explicit models of generator transitions using jointly experienced stressors can help grid planners more precisely manage their systems.

Murphy, S., Lavin, L., & Apt, J. (2020). Resource adequacy implications of temperature-dependent electric generator availability. *Applied Energy*, *262*, 114424.

**Abstract:** Current grid resource adequacy modeling assumes generator failures are both independent and invariant to ambient conditions. We evaluate the resource adequacy policy implications of correlated generator failures in the PJM Interconnection by making use of observed temperature-dependent forced outage rates. Correlated failures pose substantial resource adequacy risk, increasing PJM's required reserve margin from 15.9% to 22.9% in the 2018/2019 delivery year. However, PJM actually procured a 26.6% reserve margin in this delivery year, translating to excess capacity payments of $315 million and an implied value of lost load of approximately $700,000/MWh, a figure two orders of magnitude greater than typically used in operational contexts. Capacity requirements vary by month, with more than 95% of loss-of-load risk accruing in July. Setting monthly capacity targets could reduce annual PJM procurement by approximately 16%. We examine the resource adequacy implications of the ongoing replacement of nuclear and coal in PJM with combined-cycle gas generators, finding moderate benefits: approximately a 2% reduction in capacity requirements. We identify modest resource adequacy risks from potential future climate scenarios, modeled as temperature increases of 1 and 2 °C relative to our study period. Holding loads fixed, these scenarios increase capacity requirements by approximately 0.5% and 1.5%, respectively.

Freeman, G. M., Apt, J., & Moura, J. (2020). What causes natural gas fuel shortages at US power plants?. *Energy Policy*, *147*, 111805.

**Abstract:** Using 2012–2018 power plant failure data from the North American Electric Reliability Corporation, we examine how many fuel shortage failures at gas power plants were caused by physical interruptions of gas flow as opposed to operational procedures on the pipeline network, such as gas curtailment priority. We find that physical disruptions of the pipeline network account for no more than 5% of the MWh lost to fuel shortages over the six years we examined. Gas shortages at generators have caused correlated failures of power plants with both firm and non-firm fuel arrangements. Unsurprisingly, plants using the spot market or interruptible pipeline contracts for their fuel were somewhat more likely to experience fuel shortages than those with firm contracts. We identify regions of the Midwest and Mid-Atlantic where power plants with non-firm fuel arrangements may have avoided fuel shortage outages if they had obtained firm pipeline contracts. The volume of gas needed by power plants to fuel the lost MWh in those regions was only a small fraction of the total volume delivered to potentially non-essential commercial and industrial pipeline customers in those regions and modest prices there at the times when power plants failed indicate gas was available.

Lavin, L., Murphy, S., Sergi, B., & Apt, J. (2020). Dynamic operating reserve procurement improves scarcity pricing in PJM. *Energy Policy*, *147*(C).

**Abstract:** Competitive electricity markets can procure reserve generation through a market in which the demand for reserves is administratively established. A downward sloping or stepped administrative demand curve is commonly termed an operating reserve demand curve (ORDC). We propose a dynamic formulation of an ORDC with generator forced outage probabilities conditional on ambient temperature to implement scarcity pricing in a wholesale electricity market. This formulation improves on common existing methods used by wholesale market operators to articulate ORDCs by explicitly accounting for a large source of observed variability in generator forced outages, whereby for a fixed load, more reserves are required during times of extreme heat and cold to maintain a constant risk of reserve shortage. Such a dynamic ORDC increases social welfare by $17.1 million compared to current practice in the PJM Interconnection during a high load week in a welfare-maximizing electricity market with co-optimized procurement of energy and reserves. A dynamic ORDC increases reserve prices under scarcity conditions, but has minimal effects on total market payments. The results are directly relevant to the modeled two-settlement electricity market in PJM, which is currently undergoing enhancements to its ORDC.

Freeman, G. M., Apt, J., Blumsack, S., & Coleman, T. (2021). Could on-site fuel storage economically reduce power plant-gas grid dependence in pipeline constrained areas like New England?. *The Electricity Journal*, *34*(5), 106956.

**Abstract:** In the Northeastern United States, natural gas supply constraints have led to periods when gas shortages have caused up to a quarter of all unscheduled power plant outages. Dual fuel oil/gas generators or local gas storage might mitigate gas supply shortages. We use historical power plant operational and availability data to develop a supply curve of the costs required for generators to mitigate fuel shortage failures in New England. Based on 2012–2018 data, we find that the historical fuel shortages at approximately 2 GW worth of gas-fired capacity could be mitigated using on-site fuel storage. For comparison, New England's average reserve margin was 1.7–2.8 GW over our sample period. Oil dual fuel plants would recoup their investment if compensated with a reliability adder of $3−7/MWh during their normal operations, while $7−16/MWh would incentivize using on-site, compressed natural gas storage. We estimate that the capital expenses associated with the fuel storage options would be less expensive than installing battery backup for resource adequacy at current battery prices.

Here are three papers on assessing residential customers' willingness to pay to avoid large power outages of long duration:

Baik, S., Davis, A. L., & Morgan, M. G. (2018). Assessing the cost of large-scale power outages to residential customers. *Risk Analysis*, *38*(2), 283–296.

**Abstract:** Residents in developed economies depend heavily on electric services. While distributed resources and a variety of new smart technologies can increase the reliability of that service, adopting them involves costs, necessitating tradeoffs between cost and reliability. An important input to making such tradeoffs is an estimate of the value customers place on reliable electric services. We develop an elicitation framework that helps individuals think systematically about the value they attach to reliable electric service. Our approach employs a detailed and realistic blackout scenario, full or partial (20 A) backup service, questions about willingness to pay (WTP) using a multiple bounded discrete choice method, information regarding inconveniences and economic losses, and checks for bias and consistency. We applied this method to a convenience sample of residents in Allegheny County, Pennsylvania, finding that respondents valued a kWh for backup services they assessed to be high priority more than services that were seen as low priority ($0.75/kWh vs. $0.51/kWh). As more information about the consequences of a blackout was provided, this difference increased ($1.2/kWh vs. $0.35/kWh), and respondents' uncertainty about the backup services decreased (Full: $11 to $9.0, Partial: $13 to $11). There was no evidence that the respondents were anchored by their previous WTP statements, but they demonstrated only weak scope sensitivity. In sum, the consumer surplus associated with providing a partial electric backup service during a blackout may justify the costs of such service, but measurement of that surplus depends on the public having accurate information about blackouts and their consequences.

Baik, S., Morgan, M. G., & Davis, A. L. (2018). Providing limited local electric service during a major grid outage: A first assessment based on customer willingness to pay. *Risk Analysis, 38*(2), 272–282.

**Abstract:** While they are rare, widespread blackouts of the bulk power system can result in large costs to individuals and society. If local distribution circuits remain intact, it is possible to use new technologies including smart meters, intelligent switches that can change the topology of distribution circuits, and distributed generation owned by customers and the power company, to provide limited local electric power service. Many utilities are already making investments that would make this possible. We use customers' measured willingness to pay to explore when the incremental investments needed to implement these capabilities would be justified. Under many circumstances, upgrades in advanced distribution systems could be justified for a customer charge of less than a dollar a month (plus the cost of electricity used during outages), and would be less expensive and safer than the proliferation of small portable backup generators. We also discuss issues of social equity, extreme events, and various sources of underlying uncertainty.

Baik, S., Davis, A. L., Park, J. W., Sirinterlikci, S., & Morgan, M. G. (2020). Estimating what US residential customers are willing to pay for resilience to large electricity outages of long duration. *Nature Energy, 5*(3), 250–258.

**Abstract:** Climate-induced extreme weather events, as well as other natural and human-caused disasters, have the potential to increase the duration and frequency of large power outages. Resilience, in the form of supplying a small amount of power to homes and communities, can mitigate outage consequences by sustaining critical electricity-dependent services. Public decisions about investing in resilience depend, in part, on how much residential customers value those critical services. Here we develop a method to estimate residential willingness-to-pay for back-up electricity services in the event of a large 10-day blackout during very cold winter weather, and then survey a sample of 483 residential customers across northeast USA using that method. Respondents were willing to pay US$1.7–2.3 kWh$^{-1}$ to sustain private demands and US$19–29 day$^{-1}$ to support their communities. Previous experience with long-duration outages and the framing of the cause of the outage (natural or human-caused) did not affect willingness-to-pay.

Here are three papers describing work on how to make the power system more resilient by using distributed generators and distribution automation. While the first two papers were closely related to the work of the Centers, note that they were supported by other sources:

King, D. E., & Morgan, M. G. (2007). Customer-focused assessment of electric power microgrids. *Journal of Energy Engineering, 133*(3), 150–164.

**Abstract:** The cost-effectiveness of interconnected microgrid systems with combined heat and power is examined for 36 cases involving six microgrid applications in six different U.S. locations. In the baseline analysis, microgrids are found to be good investments in areas with relatively high "spark spreads" (electricity/natural gas price differentials). Customers with a higher value for electric reliability realize greater benefits. Results become more favorable over time if the rate of increase of electricity prices is at least 60% of the rate of increase of natural gas prices—a plausible scenario in states with substantial natural-gas fired capacity. Sensitivity analyses reveal that the choice of microgrid customer mix has a much greater impact on system economics than climate. Economies of scale are shown to be fairly modest for the systems considered here, but microgrids do show clear benefits over traditional single customer distributed generation. If performance goals of current U.S. Department of Energy research programs for IC engines and microturbines are met, rates of return for microgrid investments increase 10–20%. (Supported by Alfred P. Sloan Foundation and EPRI.)

Narayanan, A., & Morgan, M. G. (2012). Sustaining critical social services during extended regional power blackouts. *Risk Analysis, 32*(7), 1183–1193.

**Abstract:** Despite continuing efforts to make the electric power system robust, some risk remains of widespread and extended power outages due to extreme weather or acts of terrorism. One way to alleviate the most serious effects of a prolonged blackout is to find local means to secure the continued provision of critical social services upon which the health and safety of society depend. This article outlines and estimates the incremental cost of a strategy that uses small distributed generation, distribution automation, and smart meters to keep a set of critical social services operational during a prolonged power outage that lasts for days or weeks and extends over hundreds of kilometers. (Supported by John D. and Catherine T. MacArthur Foundation and the Gordon and Betty Moore Foundation.)

Bohman, A. D., Abdulla, A., & Morgan, M. G. (2021). Individual and collective strategies to limit the impacts of large power outages of long duration. *Risk Analysis.*

**Abstract:** As modern society becomes ever more dependent on the availability of electric power, the costs that could arise from individual and social vulnerability to large outages of long duration (LLD-outages) increases. During such an outage, even a small amount of power would be very valuable. This article compares individual and collective strategies for providing limited amounts of electric power to residential customers in a hypothetical New England community during a large electric power outage of long duration. We develop estimates of the emergency load required for survival and assess the cost of strategies to address outages that last 5, 10, and 20 days in either winter or summer. We find that the cost of collective solutions could be as much as 10 to 40 times less than individual solutions (less than $2 per month per home). However, collective solutions would require community-wide coordination, and if local distribution system lines are destroyed, only individual back-up systems could provide contingency power until those lines are repaired. Costs might be reduced if more robust distributed generation were employed that could be operated continuously with the ability to sell power back to the grid. Our cost-effectiveness analysis only assesses what *could* be done, developing estimates of preparedness cost. A decision about what *should* be done would require additional input from a range of stakeholders as well as some form of analytical deliberative process.

This is the paper in which Mike Rath examined the tradeoffs between cost and environmental performance of microgrids:

Rath, M., Abdulla, A., Cohon, J. L., & Morgan, M. G. (2021). Ensuring decarbonization and decentralization of energy systems are not competing goals.

**Abstract:** Many investments in microgrids are made to enhance supply reliability. However, some advocate the adoption of microgrids as a strategy to decarbonize the electricity system. Today the majority of microgrids are powered by fossil fuel. Converting such systems to generation that does not emit carbon dioxide can be difficult and expensive. We use the Distributed Energy Resources Customer Adoption Model (DER-CAM) model in a multi-objective framework that minimizes a weighted sum of cost and carbon dioxide emissions. We examine how the system configuration changes as the relative weights applied to the two objectives are changed. Using three iconic types of loads located in southern California that involve peak demand of 0.25 $MW_e$, 2.5 $MW_e$ and 12 $MW_e$, we find that only three microgrid typologies emerge as the weights on cost and emissions are varied from 0 to 1, regardless of the size of the load. When all weight is assigned to cost, the systems use natural gas. As weight begins to be applied to emissions of $CO_2$, the technology mix changes. First, there is a transformation that prioritizes increasing efficiency and a shift to combined heat and power (CHP). At even higher weights on emissions of $CO_2$ the designs shift away from NG altogether in favor of polymer electrolyte membrane (PEM) fuel cells and in some cases photovoltaics with batteries. As the size of the microgrid increases, the weight placed on $CO_2$ emissions must be higher to stimulate the transition away from fossil fueled microgrids. Parametric analysis is run to examine how results depend on natural gas prices and the cost of battery storage. Our results find serious limitations for microgrids that consist only of solar PV and storage.

**Some notable writings by authors not affiliated with our Centers:**

Cole, W., Greer, D., Ho, J., & Margolis, R. (2020). Considerations for maintaining resource adequacy of electricity systems with high penetrations of PV and storage. *Applied Energy, 279*, 115795.

Jufri, F. H., Widiputra, V., & Jung, J. (2019). State-of-the-art review on power grid resilience to extreme weather events: Definitions, frameworks, quantitative assessment methodologies, and enhancement strategies. *Applied Energy, 239*, 1049–1065.

National Academies of Science, Engineering, and Medicine. (2012). *Terrorism and the Electric Power Delivery System*, National Academies Press. Available at: www.nap.edu/catalog/12050/terrorism-and-the-electric-power-delivery-system

National Academies of Science, Engineering, and Medicine. (2017). *Enhancing the Resilience of the Nation's Electrical System*, National Academies Press. Available at: www.nap.edu/catalog/24836/enhancing-the-resilience-of-the-nations-electricity-system

National Academies of Science, Engineering, and Medicine. (2021). *The Future of Electric Power in the United States*, National Academies Press. Available at: www.nap.edu/catalog/25968/the-future-of-electric-power-in-the-united-states

Panteli, M., Trakas, D. N., Mancarella, P., & Hatziargyriou, N. D. (2017). Power systems resilience assessment: Hardening and smart operational enhancement strategies. *Proceedings of the IEEE, 105*(7), 1202–1213.

## Notes

1 A discussion of work supported by our Centers on the impacts of climate change on hurricanes can be found in Chapter 16. While not discussed in this book, center-supported work on climate change and ice storms is reported in: Klima, K., & Morgan, M. G. (2015). Ice storm frequencies in a warmer climate. *Climatic Change, 133*(2), 209–222. Work on solar storms (which are not affected by climate change) is reported in: Kirchen, K., Harbert, W., Apt, J., & Morgan, M. G. (2020). A solar-centric approach to improving estimates of exposure processes for coronal mass ejections. *Risk Analysis, 40*(5), 1020–1039.

2 See Farley, P. (2004). The unruly power grid. *IEEE Spectrum, 41*(8), 22–27.

3 Murphy, S., Apt, J., Moura, J., & Sowell, F. (2018). Resource adequacy risks to the bulk power system in North America. *Applied Energy, 212*, 1360–1376.

4 Lavin, L., Murphy, S., Sergi, B., & Apt, J. (2020). Dynamic operating reserve procurement improves scarcity pricing in PJM. *Energy Policy, 147*, 111857.

5 www.washingtonpost.com/opinions/2021/02/18/texas-power-grid-failure-weather/

6 The ICE Calculator is available at https://icecalculator.com/home

# 14

# TRANSPORTATION WITHOUT CARBON DIOXIDE

*Granger Morgan and Parth Vaishnav*

When we started our studies of global change 25 years ago, electricity was the biggest source of carbon dioxide emissions. Over the intervening years, electricity has gotten cleaner as coal plants became uneconomic with lower-cost and cleaner natural gas, solar, and wind. The fuel efficiency of the vehicle fleet in the United States has also gotten better, thanks to the periodic tightening of CAFE standards. However, we are traveling more and shipping ever greater quantities of freight. As of 2019, 29% of emissions of U.S. greenhouse gases came from transportation while emissions from electricity were a close second at 25%. Globally, the situation is different. Much of the electricity generation around the world is dirtier than in the U.S., and much of the world travels less and ships less than we do. Transportation amounts to only 16% of emissions on a global basis as compared with 27% from electricity.[1]

## Cars and Light Trucks

There are four options to reduce greenhouse gas emissions from light vehicles: convert to batteries (battery electric vehicle or BEV); add a battery but also keep a small engine (plug-in hybrid electric vehicle or PHEV); do the same thing but only charge the battery with the engine (hybrid electric vehicle or HEV); or switch to a liquid or gaseous fuel that does not produce any lifecycle greenhouse gas emissions and use it either in an internal combustion engine or in a fuel cell. Decarbonization requires that the second and third strategies (partial electrification) be married with the fourth one (a zero-carbon liquid or gaseous fuel).

The advantage of a PHEV is that it does not have the limited range of a PEV since once the battery runs down, the vehicle can refuel at gas stations and keep going on its gasoline engine. Investigators associated with our NSF-supported Centers did two analyses of PHEVs – the first in 2008 and the second eight years

DOI: 10.4324/9781003330226-14

later in 2016. The first paper titled "Life Cycle Assessment of Greenhouse Gas Emissions from Plug-in Hybrid Electric Vehicles: Implications for Policy" was published in *Environmental Science and Technology*. It drew the obvious conclusion that just how good a PHEV is for reducing greenhouse gas emissions depends very much on how much carbon dioxide is produced by the generating plants in the local electricity system. A life cycle assessment looks at the environmental impacts of a product's manufacture, use, and end-of-life. With the exception of the battery, GHG emissions from making the PHEV were essentially the same as for a conventional vehicle. How much the battery added depended on how large it is, but in all cases, it was much smaller than for making the vehicle. Results of the analysis are shown in Figure 14.1.

Like most of the analysis we have done, the paper includes a careful look at how much uncertainty in the assumptions could change the results. The overall conclusion is that, using the average carbon intensity of the U.S. electricity system at the time the study was done (it is lower now), "life cycle GHG emissions from PHEVs…[were found to] reduce GHG emissions by 32% compared to conventional vehicles, but have small reductions compared to traditional hybrids."

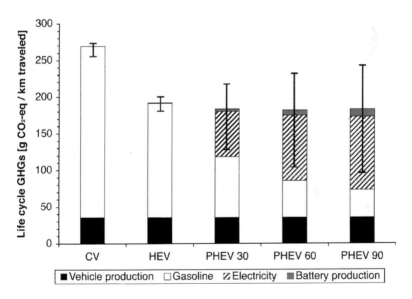

**FIGURE 14.1** Life cycle greenhouse gas emissions (in grams of $CO_2$-equivalent/kilometer traveled) of conventional vehicles (CVs), hybrid electric vehicles (HEVs), and plug-in hybrids (PHEVs) with all-electric ranges of 30, 60, or 90 km. Life cycle GHG intensity of electricity is 670 g $CO_2$-eq/kWh (186 g/MJ; U.S. average scenario). Uncertainty bars represent changes in total emissions under the carbon-intensive (950 g $CO_2$-eq/kWh) or low-carbon (200 g $CO_2$e/kWh) electricity scenarios. Figure from Samaras and Meisterling, 2008.

Samaras, C., & Meisterling, K. (2008). Life cycle assessment of greenhouse gas emissions from plug-in hybrid vehicles: Implications for policy. *American Chemical Society.*

Eight years later when Center investigators looked again at PHEVs, there were quite a few EVs and PHEVs being offered on the U.S. market. The paper reporting this analysis was published in *Environmental Research Letters*. It has a long title that pretty clearly explains what the paper is about: "Effect of regional grid mix, driving patterns and climate on the comparative carbon footprint of gasoline and plug-in electric vehicles in the United States." It begins by summarizing and providing a table that compares 12 previous studies. It then develops a county-level analysis for the U.S. based on the highest resolution data available for each factor, including emissions from electricity generation, temperatures, and driving conditions in each region. It also compares each vehicle type across each region. Figure 14.2 shows a summary of findings. The abstract for the paper is reproduced at the end of the chapter.

Of course, electric vehicles of any kind only have impacts if people buy them. Focusing on the two countries with the largest carbon emissions, China and the United States, John Helveston and several others conducted a study during 2012–13 to assess customer preferences. Because John speaks Chinese, he was able to conduct interviews in China. The paper reporting the results of this study appeared in *Transportation Research Part A* in 2015. While the situation may have changed in more recent years, at the time of the study Chinese respondents demonstrated greater willingness to adopt electric vehicles and were less concerned than their American counterparts about limited range. The authors observe that because China imports much of its oil, there are national security benefits for them associated with converting to electric vehicles. At the same time, because the Chinese grid is heavily dependent on coal, a large switch to electric vehicles in China could result in an increase in carbon dioxide emissions. There is of course an additional dimension to Chinese programs to promote the adoption of electric vehicles. Just as China has taken aggressive steps to secure a strong position in global export markets for solar power, a strong domestic market for electric vehicles could help to underpin an export market.

The batteries for electric vehicles do not last forever, and they gradually degrade over time as they are used more. PhD student Scott Peterson, working with battery expert Jay Whitaker along with Jay Apt, obtained batteries of the sort being used in electric vehicles of the time (A123 systems M1 Cell in 2010) and conducted a series of laboratory experiments in which they cycled the batteries according to a number of standard driving profiles. The batteries they studied showed good performance. They report that:

> After 2000 cycles the low rate discharge potential profile appears very similar to that collected before cycling started, and a very small fraction of the initial capacity has been lost. This observation is consistent with the hypothesis that only a minimal Solid Electrolyte Interphase (SEI) layer must be forming during cycling of these cells, and that the mechanical cycling of the electrodes does not induce loss of connection and capacity fade...

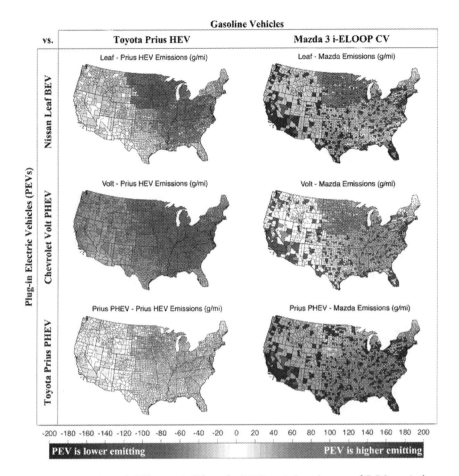

**FIGURE 14.2** Estimated difference in life cycle GHG emissions (grams of CO2 equivalent per mile) of selected plug- in electric vehicles (2013 Nissan Leaf BEV, 2013 Chevrolet Volt PHEV, and 2013 Prius PHEV) relative to selected gasoline vehicles (2010 Prius HEV and 2014 Mazda 3). In each case, blue indicates that the PEV has lower GHG emissions than the gasoline vehicle, and red indicates that the PEV has higher GHG emissions than the gasoline vehicle. Figure from Yuksel et al., 2016. To obtain full color version of image see: https://iopscience.iop.org/article/10.1088/1748-9326/11/4/044007/pdf

Yuksel, T., Tamayao, M. A. M., Hendrickson, C., Azevedo, I. M., & Michalek, J. J. (2016). Effect of regional grid mix, driving patterns and climate on the comparative carbon footprint of gasoline and plug-in electric vehicles in the United States. *Environmental Research Letters*, *11*(4), 044007.

The literature commonly indicates that the dominant mechanisms for capacity loss in Li-ion cells are (1) the formation of a resistive and progressively Li-consuming interfacial layer between the functional graphite at the anode and the electrolyte, and (2) the physical degradation of active materials and electrode structures…. Our data indicate a much lower loss of capacity as a function of cycles … a result consistent with the use of high performance nano-structured electrode (cathode) materials that are much more physically stable during use and so do not degrade…

They also considered the case where the vehicle battery is being used to supply power to the grid (V2G). They find that several thousand driving or V2G days "incur substantially less than 10% capacity loss regardless of the amount of V2G support used. However, V2G modes that are more intermittent in nature will lead to more rapid battery capacity fade and should be avoided to minimize battery capacity loss over many years of use."

In other studies, investigators associated with our Center have looked at how plug-in vehicles could impact air pollution and oil consumption (see paper by Michalek et al., 2011 at the end of the chapter) and how provisions to credit and incentivize the use of alternative fuels could render the U.S. CAFE fuel economy standards less effective and curtail the extent to which they reduced greenhouse gas emissions and gasoline consumption (see paper by Jenn et al., 2016 at the end of the chapter).

The final line of Center-related work on light vehicles has examined trade-offs between vehicle automation (smart self-driving cars) and light vehicle electrification. This work addresses the fear that the need to provide power for vehicle components – sensors and computers – required for automation will require automated vehicles to have an internal combustion engine to ensure adequate range. If this indeed turns out to be the case, that would retard progress toward electrification and light vehicle decarbonization. All of the sensors and electronics needed to automate a vehicle require electricity, which increases the load on the battery in an EV. Also, until strategies are found to integrate the sensors in a more aerodynamically streamlined way, the sensor system on the roof of the vehicle increases air resistance, and that too adds load on the battery. In a paper published in *Nature Energy* in 2020 titled "Trade-offs between automation and light vehicle electrification," the authors report that with present technology:

> …automation will likely reduce electric vehicle range by 5–10% for suburban driving and by 10–15% for city driving. The effect on range is strongly influenced by sensor drag for suburban driving and computing loads for city driving. The impact of automation on battery longevity is negligible. While some commentators have suggested that the power and energy requirements of automation mean that the first automated vehicles will be gas–electric hybrids, our results suggest that this need not be the case if automakers can implement energy-efficient computing and aerodynamic sensor stacks.

## Heavy Freight

Most large trucks and most railroad locomotives burn diesel fuel.[2] While it is relatively easy to convert passenger cars and light trucks to run on batteries, running freight trains or heavy trucks for long distances using batteries will likely be harder. In February of 2017, we organized a workshop to explore options to decarbonize heavy freight (see Figure 14.3).

We discussed a variety of strategies to make fuels that have no significant life cycle greenhouse gas emissions. While still expensive today, hydrogen and other fuels made from hydrogen are beginning to look more practical. In 2017, we concluded that the best short-term solution was to move as much truck traffic onto rail, since rail is so much more efficient than trucks. In shipping in terms of ton-mile, rail and trucks in the U.S. carry almost the same amount of freight (in 2017, it was 43% on trucks, 47% on rail). Much of the freight on rail is bulk commodities, so measured in terms of the value of the loads, trucks carry more. There is a lot of variation in the data on emissions, with different sources producing different values, but Lynn Kaack summarized many of those results in Figure 14.4. Clearly rail (the Xs) emits much less $CO_2$ per ton-kilometer than trucks, in some cases as little as 10% or less.

In an extensive review published in *Environmental Research Letters* in 2018, Lynn and others explored all the different ways that emissions from freight could be reduced and then examined the status of moving as large a portion as possible of the journey performed by a unit of freight from trucks onto rail, while acknowledging that road transport might still be the most effective way of performing the last mile of the journey. Doing that is called "intermodal shipping." While a lot of intermodal freight is already operating in the U.S., it is difficult to move more truck freight to rail because of issues of reliability, cost, and convenience.

**FIGURE 14.3** Images from the workshop we ran at Carnegie Mellon in late February 2017 in which we explored strategies to decarbonize heavy freight.

Photos taken by Granger Morgan.

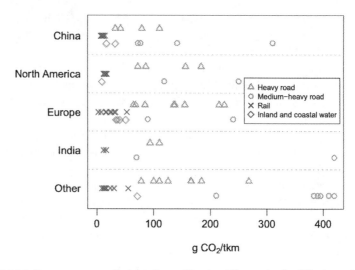

**FIGURE 14.4** Summary compiled by Lynn Kaack of how much $CO_2$ is emitted by different kinds of freight in different parts of the world. Each marker represents an average estimate for the region, or a country in that region, from a different data source. Road carbon intensity values distinguish medium- and heavy-duty vehicles. For details, see Kaack et al., 2018 at the end of the chapter.

Lynn and we concluded the review by arguing that intermodal freight has considerable potential to reduce emissions of greenhouse gases, but "today's markets will likely not lead to the levels of decarbonization that are needed to slow climate change." We argue that "additional policies that include either incentives for reductions or penalties for GHG emissions will be needed. We find that modal shift may have the potential to reduce GHG emissions, but that a systematic analysis of the possible emissions reductions and costs is yet to be found in the literature."

While there is good data on truck freight in the U.S. and some other developed countries, in most of the world the data are very poor. Because she had gotten interested in machine learning and computer pattern recognition, Lynn decided to see if she could fill some of the data gap using satellite images of trucks on highways. She built a system that she calibrated with known truck freight on the New York Throughway, where it did well. Then she applied it in Brazil. It worked, but not as well as in New York, perhaps because the look of some trucks is different in Brazil.

Much of the work discussed so far focuses on greenhouse gas emissions, the most important of which – $CO_2$ – is well-mixed and does roughly as much damage no matter where it is emitted. The extent of damage that short-lived pollutants, such as the oxides of nitrogen, fine particulate matter, and sulfur dioxide, cause depends on where they are emitted. Where these pollutants are emitted also determines *who* is hurt by them. Recent work of Tessum and colleagues[3] has shown that, in the U.S., communities of color are disproportionately exposed to fine particulate matter pollution. The work also shows that heavy-duty diesel vehicles are a significant contributor to this disparity. To show that this was true at the Census Block level, Tessum et al. considered emissions, for which estimates were available at the county level, and allocated them to Census Blocks using spatial surrogates. In the second chapter of his EPP dissertation, Priyank Lathwal improved on this approach by estimating truck emissions along individual road segments and used these values to build a bottom-up estimate of emissions at the Census Tract and county level. His analysis confirmed that areas that had higher proportions of Black or Hispanic people experienced greater emissions from trucks. As of September 2021, we are working on preparing a manuscript based on Priyank's findings and submitting it for peer review.

## Ships and Aircraft

Both ships and aircraft burn fossil fuel that releases carbon dioxide as well as air pollutants. In a series of papers published from 1997–99, then doctoral student Jim Corbett – working with Paul Fischbeck and others – showed that pollution from the bunker fuel burned by the world's ships made up a significant share of total pollution. Moreover, they found that sulfur dioxide from ships had a significant influence on the earth's radiative balance. Jim is now a Professor of Marine Policy at the University of Delaware and the leading expert on the environmental consequences of ocean shipping and on its environmental governance. Jim's work formed the basis of the International Maritime Organization's first systematic study of greenhouse gas emissions of ocean shipping and of the efficacy of different ways of cutting those emissions. The third iteration of that study was published in 2012 and a fourth iteration in 2020.

A 1999 IPCC report estimated aviation's contribution to global greenhouse gas emissions at about 2%, approximately the same as ocean shipping, although considerable uncertainty remained (and continues to remain) about the effect of aviation-induced cirrus clouds on radiative forcing. When they operate within a nation, environmental agencies can establish accounting and regulatory frameworks. Since it is extremely hard to allocate greenhouse gas emissions from international aviation or from ocean shipping, these emissions are not attributed to any particular country. In turn, no country has a responsibility to seek to reduce or eliminate them. As Parth Vaishnav wrote in a 2014 article in *Issues in Science & Technology*, "Given the nature of global transport, it is difficult to allocate its environmental impacts to any one country. Consider a ship that is registered in Liberia, operated by a Danish shipping line, and making a voyage from Shanghai to Los Angeles carrying products made in China by a European firm for sale in North America. How and to whom should the emissions from this voyage be allocated, and who should be assigned responsibility for reducing them? Questions such as these have proven to be politically intractable."

The responsibility for resolving these questions lies with the International Civil Aviation Organization (ICAO) and the International Maritime Organization (IMO). Progress on these efforts had, however, been glacial. Impatient with the lack of globally coordinated action, the European Union declared that it would include flights in and out of Europe in its emissions trading scheme (ETS). This set off some well-publicized legal wrangling and prompted Granger to wonder what could *actually* be done to reduce $CO_2$ from aviation and ocean shipping.

Attempts to answer that question have been a major component of three recent CEDM-supported PhD dissertations: those of Parth Vaishnav, Lynn Kaack, and Priyank Lathwal.

Moving freight by ship is very energy efficient. In the future, using "green" hydrogen, or ammonia made from that hydrogen, may make it possible to completely eliminate greenhouse gas and other polluting emissions from ships. While ocean shipping is responsible for ~3% of global greenhouse gas emissions, the emissions from these ships are also responsible for many premature deaths caused by short-lived pollutants that have a more immediate effect on human health. To address the climate impacts of shipping, and to reduce the health impacts of local pollution, it is possible to turn off the ship's engines when it is in port and connect it to the electric power grid. This practice is called "cold ironing." Clearly the benefits of this strategy depend on the cleanliness of the local electricity grid. In a paper published in *Environmental Science & Technology* in 2016, Parth Vaishnav obtained data on vessels calling at U.S. ports and then used two integrated assessment models to quantify the benefits of reducing the emissions from ships. This paper found that the benefits of mandating cold ironing could be as high as $150-millon per year. This paper found that, depending on the value placed on the avoided pollution, cold ironing is an attractive strategy not only in California (where the electricity grid is relatively clean and the major ports are close to densely populated cities) but also in ports along the Gulf of Mexico and in the Northeast.

Having seen the big benefit this could have in the U.S. and aware that some countries like China were also requiring the use of shore power for ships in their ports, Priyank Lathwal set out to see if doing the same thing in India might also be beneficial. Getting data on vessel calls at Indian ports turned out to be a very big challenge. However, through hard and persistent work and diplomacy, after many months Priyank was able to get data on ships calling at all the major ports in India. It turns out that because so much of India's power is made with dirty coal plants, switching to shore power is unlikely to reduce premature mortality much, and might in some places even increase it. The 2021 paper reporting this work in *Environmental Research Letters* concludes that "switching from high-sulfur fuel to shore power might avoid at most a couple of dozen premature deaths each year, whereas switching from low-sulfur fuel could lead to a slight increase in premature mortality. Therefore, policymakers must first clean up power generation for shore power to be a viable strategy to improve air quality in Indian port cities."

Airplanes are the most challenging transportation mode in terms of eliminating carbon dioxide emissions. Pound for pound, the energy density of jet fuel is about 50 times that of today's batteries. There is work being done to develop short-range, lightweight, battery-operated aircraft, but there is simply no way that batteries can be used for larger aircraft flying longer distances. Hydrogen packs about three times the energy of jet fuel per unit mass. However, hydrogen is a gas, so storing it takes up a lot of space. Hydrogen can be made more compact if it is liquified, but that requires very low (cryogenic) temperatures, so the issues of cost and practical details are yet to be worked out.

Over the course of the past decade, the International Civil Aviation Organization (ICAO) has been working to develop a set of policies to cap greenhouse gas emissions from international aviation. When ICAO issued a draft proposal, Parth worked with Annie Petsonk, who was then at the Environmental Defense Fund (EDF), to perform a comprehensive analysis of how the proposed regulation would perform over time. Since ICAO did not want to disadvantage emerging airlines from developing countries, they had proposed a fairly complicated scheme. In his analysis, Parth showed that while the proposed approach, which included a complex system of exemptions and weighting strategies, would have the desired effect in the short-term, over the longer run, as the size of the industry and traffic volumes increased, the proposed strategy would make things worse. Parth and his co-authors reported on this work in a 2016 paper in the journal *Transportation Research Part D: Transport and Environment*. In their paper, Parth and co-authors observe that "targeting such exemptions more narrowly would raise practical difficulties," which they describe. They conclude their paper by recommending that ICAO design and implement a much simpler system, and they propose one alternative. Motivated in part by the findings of this work, ICAO revised their approach.

Just as it is possible to achieve some reduction in the emissions from ships when they are in port, there are also strategies to reduce the emissions from aircraft when they are taxiing at airports. This can be done using tugs to tow the aircraft so that it does not need to run its main engines to taxi. This approach is called dispatch towing. The challenge is to do this in a way that does not sharply reduce landing gear life and ensures the safety of ground operations. In a 2014 paper in *Transportation Research Record,* Parth examined the technical and economic aspects of using this strategy. His work found that, assuming jet fuel prices of $3 per gallon, using tugs would result in cost savings of $20/ton of carbon dioxide ($CO_2$) emissions avoided if the measure were adopted for all domestic flights. Estimates of average net savings for airlines vary from $100 per flight at John F. Kennedy International Airport in New York City to a loss of $160 per flight in Honolulu, Hawaii. Dispatch towing could reduce $CO_2$ emissions from domestic flights in the United States by about 1.5 million tons each year, or about 1.1% of the total emissions in 2006. The total impact of such a strategy is clearly very modest. However, as concerns grow about the local air pollution impact from taxiing aircraft, and the world scrambles to find every way it can to reduce $CO_2$ emissions, the approach may warrant increased attention.

Finally, a 2021 *Transportation Research Part D* paper led by Lavanya Marla of the University of Illinois at Urbana-Champaign examined the cost of airline operations recovery after disruptions. The paper examined what these costs would be with and without the effects of climate change. Based on the relationship between runway length, temperature, and maximum takeoff weight, Parth worked with master's student Aradhana Gahlaut to identify times during which aircraft would have to carry a sub-optimal number of passengers to be light enough to take off. The paper showed that climate change would multiply the number of occasions in which airline operations would be disrupted and costly recovery measures (e.g., rerouting, delaying, or stranding passengers) would have to be implemented. There is some understanding (e.g., in the National Climate Assessment) that extreme events such as floods would disrupt aviation activities more frequently in the future. The Marla et al. paper demonstrated that climate change would cause costly disruptions even during times of "normal" operation simply because of increasing temperatures.

## Technical Details for Chapter 14

Here are the papers on light vehicles that we talk about in this chapter:

Samaras, C., & Meisterling, K. (2008). Life cycle assessment of greenhouse gas emissions from plug-in hybrid vehicles: Implications for policy. *Environmental Science & Technology, 42*(9), 3170–3176.

**Abstract:** Plug-in hybrid electric vehicles (PHEVs), which use electricity from the grid to power a portion of travel, could play a role in reducing greenhouse gas (GHG) emissions from the transport sector. However, meaningful GHG emissions reductions with PHEVs are conditional on low-carbon electricity sources. We assess life cycle GHG emissions from PHEVs and find that they reduce GHG emissions by 32% compared to conventional vehicles, but have small reductions compared to traditional hybrids. Batteries are an important component of PHEVs, and GHGs associated with lithium-ion battery materials and production account for 2–5% of life cycle emissions from PHEVs. We consider cellulosic ethanol use and various carbon intensities of electricity. The reduced liquid fuel requirements of PHEVs could leverage limited cellulosic ethanol resources. Electricity generation infrastructure is long-lived, and technology decisions within the next decade about electricity supplies in the power sector will affect the potential for large GHG emissions reductions with PHEVs for several decades.

Yuksel, T., Tamayao, M. A. M., Hendrickson, C., Azevedo, I. M., & Michalek, J. J. (2016). Effect of regional grid mix, driving patterns and climate on the comparative carbon footprint of gasoline and plug-in electric vehicles in the United States. *Environmental Research Letters, 11*(4), 044007.

**Abstract:** We compare life cycle greenhouse gas (GHG) emissions from several light-duty passenger gasoline and plug-in electric vehicles (PEVs) across US counties by accounting for regional differences due to marginal grid mix, ambient temperature, patterns of vehicle miles traveled (VMT), and driving conditions (city *versus* highway). We find that PEVs can have larger or smaller carbon footprints than gasoline vehicles, depending on these regional factors and the specific vehicle mode being compared. The Nissan Leaf battery electric vehicle has a smaller carbon footprint than the most efficient gasoline vehicle (the Toyota Prius) in the urban counties of California, Texas and Florida, whereas the Prius has a smaller carbon footprint in the Midwest and the South. The Leaf is lower emitting than the Mazda 3 conventional gasoline vehicle in most urban counties, but the Mazda 3 is lower emitting in rural Midwest counties. The Chevrolet Volt plug-in hybrid electric vehicle has a larger carbon footprint than the Prius throughout the continental US, though the Volt has a smaller carbon footprint than the Mazda 3 in many urban counties. Regional grid mix, temperature, driving conditions, and vehicle model all have substantial implications for identifying which technology has the lowest carbon footprint, whereas regional patterns of VMT have a much smaller effect. Given the variation in relative GHG implications, it is unlikely that blunt policy instruments that favor specific technology categories can ensure emission reductions universally.

Helveston, J. P., Liu, Y., Feit, E. M., Fuchs, E., Klampfl, E., & Michalek, J. J. (2015). Will subsidies drive electric vehicle adoption? Measuring consumer preferences in the US and China. *Transportation Research Part A: Policy and Practice, 73*, 96–112.

**Abstract:** We model consumer preferences for conventional, hybrid electric, plug-in hybrid electric (PHEV), and battery electric (BEV) vehicle technologies in China and the U.S. using data from choice-based conjoint surveys fielded in 2012–2013 in both countries. We find that with the combined bundle of attributes offered by vehicles available today, gasoline vehicles continue in both countries to be most attractive to consumers, and American respondents have significantly lower relative willingness-to-pay for BEV technology than Chinese respondents. While U.S. and Chinese subsidies are similar, favoring vehicles with larger battery packs, differences in consumer preferences lead to different outcomes. Our results suggest that with or without each country's 2012–2013 subsidies, Chinese consumers are willing to adopt today's BEVs and mid-range PHEVs at similar rates relative to their respective gasoline counterparts, whereas American consumers prefer low-range PHEVs despite subsidies. This implies potential for earlier BEV adoption in China, given adequate supply. While there are clear national security benefits for adoption of BEVs in China, the local and global social impact is unclear: With higher electricity generation emissions in China, a transition to BEVs may reduce oil consumption at the expense of increased air pollution and/or greenhouse gas emissions. On the other hand, demand from China could increase global incentives for electric vehicle technology development with the potential to reduce emissions in countries where electricity generation is associated with lower emissions.

Peterson, S. B., Apt, J., & Whitacre, J. F. (2010). Lithium-ion battery cell degradation resulting from realistic vehicle and vehicle-to-grid utilization. *Journal of Power Sources, 195*(8), 2385–2392.

**Abstract:** The effects of combined driving and vehicle-to-grid (V2G) usage on the lifetime performance of relevant commercial Li-ion cells were studied. We derived a nominal realistic driving schedule based on aggregating driving survey data and the Urban Dynamometer Driving Schedule, and used a vehicle physics model to create a daily battery duty cycle. Different degrees of continuous discharge were imposed on the cells to mimic afternoon V2G use to displace grid electricity. The loss of battery capacity was quantified as a function of driving days as well as a function of integrated capacity and energy processed by the cells. The cells tested showed promising capacity fade performance: more than 95% of the original cell capacity remains after thousands of driving days worth of use. Statistical analyses indicate that rapid vehicle motive cycling degraded the cells more than slower, V2G galvanostatic cycling. These data are intended to inform an economic model.

Michalek, J. J., Chester, M., Jaramillo, P., Samaras, C., Shiau, C. S. N., & Lave, L. B. (2011). Valuation of plug-in vehicle life-cycle air emissions and oil displacement benefits. *Proceedings of the National Academy of Sciences, 108*(40), 16554–16558.

**Abstract:** We assess the economic value of life-cycle air emissions and oil consumption from conventional vehicles, hybrid-electric vehicles (HEVs), plug-in hybrid-electric vehicles (PHEVs), and battery electric vehicles in the US. We find that plug-in vehicles may reduce or increase externality costs relative to grid-independent HEVs, depending largely on greenhouse gas and $SO_2$ emissions produced during vehicle charging and battery manufacturing. However, even if future marginal damages from emissions of battery and electricity production drop dramatically, the damage reduction potential of plug-in vehicles remains small compared to ownership cost. As such, to offer a socially efficient approach to emissions and oil consumption reduction, lifetime cost of plug-in vehicles must be competitive with HEVs. Current subsidies intended to encourage sales of plug-in vehicles with large capacity battery packs exceed our externality estimates considerably, and taxes that optimally correct for externality damages would not close the gap in ownership cost. In contrast, HEVs and PHEVs with small battery packs reduce externality damages at low (or no) additional cost over their lifetime. Although large battery packs allow vehicles to travel longer distances using electricity instead of gasoline, large packs are more expensive, heavier, and more emissions intensive to produce, with lower utilization factors, greater charging infrastructure requirements, and life-cycle implications that are more sensitive to uncertain, time-sensitive, and location-specific factors. To reduce air emission and oil dependency impacts from passenger vehicles, strategies to promote adoption of HEVs and PHEVs with small battery packs offer more social benefits per dollar spent.

Jenn, A., Azevedo, I. M., & Michalek, J. J. (2016). Alternative fuel vehicle adoption increases fleet gasoline consumption and greenhouse gas emissions under United States corporate average fuel economy policy and greenhouse gas emissions standards. *Environmental Science & Technology, 50*(5), 2165–2174.

**Abstract:** The United States Corporate Average Fuel Economy (CAFE) standards and Greenhouse Gas (GHG) Emission standards are designed to reduce petroleum consumption and GHG emissions from light-duty passenger vehicles. They do so by requiring automakers to meet aggregate criteria for fleet fuel efficiency and carbon dioxide ($CO_2$) emission rates. Several incentives for manufacturers to sell alternative fuel vehicles (AFVs) have been introduced in recent updates of CAFE/GHG policy for vehicles sold from 2012 through 2025 to help encourage a fleet technology transition. These incentives allow automakers that sell AFVs to meet less-stringent fleet efficiency targets, resulting in increased fleet-wide gasoline consumption and emissions. We derive a closed-form expression to quantify these effects. We find that each time an AFV is sold in place of a conventional vehicle, fleet emissions increase by 0 to 60 t of $CO_2$ and gasoline consumption increases by 0 to 7000 gallons (26,000 L), depending on the AFV and year of sale. Using projections for vehicles sold from 2012 to 2025 from the Energy Information Administration, we estimate that the CAFE/GHG AFV incentives lead to a cumulative increase of 30 to 70 million metric tons of $CO_2$ and 3 to 8 billion gallons (11 to 30 billion liters) of gasoline consumed over the vehicles' lifetimes – the largest share of which is due to legacy GHG flex-fuel vehicle credits that expire in 2016. These effects may be 30–40% larger in practice than we estimate here due to optimistic laboratory vehicle efficiency tests used in policy compliance calculations.

Mohan, A., Sripad, S., Vaishnav, P., & Viswanathan, V. (2020). Trade-offs between automation and light vehicle electrification. *Nature Energy, 5*(7), 543–549.

**Abstract:** Weight, computing load, sensor load and possibly higher drag may increase the energy use of automated electric vehicles relative to human-driven electric vehicles, although this increase may be offset by smoother driving. Here, we use a vehicle dynamics model to evaluate the trade-off between automation and electric vehicle range and battery longevity. We find that automation will likely reduce electric vehicle range by 5–10% for suburban driving and by 10–15% for city driving. The effect on range is strongly influenced by sensor drag for suburban driving and computing loads for city driving. The impact of automation on battery longevity is negligible. While some commentators have suggested that the power and energy requirements of automation mean that the first automated vehicles will be gas–electric hybrids, our results suggest that this need not be the case if automakers can implement energy-efficient computing and aerodynamic sensor stacks.

This is the extensive review paper Lynn and we wrote on intermodal shipping after the workshop we ran on decarbonizing heavy freight:

Kaack, L. H., Vaishnav, P., Morgan, M. G., Azevedo, I. L., & Rai, S. (2018). Decarbonizing intraregional freight systems with a focus on modal shift. *Environmental Research Letters*, *13*(8), 083001.

**Abstract:** Road freight transportation accounts for around 7% of total world energy-related carbon dioxide emissions. With the appropriate incentives, energy savings and emissions reductions can be achieved by shifting freight to rail or water modes, both of which are far more efficient than road. We briefly introduce five general strategies for decarbonizing freight transportation, and then focus on the literature and data relevant to estimating the global decarbonization potential through modal shift. We compare freight activity (in tonne-km) by mode for every country where data are available. We also describe major intraregional freight corridors, their modal structure, and their infrastructure needs. We find that the current world road and rail modal split is around 60:40. Most countries are experiencing strong growth in road freight and a shift from rail to road. Rail intermodal transportation holds great potential for replacing carbon-intense and fast-growing road freight, but it is essential to have a targeted design of freight systems, particularly in developing countries. Modal shift can be promoted by policies targeting infrastructure investments and internalizing external costs of road freight, but we find that not many countries have such policies in place. We identify research needs for decarbonizing the freight transportation sector both through improvements in the efficiency of individual modes and through new physical and institutional infrastructure that can support modal shift.

Kaack, L. H., Chen, G. H., & Morgan, M. G. (2019, July). Truck traffic monitoring with satellite images. In *Proceedings of the 2nd ACM SIGCAS Conference on Computing and Sustainable Societies* (pp. 155–164).

This is the paper in which Priyank showed that air pollution from heavy trucking disproportionately affects minority communities:

Lathwal, P., Vaishnav, P., & Morgan, M.G. (2022). Pollution from freight trucks in the contiguous United States: Public health damages and implications for environmental justice.

**Abstract:** $PM_{2.5}$ produced by freight trucking has significant adverse impacts on human health. Here we explore the spatial distribution of freight trucking emissions and demonstrate that public health impacts due to freight trucking disproportionately affect certain racial and ethnic groups. Based on the US federal government data, we build an emissions inventory to quantify heterogeneity of trucking emissions and find that ~10% of $NO_x$

and ~12% of $CO_2$ emissions from all sources in the US come from freight trucks. The costs to human health and the environment due to $NO_x$, $PM_{2.5}$, $SO_2$, and $CO_2$ from freight trucking in the US are estimated respectively to be \$11B, \$5.5B, \$100M, and \$30B (social cost of carbon of \$51 per ton). We demonstrate that more freight pollution occurs in counties and census tracts with a higher proportion of Black and Hispanic residents. Counties with a higher proportion of Black and Hispanic residents are also more likely to be net importers of pollution damages from other counties.

Here is the overview paper on emissions from international air and ships:

Vaishnav, P. (2014). Greenhouse gas emissions from international transport. *Issues in Science & Technology, 30*(2), 25–28.

Here is the paper on powering ships in U.S. ports with power from the grid:

Vaishnav, P., Fischbeck, P. S., Morgan, M. G., & Corbett, J. J. (2016). Shore power for vessels calling at US ports: Benefits and costs. *Environmental Science & Technology, 50*(3), 1102–1110.

**Abstract:** When in port, ships burn marine diesel in on-board generators to produce electricity and are significant contributors to poor local and regional air quality. Supplying ships with grid electricity can reduce these emissions. We use two integrated assessment models to quantify the benefits of reducing the emissions of $NO_X$, $SO_2$, $PM_{2.5}$, and $CO_2$ that would occur if shore power were used. Using historical vessel call data, we identify combinations of vessels and berths at U.S. ports that could be switched to shore power to yield the largest gains for society. Our results indicate that, depending on the social costs of pollution assumed, an air quality benefit of \$70−150 million per year could be achieved by retrofitting a quarter to two-thirds of all vessels that call at U.S. ports. Such a benefit could be produced at no net cost to society (health and environmental benefits would be balanced by the cost of ship and port retrofit) but would require many ships to be equipped to receive shore power, even if doing so would result in a private loss for the operator. Policy makers could produce a net societal gain by implementing incentives and mandates to encourage a shift toward shore power.

This is Priyank's paper that looked at whether connecting ships to shore power is a good thing to do in India:

Lathwal, P., Vaishnav, P., & Morgan, M. G. (2021). Environmental and health consequences of shore power for vessels calling at major ports in India. *Environmental Research Letters, 16*(6), 064042.

**Abstract:** To reduce local air pollution, many ports in developed countries require berthed ships to use shore-based electricity instead of burning diesel to meet their electricity requirement for loads such as lights, cargo-handling equipment, and air conditioning. The benefits of this strategy in developing countries remain understudied. Based on government data for all major ports in India, we find that switching from high-sulfur fuel to shore power reduces hoteling emissions of $PM_{2.5}$ by 88%; $SO_2$ by 39%; $NO_x$ by 85%; but increases $CO_2$ emissions by 12%. Switching from low-sulfur fuel reduces hoteling emissions of $PM_{2.5}$ by 46% and $NO_x$ by 84% but increases $SO_2$ emissions by 240% and $CO_2$ emissions by 17%. The lifetime cost savings from the switch to electricity are \$73M for high-sulfur fuel and \$370M for low-sulfur fuel. We estimate that switching from high-sulfur fuel to shore power might avoid at most a couple of dozen premature deaths each year, whereas switching from low-sulfur fuel could lead to a slight increase in premature mortality. Therefore, policymakers must first clean up power generation for shore power to be a viable strategy to improve air quality in Indian port cities.

These are papers Parth wrote on aviation:

Vaishnav, P. (2014). Costs and benefits of reducing fuel burn and emissions from taxiing aircraft: Low-hanging fruit?. *Transportation Research Record, 2400*(1), 65–77.

**Abstract:** While taxiing, aircraft are powered by their main engines. This paper estimates the potential reductions in costs and emissions that could be achieved with tugs or an electric motor embedded in the landing gear to propel aircraft on the ground. The use of tugs would result in the avoidance of $20/ton of carbon dioxide ($CO_2$) emissions if the measure were adopted for all domestic flights. Estimates of average net savings for airlines vary from $100 per flight at John F. Kennedy International Airport in New York City to a loss of $160 per flight in Honolulu, Hawaii. Electric taxiing would save between $30 and $240/ton of $CO_2$ emissions avoided. Either approach could reduce $CO_2$ emissions from domestic flights in the United States by about 1.5 million tons each year, or about 1.1% of the total emissions in 2006. If the switch were limited to large narrow-body aircraft on domestic service at the busiest airports in the United States, the total reduction in emissions would be 0.5 million tons of $CO_2$ annually, accompanied by savings of $100/ton. Air quality benefits associated with lower main engine use were monetized by using the air pollution emission experiments and policy model and ranged from more than $500 per flight in the New York City area to just more than $20 per flight in the Dallas–Fort Worth, Texas, area. The analysis also demonstrates that emissions reductions from different interventions (e.g., single-engine taxiing and the use of tugs) are often not independent of each other and therefore cannot be combined in a simple way.

Vaishnav, P., Petsonk, A., Avila, R. A. G., Morgan, M. G., & Fischbeck, P. S. (2016). Analysis of a proposed mechanism for carbon-neutral growth in international aviation. *Transportation Research Part D: Transport and Environment, 45*, 126–138.

**Abstract:** In October 2013, the International Civil Aviation Organization (ICAO) announced that it would put in place a market-based mechanism to cap net greenhouse gas emissions from international civil aviation at 2020 levels. This paper analyses the obligations that would be placed on real airlines under an initial draft "Strawman" proposal that was originally formulated as a starting point for discussions within ICAO, and the extent to which such a proposal would succeed in keeping emissions at or below the desired level. The provisions of the ICAO proposal were then applied to more than 100 existing airlines. In order to protect commercial sensitivities, we used hierarchical cluster analysis to identify groups of different types of airlines. We report the results for these groups rather than for individual airlines. While ambiguities in the Strawman proposal complicated the analysis, we found that, depending on their size and rate of growth, airlines will be required to offset very different proportions of their emissions from international flights. A system of *de minimis* exemptions, as currently proposed, would benefit some rich countries as well as poor ones. Targeting such exemptions more narrowly would raise practical difficulties, which we describe. We conclude by recommending that ICAO design and implement a much simpler system; and propose one alternative.

Lee, J., Marla, L., & Vaishnav, P. (2021). The impact of climate change on the recoverability of airline networks. *Transportation Research Part D: Transport and Environment, 95*, 102801.

**Abstract:** This paper studies the impact of climate change-imposed constraints on the recoverability of airline networks. We first use models that capture the modified payload-range curves for different aircraft types under multiple climate change scenarios, and the associated (reduced) aircraft capacities. We next construct a modeling and algorithmic framework that allows for simultaneous and integrated aircraft and passenger recovery that explicitly capture the above-mentioned capacity changes in aircraft at different times of day. The results, based on the RCP8.5 climate change scenario, suggest that daily total airline recovery costs increase on average by 15.7% to 49.4%; and by 10.6% to 165.0% over individual disrupted days – depending on the original source of disruption, the climate change case of interest and year of interest. Aircraft-related costs are driven by a huge increase in aircraft swaps and cancellations; and passenger-related costs are driven by increases in disrupted (misconnected) passengers who need to be rebooked on the same or a different airline. This work motivates the critical need for airlines to systematically incorporate climate change as a factor in the design of aircraft as well as in the design and operations of airline networks.

**Some notable writings by authors not affiliated with our Centers:**

Alarfaj, A. F., Griffin, W. M., & Samaras, C. (2020). Decarbonizing US passenger vehicle transport under electrification and automation uncertainty has a travel budget. *Environmental Research Letters, 15*(9), 0940c2.

Gallucci, M. (2021). The ammonia solution: Ammonia engines and fuel cells in cargo ships could slash their carbon emissions. *IEEE Spectrum, 58*(3), 44–50.

Holland, S. P., Mansur, E. T., Muller, N. Z., & Yates, A. J. (2016). Are there environmental benefits from driving electric vehicles? The importance of local factors. *American Economic Review, 106*(12), 3700–3729.

Milovanoff, A., Minet, L., Cheah, L., Posen, I. D., MacLean, H. L., & Balasubramanian, R. (2021). Greenhouse gas emission mitigation pathways for urban passenger land transport under ambitious climate targets. *Environmental Science & Technology, 55*(12), 8236–8246.

Moultak, M., Lutsey, N., & Hall, D. (2017). Transitioning to zero-emission heavy-duty freight vehicles. *Int. Counc. Clean Transp.*

Sripad, S., & Viswanathan, V. (2017). Performance metrics required of next-generation batteries to make a practical electric semi truck. *ACS Energy Letters, 2*(7), 1669–1673. https://doi.org/10.1021/acsenergylett.7b00432

Sripad, S., Viswanathan, V. (2019). Quantifying the economic case for electric semi-trucks. *ACS Energy Letters, 4*, 149–155. https://doi.org/10.1021/acsenergylett.8b02146

## Notes

1 If we convert these percentages into tons of carbon dioxide and equivalent greenhouse gases for the U.S., which has total emissions of about 6.6 billion tons, transportation contributes just under 2 billion tons and electricity contributes over 1.5 billion tons. Global emissions are approximately 51 billion tons, to which transportation contributes a little over 8 billion tons and electricity contributes almost 14 billion tons.

2 Some railroad lines in the east are electrified, but long cross-country routes are not and adding overhead wires to convert them to electricity would be very expensive.

3 Tessum, C. W., Paolella, D. A., Chambliss, S. E., Apte, J. S., Hill, J. D., & Marshall, J. D. (2021). $PM_{2.5}$ polluters disproportionately and systemically affect people of color in the United States. *Science Advances, 7*(18), eabf4491.

# 15

# UNCERTAINTY IN ENERGY AND OTHER FORECASTS

*Granger Morgan*

We named our third NSF-supported center the "Center for Climate and Energy Decision Making" because understanding, choosing, and managing the sources and uses of energy lies at the heart of addressing the problems of climate change. Especially important to understanding how the climate system is likely to evolve in the future is anticipating how society's future use of different sources of energy may evolve.

The Shell Oil Company has long been a proponent of using scenarios – stories about different possible futures. They argue that these relatively detailed stories about different ways in which the future may unfold "are not forecasts, projections or predictions of what is to come. Nor are they preferred views of the future. Rather, they are plausible alternative futures: they provide reasonable and consistent answers to the 'what if?' questions relevant to business."[1] The Intergovernmental Panel on Climate Change (IPCC) has followed their lead and used scenarios that spell out a variety of ways in which future human activities and emissions of greenhouse gas may evolve. The IPCC, like Shell, argues that they are not forecasts or projections and that no probability should be attached to them.

There are several problems with the use of scenarios. Because they often provide quite a bit of detail, they can be very compelling. Indeed, experimental psychologists have conducted studies in which they have shown that the more detail that is added, the higher the probability people attach to that scenario being true. In fact, of course, if a more detailed story is a subset of a more general one, the probability that it will happen exactly in that way goes down, not up. The issue is well illustrated by results from a classic study conducted by experimental psychologists Amos Tversky and Daniel Kahneman.[2] They provided people with a personality sketch of a young man they called "Tom W." who sounds like a "nerd." Then they recruited three separate groups of people and asked the first group the simple question: "What is the probability that Tom W. will select journalism as his major in college?" They asked

DOI: 10.4324/9781003330226-15

the second group a slightly more detailed question: "What is the probability that Tom W. will select journalism as his major in college but decides he does not like it and decides to change his major?" Finally, they asked the third group an even more detailed question: "What is the probability that Tom W. will select journalism as his major in college but decides he does not like it and decides to change his major to engineering?" The probabilities given by the groups that got more detailed stories were higher. Figure 15.1 shows why they should go down.

There are a number of other issues that raise problems and limit the usefulness of scenarios. In 2008, David Keith and I published a paper in the journal *Climatic Change* titled "Improving the Way We Think About Projecting Future Energy Use and Emissions of Carbon Dioxide" in which we described many of the problems with using scenarios and then, in place of telling stories that only define a very narrow line out into the future, showed how to define a "cone" of future possible outcomes sort of like the cone of uncertainty that hurricane forecasters use to describe how a hurricane may evolve in the future.[3]

Of course, if, despite their limitations, someone really wants to use scenarios, they at least should be careful to include a set of futures that are among the most likely to occur and are internally consistent. In order to assess how well the scenarios developed and used by the IPCC are internally consistent, EPP PhD student Vanessa Schweizer worked with Elmer Kriegler to apply an approach called the cross-impact balance (CIB) method. They found that some IPCC scenarios had some serious problems with respect to internal consistency. After completing her PhD, Vanessa went on to do a variety of other analyses looking at how best to improve the coverage and the internal consistency of scenarios.

"Bounding analysis" is a strategy that engineers and scientists often use when they are thinking about something that is not understood very well, such as how much energy the country will use at a date in the far future. The idea is to consider

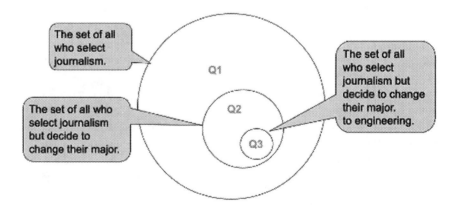

**FIGURE 15.1** Illustration of why the probability of the three scenarios described in the text should go down, not up, as more details are added to the scenario.

Drawn by Granger Morgan.

all the reasons the number must be at least as big as a lower value or "lower bound." Then one considers all the reasons the number can't be larger than an upper value or "upper bound." The resulting range "bounds" the answer. While common in technical communities, such an approach has not seen wide use in the area of public policy.

After thinking a lot about how to do climate analysis that goes far out into the future, in 1999, Liz Casman, Hadi Dowlatabadi, and I wrote a paper titled "Mixed Levels of Uncertainty in Complex Policy Models" in which we made extensive use of bounding analysis. We argued that as an analysis moved further and further out into the uncertain future, the models that are used to do analysis should gradually become simpler, and that finally they should just switch over to simple bounding analysis. Unfortunately, we have not been very successful in persuading people to adopt such a strategy. Many analysts continue to run models far out into the future, far beyond the point at which the assumptions on which they are based remain valid.

Applying the idea of bounding analysis to energy forecasting, Vanessa Schweitzer and I did an analysis to bound the range of U.S. electricity demand in the year 2050. We broke up future electricity demand into two parts: an expected, or "business-as-usual," part and a "new demand" part in which we included possible technological changes in response to climate change. We concluded that under a variety of aggressive adaptation and mitigation conditions, low or high growth in GDP, and modest or substantial improvements in energy intensity, U.S. electricity demand could be as little as 3100 TWh or as much as 17,000 TWh in 2050. We estimated that electrification of the U.S. transportation sector could introduce the largest share of new electricity demand. Projections for expected electricity demand are most sensitive to assumptions about the rate of reduction of U.S. electricity intensity per unit GDP. A result from this study is shown in Figure 15.2.

We had quite a bit of difficulty getting this paper accepted for publication because most reviewers were not familiar with the idea of bounding analysis and kept insisting that they wanted to see detailed models projected out for many decades. After quite a bit of back-and-forth, we finally persuaded the journal editor that projecting very detailed models far out into the future did not make very much sense, and the paper was accepted.

In the balance of this chapter, we discuss how others have engaged in forecasting U.S. energy demand and energy prices and some of the strategies that we believe could be used to better characterize the uncertainty in those forecasts.

Each year, the U.S. Energy Information Agency (EIA), an agency within the U.S. Department of Energy, publishes an "Annual Energy Outlook" that "provides modeled projections of domestic energy markets." The EIA is careful to make clear that these are not predictions of what will happen. Rather, they say they show "…what may happen given the assumptions in the underlying National Energy Modeling System (NEMS)." Despite the caveats, many organizations and analysts use the EIA forecasts as predictions. Over the years, lots of other groups in the U.S. and around the world have also made similar forecasts.

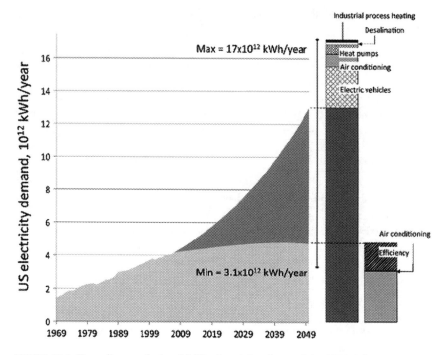

**FIGURE 15.2** Bounding analysis of U.S. electricity demand in 2050. The upper case assumes sustained GDP growth of about 3.3% per year with modest improvement in the efficiency with which the country uses energy ($-2.5$ Wh/$GDP/year). Longer-range (60-km, or 40-mile, range) batteries were assumed for plug-in hybrid electric vehicles. The lower case assumes GDP growth of 1.8% per year with sustained electricity intensity improvement ($-4.5$ Wh/$GDP/year). Figure from Schweizer & Morgan, 2016.

Schweizer, V. J., & Morgan, M. G. (2016). Bounding US electricity demand in 2050. *Technological Forecasting and Social Change, 105*, 215–223.

As we began to examine how those various forecasts had performed in light of subsequent developments, it became clear that making good forecasts is a *really* tough problem, and nobody is very good at it. For example, Figure 15.3 shows forecasts that EIA made of the price that U.S. power plants would pay for coal over the years of 1984 through 2004. These forecasts are compared with actual prices (black line). Notice that for over a decade EIA was forecasting prices to rise more rapidly than they actually did. Figure 15.4 compares a number of past forecasts with how things actually worked out.

In order to better understand the problems of making forecasts of energy and other important social quantities, such as future population, we organized an invitational workshop that we ran in Washington, DC in March of 2013. Participants who talked about energy included a number of experts like Paul Craig, who had done retrospective analyses of past energy forecasts, and Howard Gruenspecht, who ran the EIA. We also heard from several folks who had worked on scenario analysis.

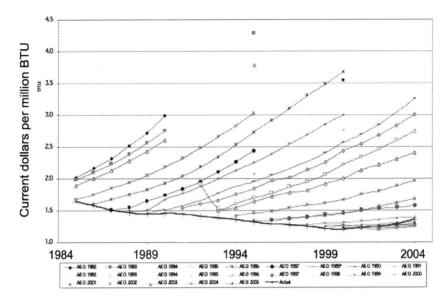

**FIGURE 15.3** Modeled projections of coal prices to electric generating plants from the EIA Annual Energy Outlook (rising curves) compared with actual historic prices (bold curve at the bottom) over the period 1984–2004. Data from EIA plotted by Adam Newcomer.

Data from EIA plotted by Adam Newcomer.

These included Nebojsa Nakicenovic and Richard Moss, who had led the development of scenarios for the IPCC, and Vanessa Schweizer, who, as noted above, had done some very inventive work on assessing the EIA scenarios and had worked with me on applying bounding analysis to a longer-term forecast of U.S. electricity demand.

The discussions and presentations in that workshop stimulated a number of new ideas, and we recruited two new EPP PhD students to follow up. Lynn Kaack worked on developing strategies to add uncertainty to EIA forecasts, and Evan Sherwin explored whether the frequency and magnitude of surprises related to energy forecasts had increased in recent years.

One way the EIA might add uncertainty to its projections would be to go back and see how well past forecasts had done and then use those results to put uncertainty ranges on the most recent projections. Lynn set out to do this, working with me, Jay Apt, and Patrick McSharry. Lynn applied this method to 18 core quantities being projected by the EIA and produced probabilistic uncertainties for their 2016 Annual Energy Outlook. An example of the sort of results she produced is shown in Figure 15.5. Lynn suggested that the approach she had developed could be used to "give guidance on how to evaluate and communicate uncertainty in future energy outlooks" and that the standard deviations of past forecasting errors should be made public.

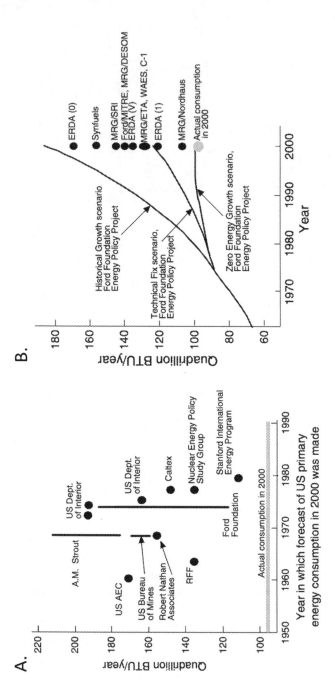

**FIGURE 15.4  A** (left) Summary of forecasts of U.S. primary energy consumption for the year 2000 compiled by Smil[4] as a function of the date on which they were made compared with the actual value that occurred in that year (line at the bottom). **B** (right) Forecasts of U.S. primary energy consumption for the year 2000 compiled by Greenberger[5] in the early 1980s compared with three scenarios developed by the Ford Foundation Energy Project compared with the actual value that occurred in that year (large dot).

Smil, V. (2000). Perils of long-range energy forecasting: Reflections on looking far ahead. *Technological Forecasting and Social Change, 65*(3), 251–264.

Morgan, M. G., & Keith, D. W. (2008). Improving the way we think about projecting future energy use and emissions of carbon dioxide. *Climatic Change, 90*(3), 189–215.

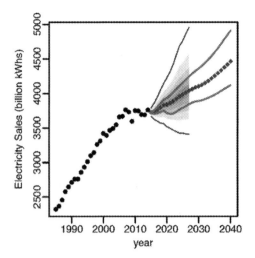

**FIGURE 15.5** An example of the type of uncertainty bounds that Lynn Kaack was able to add to the EIA's projections (in this case, for future electricity sales) by performing statistical analysis of past projections. Historical values are indicated by black circles, the AEO 2016 reference case by green diamonds, and the probability density forecast by blue shaded areas. AEO 2016 envelope scenarios are in green.

Kaack, L. H., Apt, J., Morgan, M. G., & McSharry, P. (2017). Empirical prediction intervals improve energy forecasting. *Proceedings of the National Academy of Sciences*, *114*(33), 8752–8757.

After her paper was completed, Lynn went down to Washington and gave a seminar to the staff at the EIA. They found her work very interesting and later adopted her recommendation in their 2018 Annual Energy Outlook (AEO) Retrospective Review (see technical details at the end of the chapter).

In his analysis, EPP PhD student Evan Sherwin worked with Max Henrion and Inês Azevedo to explore whether the energy system of today has become harder to predict than it was in the past. He did this by quantifying the unpredictability in terms of the frequency of extreme errors he found when he looked back at past AEO energy projections. He found that very large projection errors and very large year-to-year changes both became more common in recent decades.

As these examples illustrate, uncertainty in forecasts has only recently entered the conversation. Our centers have done a lot of work to improve how it is understood and communicated, but there is still much more work to be done.

## Technical Details for Chapter 15

Here is the paper in which we argued that while policy models might start out being pretty complicated, as we go further and further out into an uncertain future, they should gradually get simpler and should ultimately switch over to simple bounding analysis:

Casman, E. A., Morgan, M. G., & Dowlatabadi, H. (1999). Mixed levels of uncertainty in complex policy models. *Risk Analysis, 19*(1), 33–42.

**Abstract:** The characterization and treatment of uncertainty poses special challenges when modeling indeterminate or complex coupled systems such as those involved in the interactions between human activity, climate and the ecosystem. Uncertainty about model structure may become as, or more important than, uncertainty about parameter values. When uncertainty grows so large that prediction or optimization no longer makes sense, it may still be possible to use the model as a "behavioral test bed" to examine the relative robustness of alternative observational and behavioral strategies. When models must be run into portions of their phase space that are not well understood, different submodels may become unreliable at different rates. A common example involves running a time stepped model far into the future. Several strategies can be used to deal with such situations. The probability of model failure can be reported as a function of time. Possible alternative "surprises" can be assigned probabilities, modeled separately, and combined. Finally, through the use of subjective judgments, one may be able to combine, and over time shift between models, moving from more detailed to progressively simpler order-of-magnitude models, and perhaps ultimately, on to simple bounding analysis.

Here is the paper David Keith and I wrote that is critical of using scenario analysis in studies of energy and climate change and lays out what we believe are some better strategies:

Morgan, M. G., & Keith, D. W. (2008). Improving the way we think about projecting future energy use and emissions of carbon dioxide. *Climatic Change, 90*(3), 189–215.

**Abstract:** A variety of decision makers need projections of future energy demand, $CO_2$ emissions and similar factors that extend many decades into the future. The past performance of such projections has been systematically overconfident. Analysts have often used scenarios based on detailed story lines that spell out "plausible alternative futures" as a central tool for evaluating uncertainty. No probabilities are typically assigned to such scenarios. We argue that this practice is often ineffective. Rather than expanding people's judgment about the range of uncertainty about the future, scenario-based analysis is more likely to lead to systematic overconfidence, to an underestimate of the range of possible future outcomes. We review relevant findings from the literature on human judgment under uncertainty and discuss their relevance to the task of making probabilistic projections. The more detail that one adds to the story line of a scenario, the more probable it will appear to most people, and the greater the difficulty they likely will have in imagining other, equally or more likely, ways in which the same outcome could be reached. We suggest that scenario based approaches make analysts particularly prone to such cognitive biases, and then outline a strategy by which improved projections, tailored to the needs of specific decision makers, might be developed.

Of course, if one is going to use scenario analysis, you should work to make sure the scenarios you use include some of the more likely future developments and are internally consistent. In this paper, Vanessa Schweizer and Elmer Krielger outline a set of strategies to do that and show that some of the scenarios that the IPCC had been using were not very internally consistent:

Schweizer, V. J., & Kriegler, E. (2012). Improving environmental change research with systematic techniques for qualitative scenarios. *Environmental Research Letters, 7*(4), 044011.

**Abstract:** Scenarios are key tools in analyses of global environmental change. Often they consist of quantitative and qualitative components, where the qualitative aspects are expressed in narrative, or storyline, form. Fundamental challenges in scenario development and use include identifying a small set of compelling storylines that span a broad range of policy-relevant futures, documenting that the assumptions embodied in the storylines are internally consistent, and ensuring that the selected storylines are sufficiently comprehensive, that is, that descriptions of important kinds of future developments are not left out. The dominant approach to scenario design for environmental change research has been criticized for lacking sufficient means of ensuring that storylines are internally consistent. A consequence of this shortcoming could be an artificial constraint on the range of plausible futures considered. We demonstrate the application of a more systematic technique for the development of storylines called the cross-impact balance (CIB) method. We perform a case study on the scenarios published in the IPCC Special Report on Emissions Scenarios (SRES), which are widely used. CIB analysis scores scenarios in terms of internal consistency. It can also construct a very large number of scenarios consisting of combinations of assumptions about individual scenario elements and rank these combinations in terms of internal consistency. Using this method, we find that the four principal storylines employed in the SRES scenarios vary widely in internal consistency. One type of storyline involving highly carbon-intensive development is underrepresented in the SRES scenario set. We conclude that systematic techniques like CIB analysis hold promise for improving scenario development in global change research.

Here is the paper in which Vanessa Schweizer and I applied bounding analysis to estimate the (very large) range of ways in which future electricity demand might evolve in the U.S. by the year 2050:

Schweizer, V. J., & Morgan, M. G. (2016). Bounding US electricity demand in 2050. *Technological Forecasting and Social Change, 105,* 215–223.

**Abstract:** Limiting climate change requires a radical shift in energy supply and use. Because of time lags in capital investments, the political process, and the climate system, potential developments decades from now must be considered for energy policy decisions today. Traditionally, scenario analysis and forecasting are used to conceptualize the future; however, past energy demand forecasts have performed poorly displaying overconfidence, or a tendency to overly discount the tails of a distribution of possibilities under uncertainty. This study demonstrates a simple analytical approach to bound US electricity demand in 2050. Long-term electricity demand is parsed into two terms — an expected, or "business-as-usual," term and a "new demand" term estimated explicitly to account for possible technological changes in response to climate change. Under a variety of aggressive adaptation and mitigation conditions, low or high growth in GDP, and modest or substantial improvements in energy intensity, US electricity demand could be as little as 3100 TWh or as much as 17,000 TWh in 2050. Electrification of the US transportation sector could introduce the largest share of new electricity demand. Projections for expected electricity demand are most sensitive to assumptions about the rate of reduction of US electricity intensity per unit GDP.

Finally, on the topic of bounding analysis, here is an opinion piece I wrote arguing for the use of bounding analysis in environmental policy analysis when uncertainty is very high:

Morgan, M. G. (2001). The neglected art of bounding analysis. Viewpoint, *Environmental Science & Technology, 35,* 162A–164A, April 1, 2001.

The next papers move on from bounding analysis to examine forecasts and performance of the U.S. energy system.

Here is the paper that reports the work that Lynn Kaack did on adding uncertainty to forecasts made by the EIA:

> Kaack, L. H., Apt, J., Morgan, M. G., & McSharry, P. (2017). Empirical prediction intervals improve energy forecasting. *Proceedings of the National Academy of Sciences, 114*(33), 8752–8757.

**Abstract:** Hundreds of organizations and analysts use energy projections, such as those contained in the US Energy Information Administration (EIA)'s Annual Energy Outlook (AEO), for investment and policy decisions. Retrospective analyses of past AEO projections have shown that observed values can differ from the projection by several hundred percent, and thus a thorough treatment of uncertainty is essential. We evaluate the out-of-sample forecasting performance of several empirical density forecasting methods, using the continuous ranked probability score (CRPS). The analysis confirms that a Gaussian density, estimated on past forecasting errors, gives comparatively accurate uncertainty estimates over a variety of energy quantities in the AEO, in particular outperforming scenario projections provided in the AEO. We report probabilistic uncertainties for 18 core quantities of the AEO 2016 projections. Our work frames how to produce, evaluate, and rank probabilistic forecasts in this setting. We propose a log transformation of forecast errors for price projections and a modified nonparametric empirical density forecasting method. Our findings give guidance on how to evaluate and communicate uncertainty in future energy outlooks.

The retrospective study EIA did which adopted Lynn Kaack's method is cited at the end of this chapter.

This is the paper describing the work in which Evan Sherwin explored surprises in the U.S. energy system:

> Sherwin, E. D., Henrion, M., & Azevedo, I. M. (2018). Estimation of the year-on-year volatility and the unpredictability of the United States energy system. *Nature Energy, 3*(4), 341.

**Abstract:** Long-term projections of energy consumption, supply and prices heavily influence decisions regarding long-lived energy infrastructure. Predicting the evolution of these quantities over multiple years to decades is a difficult task. Here, we estimate year-on-year volatility and unpredictability over multi-decade time frames for many quantities in the US energy system using historical projections. We determine the distribution over time of the most extreme projection errors (unpredictability) from 1985 to 2014, and the largest year-over-year changes (volatility) in the quantities themselves from 1949 to 2014. Our results show that both volatility and unpredictability have increased in the past decade, compared to the three and two decades before it. These findings may be useful for energy decision-makers to consider as they invest in and regulate long-lived energy infrastructure in a deeply uncertain world.

On the topic of forecasting, here is a commentary that appeared in *PNAS*:

> Morgan, M. G. (2018). Uncertainty in long-run forecasts of quantities such as per capita gross domestic product. *Proceedings of the National Academy of Sciences, 115*(21), 5314–5316.

**Some notable writings by authors not affiliated with our Centers:**

> Fischer, C., Herrnstadt, E., & Morgenstern, R. (2009). Understanding errors in EIA projections of energy demand. *Resource and Energy Economics, 31*(3), 198–209.

Gilbert, A. Q., & Sovacool, B. K. (2016). Looking the wrong way: Bias, renewable electricity, and energy modelling in the United States. *Energy*, *94*, 533–541.

Linderoth, H. (2002). Forecast errors in IEA-countries' energy consumption. *Energy Policy*, *30*(1), 53–61.

Milford, J., Henrion, M., Hunter, C., Newes, E., Hughes, C., & Baldwin, S. F. (2022). Energy sector portfolio analysis with uncertainty. *Applied Energy*, *306*, 117926.

Smil, V. (2000). Perils of long-range energy forecasting: Reflections on looking far ahead. *Technological Forecasting and Social Change*, *65*(3), 251–264.

Smil, V. (2008). Long-range energy forecasts are no more than fairy tales. *Nature*, *453*(7192), 154.

U.S. Energy Information Agency. (2018). Annual Energy Outlook (AEO) Retrospective Review: Evaluation of AEO2018 and Previous Reference Case Projections, U.S. Department of Energy, 36pp.

Winebrake, J. J., & Sakva, D. (2006). An evaluation of errors in US energy forecasts: 1982–2003. *Energy Policy*, *34*, 3475–3483.

# Notes

1 Quote is from the Executive Summary of "Shell Global Scenarios to 2025" (2005).

2 For details on these experimental studies, see: Kahneman, D., & Tversky, A. (1973). On the psychology of prediction. *Psychological Review*, *80*(4), 237.

3 Here is a more detailed explanation for more technical readers. The folks who develop the scenarios for IPCC argue that people should not attach probabilities to them, that they are simply provided as a means of exploring a range of future possibilities (although some investigators have done that anyway). Because these scenarios are described as "feasible" or "plausible," we argued that this must mean they are more *probable* than alternatives. At the same time, for entirely different reason, we agree that a probability should not be attached to them. Such scenarios can be thought of as a line out through N-dimensional space. No probability can be attached to a point or line in a continuous space. However, if the line is expanded into a cone that spans a region of that space, then it does make sense to attach a probability. This is the basic approach our paper outlines and recommends.

4 See: Smil, V. (2000). Perils of long-range energy forecasting: Reflections on looking far ahead. *Technological Forecasting and Social Change*, *65*(3), 251–264.

5 Ford Foundation Energy Project. (1974). *A Time to Choose: America's Energy Future*, Ballenger, 511pp.

# 16

# WE HAVE NO CHOICE BUT TO ADAPT

*Granger Morgan and Hadi Dowlatabadi*

Even if the world had been reducing emissions of greenhouse gas over 50 years ago when, as a graduate student, Granger first started going to sessions on climate science at the annual meeting of the American Geophysical Union together with Steve Schneider, we'd need to be adapting today to some of the effects of a changing climate. But because the world has still taken only very limited steps to reduce emissions of greenhouse gases, today the impacts from climate change to which we have committed the world are very large and growing. Even if we move much more aggressively to reduce the emission of greenhouse gases to zero, and develop ways to take some carbon dioxide out of the atmosphere, so much long-lived climate warming gas has been added to earth's atmosphere that substantial climate change is inevitable. There is simply no way to avoid the fact that we will have to adapt to many of the consequence of warmer planet that has a changed climate.

We can start with sea level. As water warms, it expands a little bit. So as the planet warms, the oceans expand. This along with melting glaciers and icecaps causes sea level to rise. In some of the first studies of the impacts of sea level rise, economists estimated future damage from sea level by projecting the value of coastal land that would be submerged under the sea. Coastal engineers argued that wherever the value of land and property was high enough, they would be building sea walls and there would be no retreat from sea level rise – this lowered the estimated damage from sea level rise, especially in high-density population centers. Economists then stepped back into the debate, noting that sea level rise is a slow and predictable process. This means that long before there is inundation, people will know it is going to happen so the premium land values for shorefront properties will migrate upland (away from the rising seas) in step with the sea level rise. Under this model, damage estimates were reduced even more. Of course, these analysts argued, if it was not possible to limit damages by building sea walls, the value of land lost to the sea would be that of land far inland as whole developments move upland.

DOI: 10.4324/9781003330226-16

This process of "refining" assessments of the cost of sea level rise effectively reduced estimates of future damages by a factor of 100. However, Hadi was not convinced. He noted that foresight about inundation by sea level rise did not account for how, long before inundation, storms damage coastal developments far inland from the shoreline. Together with Jason West and Mitch Small, they developed a damage assessment model that included the impact from storms. The model was based on data from the coastal town of Duck, North Carolina. It included spatial data on homes, property tax data on their value, and historic data from Corp of Engineers data on wave height, tide gauges, and storm reach. They ran their model for a time horizon of 2000 to 2100, over which there would be 40 cm of sea level rise. If they only made one run of their model with 40 cm of sea level rise and no serious storms, the homes closest to the shoreline were all just fine. On the other hand, when they ran the model hundreds of times, with different simulations of storms over 100 years and no change in storm intensity of frequency, their model showed significant damage from sea level rise and storms combined. It also demonstrated that simple retreat policies, such as a ban on rebuilding in the same risky location if damage exceeded 40% of the property's assessed value, would help the community adapt to sea level rise and dramatically reduce the long-term damage that could be suffered.

The model also showed that there would have been significant damage from storms regardless of sea level rise. This highlights the general challenge of disentangling impacts and adaptation from present conditions that, regardless of climate change, can place people and property at predictable risk. In the case of our Duck Island model, concern about sea level rise and introduction of a "retreat" policy reduces discounted damage to properties over the simulation period (regardless of sea levels rising). One of the more counterintuitive findings from this work was that, with retreat policies in place, a future timeline of small storms would generate the greatest damage (because homeowners kept rebuilding and not retreating) while one early large storm that caused people to retreat would lead to lower long-term damages. More details on this work can be found in a paper titled "Storms, investor decisions, and the economic impacts of sea level rise" that was published in *Climatic Change* in 2001.

More recently, our team worked on whether the storm models used in such simulations should grow in severity or frequency over time. In addition to sea level rise, another consequence of a warmer planet is warmer sea surface temperatures that will give rise to more intense (and perhaps more frequent) hurricanes. Iris Grossmann did an extensive review of the available evidence on hurricanes[1] and published a paper in the journal *Climatic Change* in 2011. She found clear evidence that hurricanes in the Atlantic were becoming more intense.

Between 1962 and 1983, the U.S. government ran a program called Project Stormfury that tried to modify hurricanes. The idea was to seed the clouds with silver iodide, which was thought would disrupt the inner structure of the storm. Several Atlantic hurricanes were seeded. However, the project never produced promising results and more recent research showed that the basic approach was not scientifically sound. Years later, the Department of Homeland Security began to discuss the possibility of restarting work on modifying hurricanes. PhD student Kelly Klima and Granger thought they were nuts but decided to do an analysis to examine the issue. We teamed up with MIT hurricane expert Kerry Emanuel, some of his MIT colleagues, and Iris Grossmann at CMU and carried out a series of analyses. The basic idea was that the intensity of a storm could be decreased by lowering the temperature of the ocean surface, which is what drives the intensity of the storm.

We did this by assuming that a series of wind-wave pumps, which bring cold water from lower depths up to the surface, were placed in front of the storm. Using Kerry's state-of-the-art hurricane model, we used historical hurricane data to examine how much storm intensity and the associated damage to buildings might be reduced. We published the results in a paper in *Environmental Science & Technology* in 2011. While the approach looked more promising than we had expected, we concluded that other strategies like hardening buildings (secure tie-downs of roofs to foundations, storm shutters) were more cost-effective and reliable.

This first analysis examined a region in which flooding from storm surge was not a major issue. A year later, in a second paper, also in *Environmental Science & Technology*, we added a consideration of storm surge. For storm surge adaptation, we found that in most regions a surge barrier performed best, combined in some areas, with raising buildings.

While it looks like wind-wave pumps could reduce the surface temperature and in that way reduce a hurricane's intensity, it is not clear that there is a technology that could steer a hurricane. Of course, even if it were possible to steer a storm away from a heavily populated area to a less populated area, one faces many ethical and legal issues. In 2012, Kelly and Granger published a paper in *Journal of Risk Research* that explored these issues. We also teamed up with Wändi Bruine de Bruin to do a study of what the public thinks about hurricane modification techniques. We published the results of this work in *Risk Analysis* in 2012.

In addition to hurricanes, wildfire and drought are two other consequences of climate change. Every time he hears about the next major western wildfire, Granger is vividly reminded that when he interviewed climate scientist Steve Schneider in his second expert elicitation in 2009, Steve talked at great length about how wildfires caused by drought would soon be devastating the west.

Australia is another part of the world in which drought and wildfires are becoming ever more serious. Anticipating this growing problem, Hadi, together with James Risbey and Milind Kandlikar, arranged to team with Dean Graetz of the Australian Commonwealth Scientific and Industrial Research Organization (CSIRO). Dean was keen on us sharing our integrated assessment methods. We were keen on learning about adaptation to droughts in a developed agricultural sector. Everyone agreed that Australia, with a great deal of rain-fed agriculture and frequent droughts, was an ideal place to study adaptation policies. So, our team of Milind, James, and Hadi held a weeklong workshop with Dean and various master-farmers and agricultural scientists from CSIRO. During that week, we conducted informal elicitations with the workshop participants to develop an integrated assessment framework for CSIRO. The most important finding of the workshop is summarized in the table below. There are many different and simultaneous impacts to consider at any given time. These have different magnitudes and frequencies, each requiring different types of adaptive responses. These impacts can work to amplify or ameliorate one another. Meanwhile, there are many different adaptation measures that can be taken. These too have different timescales and efficacies in addressing a specific impact, but they too can amplify or ameliorate other impacts as well as the efficacy of other adaptation measures. Table 16.1 provides a summary of impact-and-adaptation interactions. Of course, there should also be careful assessments of impact-and-impact and adaptation-and-adaptation interactions before we can be confident that adaptation interventions will lead to greater resilience rather than inevitable fragility.

If there's any group you would think would benefit from more information about how climate change will affect hurricanes, it's the insurance industry. With that in mind, we organized a workshop with representatives from insurance companies and also the big companies that backed them up (called reinsurers). Perhaps we were naïve, but the insurance companies basically told us "unless you could tell us how many hurricanes there will be next season, how strong they will be, and

**TABLE 16.1** Summary of impact-and-adaptation interactions.

| Adaptations | Australian Agriculture Impacts and Adaptations | | | | | |
| --- | --- | --- | --- | --- | --- | --- |
| | Impacts | | | | | |
| | Weather/climate | Internal markets | Export markets | C and N cycles | Pests | Soil & water |
| Storage | low | low | low | – | low | – |
| Insurance | low | low | low | – | low | – |
| Engineering | low | – | – | – | – | high |
| Practice | high | low | low | high | high | high |
| Land Use | high | – | – | low | low | low |
| Decision Support | high | low | low | – | low | low |
| R & D | low | low | low | low | low | low |
| Incentives | low | high | high | – | – | high |
| Disaster Aid | low | – | – | – | – | – |
| Relative Impact | 0.3 | 1 | 1 | 0.2 | 0.5 | 1 |
| Freq. Impact (yr.) | 1 | 10 | – | 20 | 1 | >100 |

Source: Risbey, Kandlikar, Dowlatabadi, & Graetz, 1999.

This table was constructed by elicitation of the panel of experts participating in the workshop on Integrated Assessment of Agriculture Adaptation for Australia. The columns show a variety of impact categories on agriculture and the rows show a range of potential adaptations to those impacts. The subjective estimates of the efficacy of adaptive measures were ranked as 'high' or 'low.' Relative sizes of impacts on agriculture were ranked on a 0–1 scale, where 0 is no impact and 1 is the maximum relative impact.

where they will make landfall, we are not interested." While the big reinsurance companies, like SwissRe, do have groups that track climate science, the frontline insurance companies basically write coverage one season at a time and base their policies on the experience of the last few years.

Insurance companies also get advice from several private firms that do risk assessments for them. The models that those private firms use are confidential, and they have generally not been willing to share many details of how they work. About a decade after our workshop, Howard Kunreuther, in the Wharton School at the University of Pennsylvania, convened a similar workshop. The results were pretty much the same. This was just a few years after a couple of major hurricanes had hit the Gulf Coast. Representatives from one of the frontline insurance companies described the impact and losses from those storms as a "black swan"—an event that was entirely unpredictable. Of course, any geophysicist could have told them that large storms hit the Gulf Coast on a very regular basis. The simple fact that there hadn't been such a storm in recent years doesn't mean that such events couldn't be easily anticipated by anyone who looked at the historical and geological record.

With Christina Cook of the University of British Columbia, Hadi wrote a paper in 2008 and a book chapter in 2011 that explored key issues in insurance and climate change. They explored three ways in which the insurance industry might learn about, and better adjust to, the growing risks posed by climate change. These are: (1) learning from recent events; (2) using models of how risks may change in the future; (3) learning by getting sued (i.e., from litigation). In concluding the 2008 article, we wrote:

> So far, extreme event losses have only dealt a small blow to insurance industry coffers (less than 5% of their revenues). The evidence from rising exposures and new probabilistic models estimating the risks quantitatively are growing in their influence over underwriting decisions. It is tempting to assume that this new knowledge is prompting the industry to offer better-informed terms for underwriting. It would be more accurate to say that the industry has learned that some former underwritings are not insurable at rates that are acceptable to consumers. This will eventually lead to a renegotiation of risk management through a coordination of private and public entities—best achieved in an atmosphere of cooperation.
>
> In countries where the insurance industry is well established, climate concerns are forcing a careful re-examination of underwriting and risk mitigation practices. The outcome will more clearly recognize the inadvisability of property developments in hazardous areas, transfer some risks and burdens of mitigation to property owners, and engage the government and new instruments for provision of risk coverage. The recognition that many such risks are created by our own lack of foresight will entail significant public benefits. In the interim, the costs of transition to a more enlightened pattern of land-use is likely to fall on the shoulders of the poor, who cannot afford

insurance, and on the government, which effectively acts as an insurer of last resort while fumbling to find the right mix of policies to mitigate risks.

One consequence of climate change causing more, and more extreme, events is more frequent disruptions of electric power. This is a serious problem because modern society has become critically dependent on reliable electricity. Accordingly, Center investigators have undertaken a number of studies to develop strategies that could limit social vulnerability to power disruptions. In 2011, EPP PhD student Anu Narayanan published a paper in *Risk Analysis* that laid out a strategy by which electric power distribution systems could be reconfigured to operate as an isolated microgrid during a large outage of long duration (LLD-outage). She argued that the arrangements she proposed could be affordable, however, because the benefits accrued to all of society, it was probably not appropriate to cover their cost only through the traditional rate-base (i.e., requiring electricity users to cover all the cost). In 2021, EPP PhD student Angelena Bohman published a second paper in *Risk Analysis* that took a more detailed look at how service to customers on a hypothetical distribution feeder in the upper Connecticut River Valley[2] might be made more robust in the face of LLD-outages. Her work showed that collective strategies (as opposed to individual generators at each home) could offer a far more affordable solution.

Finally, several PhD students in EPP, including Mike Rath, Doug King, Neil Strachen, and Anu Narayanan, have also explored the possible environmental and resiliency benefits of distributed generation sources and microgrids.

While not directly a part of our centers' work, insights from center work helped to inform three consensus studies performed by the National Academies that Granger chaired:

- *Terrorism and the Electric Power Delivery System*, 2012, 146 pp.
- *Enhancing the Resilience of the Nation's Electricity System*, 2017, 156pp.
- *The Future of Electric Power in the U.S.*, 2021, 337 pp.

These reports can be downloaded for free from the website of the National Academies Press.

In 1999, Hadi teamed with economist Gary Yohe of Wesleyan University to write a paper titled "Risk and uncertainties, analysis and evaluation: lessons for adaptation and integration." We think the lessons they drew then are still highly relevant today. Here is what they wrote in the abstract of their paper:

This paper draws ten lessons from analyses of adaptation to climate change under conditions of risk and uncertainty: (1) Socio-economic systems will likely respond most to extreme realizations of climate change. (2) Systems have been responding to variations in climate for centuries. (3) Future change will affect future citizens and their institutions. (4) Human systems can be the sources of surprise. (5) Perceptions of risk depend upon welfare valuations that depend upon expectations. (6) Adaptive decisions will be made in response to climate change *and* climate change policy. (7) Analysis of adaptive decisions should recognize the second-best context of those decisions. (8) Climate change offers opportunity as well as risk. (9) All plausible futures should be explored. (10) Multiple methodological approaches should be accommodated. These lessons support two pieces of advice for the Third Assessment Report: (1) Work toward consensus, but not at the expense of thorough examination and reporting of the "tails" of the distributions of the future. (2) Integrated assessment is only *one* unifying methodology; others that can better accommodate those tails should be encouraged and embraced.

## Technical Details for Chapter 16

Here is the paper on retreating or rebuilding after hurricanes and storm surge:

West, J. J., Small, M. J., & Dowlatabadi, H. (2001). Storms, investor decisions, and the economic impacts of sea level rise. *Climatic Change, 48*(2), 317–342.

**Abstract:** Past research on the economic impacts of a climate-induced sea level rise has been based on the gradual erosion of the shoreline, and human adaptation. Erosion which is accelerated by sea level rise may also increase the vulnerability to storm damage by decreasing the distance between the shore and structures, and by eroding protective coastal features (dunes). We present methods of assessing this storm damage in coastal regions where structural protection is not pursued. Starting from the bounding cases of *no foresight* and *perfect foresight* of Yohe et al. (1996), we use a disaggregated analysis which models the random nature of storms, and models market valuation and private investor decisions dynamically. Using data from the National Flood Insurance Program and a hypothetical community, we estimate that although the total storm damage can be large, the increase in storm damage attributable to sea level rise is small (<5% of total sea level rise damages). These damages, however, could become more significant under other reasonable assumptions or where dune erosion increases storm damage.

Here is Iris' review of what we know about how climate change is affecting hurricanes:

Grossmann, I., & Morgan, M. G. (2011). Tropical cyclones, climate change, and scientific uncertainty: What do we know, what does it mean, and what should be done?. *Climatic Change, 108*(3), 543–579.

**Abstract:** The question of whether and to what extent global warming may be changing tropical cyclone (TC) activity is of great interest to decision makers. The presence of a possible climate change signal in TC activity is difficult to detect because interannual variability necessitates analysis over longer time periods than available data allow. Projections of future TC activity are hindered by computational limitations and uncertainties about changes in regional climate, large scale patterns, and TC response. This review discusses the state of the field in terms of theory, modeling studies and data. While Atlantic TCs have recently become more intense, evidence for changes in other basins is not persuasive, and changes in the Atlantic cannot be clearly attributed to either natural variability or climate change. However, whatever the actual role of climatic change, these concerns have opened a "policy window" that, if used appropriately, could lead to improved protection against TCs.

Here are the two main papers on modifying hurricanes:

Klima, K., Morgan, M. G., Grossmann, I., & Emanuel, K. (2011). Does it make sense to modify tropical cyclones? A decision-analytic assessment. *Environmental Science & Technology*, 45, 4242–4248.

**Abstract:** Recent dramatic increases in damages caused by tropical cyclones (TCs) and improved understanding of TC physics have led DHS to fund research on intentional hurricane modification. We present a decision analytic assessment of whether it is potentially cost-effective to attempt to lower the wind speed of TCs approaching South Florida by reducing sea surface temperatures with wind-wave pumps. Using historical data on hurricanes approaching South Florida, we develop prior probabilities of how storms might evolve. The effects of modification are estimated using a modern TC model. The FEMA HAZUS-MH MR3 damage model and census data on the value of property at risk are used to estimate expected economic losses. We compare wind damages after storm modification with damages after implementing hardening strategies protecting buildings. We find that if it were feasible and properly implemented, modification could reduce net losses from an intense storm more than hardening structures. However, hardening provides "fail safe" protection for average storms that might not be achieved if the only option were modification. The effect of natural variability is larger than that of either strategy. Damage from storm surge is modest in the scenario studied but might be abated by modification.

Klima, K., Lin, N., Emanuel, K., Morgan, M. G., & Grossmann, I. (2012). Hurricane modification and adaptation in Miami-Dade County, Florida. *Environmental Science & Technology*, 46(2), 636–642.

**Abstract:** We investigate tropical cyclone wind and storm surge damage reduction for five areas along the Miami-Dade County coastline either by hardening buildings or by the hypothetical application of wind-wave pumps to modify storms. We calculate surge height and wind speed as functions of return period and sea surface temperature reduction by wind-wave pumps. We then estimate costs and economic losses with the FEMA HAZUS-MH MR3 damage model and census data on property at risk. All areas experience more surge damages for short return periods, and more wind damages for long periods. The return period at which the dominating hazard component switches depends on location. We also calculate the seasonal expected fraction of control damage for different scenarios to reduce damages. Surge damages are best reduced through a surge barrier. Wind damages are best reduced by a portfolio of techniques that, assuming they work and are correctly deployed, include wind-wave pumps.

And here are the other papers we did on modifying hurricanes:

Klima, K., & Morgan, M. G. (2012). Thoughts on whether government should steer a tropical cyclone if it could. *Journal of Risk Research*, 15(8), 1013–1020.
Klima, K., Bruine de Bruin, W., Morgan, M. G., & Grossmann, I. (2012). Public perceptions of hurricane modification. *Risk Analysis*, 32(7), 1194–1206.

Here are the CSIRO report and the paper on climate change impacts on agriculture in Australia:

Graetz, D., Dowlatabadi, H., Risbey, J., & Kandlikar, M. (1997). Applying Frameworks for Assessing Agricultural Adaptation to Climate Change in Australia. *Earth Observing Center, CSRIO Report*, 97(1).

Risbey, J., Kandlikar, M., Dowlatabadi, H., & Graetz, D. (1999). Scale, context, and decision making in agricultural adaptation to climate variability and change. *Mitigation and Adaptation Strategies for Global Change, 4*(2), 137–165.

**Abstract:** This work presents a framework for viewing agricultural adaptation, emphasizing the multiple spatial and temporal scales on which individuals and institutions process information on changes in their environment. The framework is offered as a means to gain perspective on the role of climate variability and change in agricultural adaptation, and developed for a case study of Australian agriculture. To study adaptation issues at the scale of individual farms we developed a simple modelling framework. The model highlights the decision making element of adaptation in light of uncertainty, and underscores the importance of decision information related to climate variability. Model results show that the assumption of perfect information for farmers systematically overpredicts adaptive performance. The results also suggest that farmers who make tactical planting decisions on the basis of historical climate information are outperformed by those who use even moderately successful seasonal forecast information. Analysis at continental scales highlights the prominent role of the decline in economic operating conditions on Australian agriculture. Examples from segments of the agricultural industry in Australia are given to illustrate the importance of appropriate scale attribution in adapting to environmental changes. In particular, adaptations oriented toward short time scale changes in the farming environment (droughts, market fluctuations) can be limited in their efficacy by constraints imposed by broad changes in the soil/water base and economic environment occurring over longer time scales. The case study also makes the point that adaptation must be defined in reference to some goal, which is ultimately a social and political exercise. Overall, this study highlights the importance of allowing more complexity (limited information, risk aversion, cross-scale interactions, mis-attribution of cause and effect, background context, identification of goals) in representing adaptation processes in climate change studies.

Here are a paper and chapter that Hadi and Christina wrote on insurance:

Dowlatabadi, H., & Cook, C. (2008). Climate risk management & institutional learning. *Integrated Assessment, 8*(1).

**Abstract:** Insurance is a prominent mechanism for risk transfers. Many initiatives are looking towards private-public partnerships and new risk management instruments to provide a cushion for climate change related impacts.

In order for this aspiration to be fulfilled, the insurers and institutions within which they operate need to learn about emergent risks and develop workable strategies. We explore three factors shaping the evolution of insurance practices: quantitative models of catastrophic loss, experience of catastrophic loss and outcomes of litigated cases. We use the available evidence from the USA to assess the importance of each of these factors in how the industry is evolving and hence what actual risk reductions and transfers are more likely in the USA for the foreseeable future.

Cook, C., & Dowlatabadi, H. (2011). Learning adaptation: Climate-related risk management in the insurance industry. In *Climate Change Adaptation in Developed Nations*, 255–265. Springer, Dordrecht.

**Abstract:** Insurance is a prominent, well-established mechanism for risk transfer in developed countries. While North American governments have stalled on both mitigation of and adaptation to climate change, the insurance industry (globally and in North

America) is already viewing recent catastrophic events as being partially climate change related and exploring new adaptation initiatives. In general, the intent of these initiatives is to assure the prosperity of the insurance sector, not to prevent damage to life and property. Reliance by insurers on predictive risk modeling continues to be limited, as new initiatives are prompted by extreme events rather than modeled projections of damage. As another example of reactive behavior, insurers rely on legal judgments to determine the extent of their liabilities. This pattern of learning and response has two implications. First, opportunities for anticipatory adaptation prompted by insurer initiatives are very limited, which guarantees continued large losses from extreme events into the future. Second, proactive risk mitigation will have to be pursued and implemented on behalf of public welfare by the relevant branches of government and cannot be left to market forces.

Here are two papers on improving the resilience of the electric power system:

Narayanan, A., & Morgan, M. G. (2012). Sustaining critical social services during extended regional power blackouts. *Risk Analysis, 32*(7), 1183–1193.

**Abstract:** Despite continuing efforts to make the electric power system robust, some risk remains of widespread and extended power outages due to extreme weather or acts of terrorism. One way to alleviate the most serious effects of a prolonged blackout is to find local means to secure the continued provision of critical social services upon which the health and safety of society depend. This article outlines and estimates the incremental cost of a strategy that uses small distributed generation, distribution automation, and smart meters to keep a set of critical social services operational during a prolonged power outage that lasts for days or weeks and extends over hundreds of kilometers.

Bohman, A. D., Abdulla, A., & Morgan, M. G. (2021). Individual and collective strategies to limit the impacts of large power outages of long duration. *Risk Analysis, 42*(3), 544–560.

**Abstract:** As modern society becomes ever more dependent on the availability of electric power, the costs that could arise from individual and social vulnerability to large outages of long duration (LLD-outages) increases. During such an outage, even a small amount of power would be very valuable. This paper compares individual and collective strategies for providing limited amounts of electric power to residential customers in a hypothetical New England community during a large electric power outage of long duration. We develop estimates of the emergency load required for survival and assess the cost of strategies to address outages that last 5, 10, and 20 days in either winter or summer. We find that the cost of collective solutions could be as much as 10 to 40 times less than individual solutions (less than $2 per month per home). However, collective solutions would require community-wide coordination, and if local distribution system lines are destroyed, only individual back-up systems could provide contingency power until those lines are repaired. Costs might be reduced if more robust distributed generation were employed that could be operated continuously with the ability to sell power back to the grid. Our cost-effectiveness analysis only assesses what *could* be done, developing estimates of preparedness cost. A decision about what *should* be done would require additional input from a range of stakeholders as well as some form of analytical deliberative process.

Here is the paper on lessons for adaptation that is quoted at the end of this chapter:

Yohe, G., & Dowlatabadi, H. (1999). Risk and uncertainties, analysis and evaluation: Lessons for adaptation and integration. *Mitigation and Adaptation Strategies for Global Change, 4*(3), 319–329.

**Some notable writings by authors not affiliated with our Centers:**

Bierbaum, R., Smith, J. B., Lee, A., Blair, M., Carter, L., Chapin, F. S., Fleming, P., Ruffo, S., Stults, M., McNeeley, S., & Wasley, E. (2013). A comprehensive review of climate adaptation in the United States: More than before, but less than needed. *Mitigation and Adaptation Strategies for Global Change, 18*(3), 361–406.

IPCC WG. (2022). Climate change 2022: Impacts, Adaptation and Vulnerability.

Kane, S., & Yohe, G. (2000). Societal adaptation to climate variability and change: An introduction. In *Societal Adaptation to Climate Variability and Change* (pp. 1–4). Springer, Dordrecht.

Reilly, J., & Schimmelpfennig, D. (2000). Irreversibility, uncertainty, and learning: Portraits of adaptation to long-term climate change. *Climatic Change, 45*(1), 253–278.

Sovacool, B. K., Linnér, B. O., & Goodsite, M. E. (2015). The political economy of climate adaptation. *Nature Climate Change, 5*(7), 616–618.

Yohe, G., & Toth, F. L. (2000). Adaptation and the guardrail approach to tolerable climate change. *Climatic Change, 45*(1), 103–128.

## Notes

1 Scientists call these storms *tropical cyclones* (TCs). They are referred to as *hurricanes* when they happen in the Atlantic and *typhoons* when they happen in the (western) Pacific.

2 We chose this region because in the past it has experienced both hurricanes and ice storms. On the issue of how climate change might affect ice storms, see: Klima, K., & Morgan, M. G. (2015). Ice storm frequencies in a warmer climate. *Climatic Change, 133*(2), 209–222.

# 17

# SCRUBBING CARBON DIOXIDE OUT OF THE ATMOSPHERE

*Joshuah Stolaroff*

My graduate advisor, Greg Lowry, once remarked that his diploma should have read "PhD in Plumbing" because that's what he spent most of his time doing during his graduate studies in Civil & Environmental Engineering at Stanford University. True to form, it was plumbing on my mind in the summer of 2005 while I ran an early test of the world's first prototype device to capture $CO_2$ from the atmosphere. I had moved for the summer to Calgary, Canada, to live with my other graduate advisor, David Keith, who had recently moved to the University of Calgary from CMU. We had just the summer to build and test the 4-meter-tall spray tower before I had to go back to Pittsburgh for classes and before the unoccupied high bay where we had built the device had to be turned over to whatever civil engineering uses it was built for. And much of that summer had been spent rush-ordering equipment and waiting for it to arrive. The project was perpetually in a state of "if this doesn't work, we might not be able to get all the data we need in time." However, most of what we had planned was actually working out pretty well. In our 1-meter-diameter tower, we had nozzles spraying highly caustic 5 molar sodium hydroxide solution, a blower pushing air through, and a hand-built assembly to collect the solution at the bottom and filter exhaust air to make it abundantly safe to breathe. The cone-shaped collector at the bottom was important because it recirculated the caustic solution to the pump. We didn't have much extra solution, and ordering more could set us back a week. So it was concerning to see a dripping leak from somewhere around the bottom of the cone.

I had selected most of the plumbing connections myself; I was not afraid to get under the device while it was running to look for the source of the leak. With several meters of hand-epoxied seams to inspect, it wasn't all that easy to find. That was why, in an act of profound stupidity, I removed my safety goggles "for a closer look." It was then, of course, that the large hose at the bottom of the collector, the one that carried the entire flow of caustic through the prototype, popped free from

DOI: 10.4324/9781003330226-17

its pipe clamp and gushed 5 molar sodium hydroxide all over my face and shoulder. I don't remember what I thought in the moment, but somehow I managed to re-secure the hose and save our reservoir of sodium hydroxide before running upstairs to wash off, which may have been fortunate for the history of air capture research. (I was very lucky not to get any in my eye and walk away that day with only a mild rash. PhD plumbing lessons: tighten your hose clamps like you friggin' mean it, and always wear your goggles.)

Bench-scale (let alone high-bay-scale) experiments like these were unusual in our Climate Decision Making Center, but my advisors and I were convinced that this little demonstration could have a big influence on global climate policy. Much of our discussion in the Center and in the literature was about economics and, in particular, the "price on carbon." We often assumed that there would be a regula-tory incentive to reduce $CO_2$ emissions at a certain price point, and much of the climate policy research was concerned with estimating the cost of various carbon reduction methods relative to that price point and relative to each other.

One challenge to this approach to climate policy is that climate models were telling us that greenhouse gas emissions would need to be reduced nearly to zero in the second half of the century to maintain a safe climate (models today tend to say even that is not good enough – that we'll actually need some negative emissions). However, some sources of greenhouse gas emissions are hard to get rid of, either because there is no obvious replacement process (e.g., combustion from airplanes,[1] methane emissions from livestock, $N_2O$ emissions from agriculture), or because changing the process is expensive relative to the amount of emissions avoided (e.g., gas cooking, heavy construction vehicles, and certain specialized industries). Even a more straightforward solution, like switching from gasoline to hydrogen for pas-senger cars, could be very expensive when the infrastructure costs are included, especially if done quickly. Altogether, these hard-to-mitigate sectors could make achieving climate goals much more expensive. In this economic framing of climate policy, some people were arguing that the expensive options shouldn't be pursued at all and that it's better to make our climate goals less ambitious.

A potential solution to the challenge of hard-to-mitigate sectors is the idea of "negative emissions," also known as carbon dioxide removal (CDR). There are various ways to pull $CO_2$ out of the air. Plants do it as part of photosynthesis, so many of the negative emissions methods proposed at the time were biologic-ally based: afforestation, ocean fertilization, and burning biomass in power plants equipped with conventional CCS. All of these arguably had natural limits on the scale of emissions they could remove. However, the method with the largest poten-tial scale was industrial capture of $CO_2$ from the atmosphere, followed by seques-tration underground. We called this "air capture" to distinguish it from point-source capture of $CO_2$. (The process is now commonly referred to as "direct air capture," a phrase that gives it a more pronounceable acronym – DAC – but is even less helpful to outsiders.)

It was already known that $CO_2$ could be physically removed from air. Industrial means to scrub $CO_2$ from air already operated in the 1930s as part of gas production plants. Other systems were used on submarines and space shuttles to keep air breathable. Whether this could be done at large scale – gigaton scale,[2] as we would say – at a reasonable price for climate mitigation was another question.

However, if you *could* take a gigaton of $CO_2$ out of the air, all sorts of climate policy issues would become simpler. The whole idea of underground $CO_2$ sequestration from point sources became less risky because a slow leak rate, which could otherwise invalidate the whole endeavor over time, would become simply a cost to manage. And those hard-to-mitigate sectors would only be as costly as air capture, because if you had any emissions that were more expensive to reduce than air capture, you could just do air capture instead.

Air capture also had this amazing feature of being able to remove past emissions. Perhaps unique among the negative emissions technologies, air capture could be envisioned to operate at the scale of the energy system, removing gigatons of $CO_2$. That would be enough to meaningfully turn back the clock on $CO_2$ emissions, which are otherwise stuck in the atmosphere, once emitted, for hundreds of years.

All of these features of air capture would be remarkable indeed, but they required a technology to implement air capture at a reasonable cost and modest demand for resources. That such a technology could be feasible was a proposition that many in the climate policy community did not accept. One line of argument for why air capture would be way too expensive used something known as the Sherwood plot. Going back to 1959, Thomas Sherwood observed that the market prices of some pure materials were related to the concentrations of those materials in the sources they were separated from. So, for example, gold was really expensive because there was only a little gold in the rock it was extracted from, whereas iron was much cheaper because there is a lot of iron in iron ore. A number of researchers have since followed up on Sherwood's work and found that when you plot a lot of these commodities on a logarithmic plot, they fall roughly on a straight line, as shown in Figure 17.1. The implication of this line is that, for a given class of commodities, the purified price is inversely proportional to the starting concentration.

If you follow this logic for air capture, you would look to $CO_2$ capture from power plants as a reference point. Ed Rubin and others in our Center had been doing important work on that front. In 2004, Anand Rao and Ed published what would become the most cited paper in carbon capture and estimated the cost of capture from power plants at \$59/ton-$CO_2$ (uncertainty range: \$21 to 79/ton-$CO_2$). $CO_2$ in air is about 300 times more dilute than $CO_2$ in the smokestack of a coal power plant. Thus, following the Sherwood logic, air capture should cost tens of thousands of dollars per ton of $CO_2$. This figure was wildly at odds with the few bottom-up estimates of the cost of air capture that had previously been attempted, mostly by Klaus Lackner at Columbia University. It was also at odds with the basic thermodynamics of the process, which said that air capture should only be about three times as hard as capture from power plants. Nonetheless, the Sherwood plot argument was influential at the time and remained so for years afterward.[3]

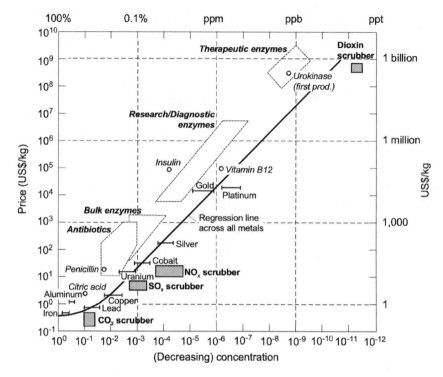

**FIGURE 17.1** This figure, called a Sherwood plot, shows the relationship between the concentration of a target material in a feed stream and the cost of removing the target material. Reproduced with permission from Arnulf Grübler, *Technology and Global Change*, Cambridge University Press, 1998.

Grübler, A. (1998). *Technology and global change.* Cambridge University Press.

If the climate policy community was going to take the idea of air capture seriously, we felt that a proof-of-concept demonstration was needed. If we could construct (at least conceptually) a plausible air capture system and make a reasonable estimate of its cost, that would set an upper bound on not only the cost of air capture but all of climate mitigation.

The approach was a gamble. If we devised a system that turned out to cost a lot (which we tended to think of as greater than $500/ton-CO$_2$), it could mean that air capture was a bad idea, or it could mean that we just didn't design the right system. Basically, no information would be gained. However, if we threw a dart in the dark and landed on an inexpensive system, we would have an upper bound on climate mitigation and a lever on the whole climate policy debate.

One thing that was clear from the Sherwood plot arguments was that if you tried to take the *same* technology developed for power plants and apply it to the air, it would be way too expensive. This point was fairly easy to make with an understanding of conventional carbon capture systems and rules of thumb about mass transfer.

The conventional method to capture $CO_2$ from points sources consisted of three big components:

1. *The absorber*, where $CO_2$ is absorbed into a chemical solvent from a flue gas – this is typically called the "contactor" in air capture literature because it contacts the capture agent with air.
2. *The stripper*, where $CO_2$ is "stripped" from the solvent with heat and steam – in air capture literature, this is often called the "regenerator" (or the regeneration step when it isn't a separate component).
3. *The compressor*, where $CO_2$ is dried and pressurized to a liquid for transport and, later, injection underground.

We knew from techno-economic analyses like Ed Rubin's that the absorber was the most expensive piece of capital equipment in the system. The reason is that it is a huge tower, several stories tall and more than 10 meters in diameter, filled with expensive, stainless steel structures called packing. The point of the packing is to get wet with solvent and provide a place for the gas that flows through the tower to interact with the liquid.

The concentration of $CO_2$ in the air is about 300 times lower than in power plant exhaust. The standard mass transfer models suggested that, if you didn't change anything else about the process, you would need a contactor at least 300 times bigger to make the conventional technology work with air (although the regenerator and compressor could stay more or less the same). That larger contactor would cost 50 to 300 times more, depending on how well economies of scale would help. Since the contactor was already the most expensive piece of the system, you'd end up with total costs well over $1000/ton-$CO_2$.

The cost of all that stainless steel packing would turn out to be a linchpin of the air capture debate for years to come, and it remains contentious even today. We knew it was something to avoid in the design we chose to assess.

The goal of our project was to describe a proof-of-concept air capture system for which we could estimate believable costs. To achieve this, our design philosophy was to use well-known technology and off-the-shelf components wherever possible. As a start, we had the long history of caustic solutions used to remove $CO_2$ from air: sodium hydroxide or potassium hydroxide. We strongly considered calcium hydroxide, and that was even the subject of our first related paper (see Stolaroff et al., 2005 at the end of this chapter), but calcium hydroxide solution ended up being too slow to absorb $CO_2$ compared to the stronger caustics.

When sodium hydroxide solution, say, reacts with $CO_2$ from the air, it transitions to a sodium carbonate solution. To run a continuous air capture system, you need to regenerate that sodium carbonate to sodium hydroxide and recover the $CO_2$ in pure form. Fortunately, there is a widely used industrial process for converting sodium carbonate to sodium hydroxide. It is part of the Kraft process, which is used by paper mills to separate cellulose from lignin. It is a complicated and thermodynamically clumsy process, but it works at large scale and we felt we could find costs for it. We also reasoned that, if there were a significantly better way to regenerate sodium hydroxide, probably someone in the paper industry would have it by then, so maybe the process was close to as efficient as it gets.

The first cost estimate we made for air capture based on the sodium carbonate system appeared as an appendix in our 2005 paper "Climate Strategy with $CO_2$ Capture from the Air." The paper was not originally about air capture technology itself but rather about the implications for global climate policy of future negative emissions. In it, Minh Ha-Duong, David Keith, and I used a coupled global climate and global economic model to estimate the lowest-cost pathways for limiting climate change by the end of the century. When negative emissions are available, we found, the best policy is to allow more emissions in the near term (because some worst-case climate scenarios become less risky with negative emissions) but cut more aggressively in the long term (because those expensive-to-mitigate sectors are suddenly easier to address). When the paper went out for review, one of the reviewers said, effectively, "It's a nice model, but I don't believe air capture is a real option." We then conceived of the appendix, which became a sort of preview of my thesis work, to outline a plausible air capture system so the reviewers would take the idea seriously. A diagram of the system presented in that paper is shown in Figure 17.2.

For the estimate, we assembled the re-causticizer components of the Kraft process with a conceptual contactor based on a hyperboloid cooling tower, those giant concrete chimneys iconic of nuclear power plants. We knew that the contactor had to be giant, and we knew that those cooling towers were some of the largest structures that were also relatively cheap for their size.

To get around the cost of lots of stainless steel packing, we simply made the contactor empty and relied on spraying the sodium hydroxide into tiny drops to get the surface area needed, using a so-called spray tower. There was good precedent for this, since some cooling towers operated by spraying water for evaporative cooling. Spray towers were also typically used to remove sulfur dioxide from power plant

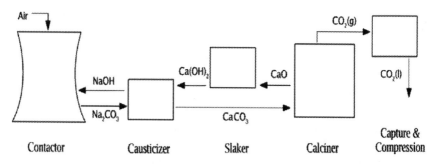

**FIGURE 17.2** Top-level process diagram of a sodium hydroxide-based air capture system, as presented in Keith et al., 2005.

Keith, D. W., Ha-Duong, M., & Stolaroff, J. K. (2006). Climate strategy with $CO_2$ capture from the air. *Climatic Change*, 74(1), 17–45.

flue gas, a close analogue of $CO_2$ removal. The details of the spray-based contactor were too much to work out for the paper appendix, but we made some rough guesses about performance to produce our first estimate of air capture system cost. We estimated the process would cost \$140/t-$CO_2$.

The contactor was the crucial component of the system for two reasons. First, there was no industrial analogue for the contactor in the way there were for the various regeneration and compression components of the process. Second, the contactor is inherently the biggest piece of the system because the low concentration of $CO_2$ in air means a huge amount of air has to be passed across a huge amount of surface area relative to point-source capture and other industrial processes. In order to estimate the cost of the air capture, we had to estimate, in turn, two basic properties of the contactor: (1) how big did it need to be for a given throughput, and (2) how much energy did it use for pumps and fans?

In order to estimate those two properties for a spray tower, we needed to model the system, including the movement of air and droplets and the transfer of $CO_2$ between them. To do that, we found, we needed to know more about the rate of $CO_2$ absorption of a drop of caustic solution.

We started by trying to measure that rate in a bench-scale experiment. I set up a nozzle that would make drops to fall about 1.5 meters through air. The nozzle had a ring below it, charged up to 10,000 volts, that would pull electrostatically on the drops so they would come out of the nozzle consistently small. It turned out to be a difficult experiment to get data from, largely because the amount of $CO_2$ each drop could pull from the air on that short flight was hard to measure above the background in solution.

It would be easier, we reasoned, to set up a system where we have a lot of drops and then measure the change in $CO_2$ concentration in the air. Conveniently, spray nozzles could produce a lot of small drops easily, and without applying 10,000 volts. This basically led to the idea of a spray-tower experiment. That's why I found myself a year later in Calgary, watching my experimental apparatus be hoisted up with a crane.

Figure 17.3 shows the prototype spray tower and some of its construction. The size was chosen to be wide enough to accommodate the spray pattern of a single nozzle and tall enough compared with its width to reduce edge effects. The basic idea was that we would blow air through the tower, spray sodium hydroxide from the top, and measure the $CO_2$ concentration in the air going in and coming out. After two months of assembling and getting it hooked up, and dealing with the odd hose clamp disaster, the device worked like a charm.

Some sample data we collected are shown in Figure 17.4. We would turn the spray on and off periodically to separate the effect of the $CO_2$ absorbed by the spray from the $CO_2$ absorbed by solution clinging to the walls of the tower. As you can see, we absorbed a significant amount of $CO_2$ from the air that passed through the system. The concept worked! You might also notice from the data plot that the amount of $CO_2$ absorbed by the wall and the spray was about the same. In later analysis, this made sense because the wall and the spray had similar amounts of surface

**FIGURE 17.3** Prototype air capture device and assembly. **A**. Leif Menezes and Kenton Heidel glue PVC sheets to the inside of the tower. **B**. David Keith and Joshuah Stolaroff watch as the main tower is lifted by crane for placement. **C**. Kenton Heidel (who later became the chief engineer for Carbon Engineering) makes adjustments on the assembled prototype.

Photos taken by Joshua Stolaroff.

area. This also had a fateful implication: if you made even a rudimentary packed tower, say, with spaced sheets of PVC like we used to line the tube, you could out-perform the spray system. Later, when David Keith and Carbon Engineering went on to build the next-generation contactor, they heeded this insight and went with a packed tower design over a spray tower. The packing in that second tower was, in fact, PVC and not expensive stainless steel.

Overall, the spray tower did its job, and we measured the $CO_2$ absorption performance of sodium hydroxide spray of varying size. However, when we went to extrapolate these measurements from the 4-meter-tall prototype data to the 120-meter-tall, full-size contactor we had envisioned in Keith et al., 2006, we ran into another challenge.

When a dense collection of droplets falls through the air, the droplets tend to collide with each other and coalesce into larger drops. The larger drops have less surface area and absorb $CO_2$ much more slowly. The coalescence process is the same one by which clouds become rain, and coincidentally, one of my committee members, Peter Adams, had studied this process for *his* PhD thesis. Peter provided

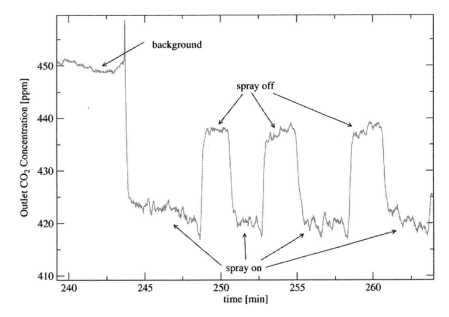

**FIGURE 17.4** Measured $CO_2$ concentration during a typical trial of the spray tower prototype.

Source: Stolaroff, J., Keith, D., & Lowry, G. (2006, June). A pilot-scale prototype contactor for $CO_2$ capture from ambient air: Cost and energy requirements. In Proc. GHGT-8, the 8th Int. Conf. on Greenhouse Gas Control Technologies, Trondheim, Norway.

me with a piece of code he had written to calculate drop coalescence in a field of falling drops, which I then incorporated into my model of the spray tower.

As it turned out, our simulation indicated that the drops would go from mist to heavy rain well before they reached the bottom of a tall tower. (Even in the 4-meter prototype, the code predicted the surface area of the spray had dropped by nearly half in the course of the fall, which helped explain some of our measurements.) For this reason, we changed the shape of the contactor from a tall tower to a wide, short structure, more analogous to a warehouse or airplane hangar. This is the shape we assessed in my thesis and in our 2008 paper "Carbon Dioxide Capture from Atmospheric Air Using Sodium Hydroxide Spray."

For my thesis, I put together the various industrial analogies with the contactor model we had developed, and I estimated the cost of air capture with the sodium hydroxide system would be between $80 and 250/ton-$CO_2$. Refining the estimate with values from our 2008 paper on the contactor would put this range from $180 to 290/ton-$CO_2$. These values were well above the estimates for point-source capture and a lot of other conventional mitigation, but they were still well within the costs that were interesting for climate mitigation. With improvements to the technology, we thought the concept could even be commercially interesting. David

Keith had enough confidence in the commercial potential of air capture that he went on to start a company called Carbon Engineering to develop the technology.[4] After completing my PhD, I joined the scientific staff at Lawrence Livermore National Laboratory, where I have continued to work on a variety of issues related to reducing the carbon intensity of the energy system.

## Technical Details for Chapter 17

Here are three papers from Josh Stolaroff's thesis work on directly capturing $CO_2$ from the atmosphere:

Stolaroff, J. K., Lowry, G. V., & Keith, D. W. (2005). Using CaO-and MgO-rich industrial waste streams for carbon sequestration. *Energy Conversion and Management*, *46*(5), 687–699.

**Abstract:** To prevent rapid climate change, it will be necessary to reduce net anthropogenic $CO_2$ emissions drastically. This likely will require imposition of a tax or tradable permit scheme that creates a subsidy for negative emissions. Here, we examine possible niche markets in the cement and steel industries where it is possible to generate a limited supply of negative emissions (carbon storage or sequestration) cost-effectively.

$Ca(OH)_2$ and CaO from steel slag or concrete waste can be dissolved in water and reacted with $CO_2$ in ambient air to capture and store carbon safely and permanently in the form of stable carbonate minerals ($CaCO_3$). The kinetics of Ca dissolution for various particle size fractions of ground steel slag and concrete were measured in batch experiments. The majority of available Ca was found to dissolve on a time scale of hours, which was taken to be sufficiently fast for use in an industrial process.

An overview of the management options for steel slag and concrete waste is presented, which indicates how their use for carbon sequestration might be integrated into existing industrial processes. Use of the materials in a carbon sequestration scheme does not preclude subsequent use and is likely to add value by removing the undesirable qualities of water absorption and expansion from the products.

Finally, an example scheme is presented which could be built and operated with current technology to sequester $CO_2$ with steel slag or concrete waste. Numerical models and simple calculations are used to establish the feasibility and estimate the operating parameters of the scheme. The operating cost is estimated to be US\$8/t-$CO_2$ sequestered. The scheme would be important as an early application of technology for capturing $CO_2$ directly from ambient air.

Keith, D. W., Ha-Duong, M., & Stolaroff, J. K. (2006). Climate strategy with $CO_2$ capture from the air. *Climatic Change*, *74*(1), 17–45.

**Abstract:** It is physically possible to capture $CO_2$ directly from the air and immobilize it in geological structures. Air capture differs from conventional mitigation in three key aspects. First, it removes emissions from any part of the economy with equal ease or difficulty, so its cost provides an absolute cap on the cost of mitigation. Second, it permits reduction in concentrations faster than the natural carbon cycle: the effects of irreversibility are thus partly alleviated. Third, because it is weakly coupled to existing energy infrastructure, air capture may offer stronger economies of scale and smaller adjustment costs than the more conventional mitigation technologies.

We assess the ultimate physical limits on the amount of energy and land required for air capture and describe two systems that might achieve air capture at prices under 200 and 500 \$/tC using current technology.

Like geoengineering, air capture limits the cost of a worst-case climate scenario. In an optimal sequential decision framework with uncertainty, existence of air capture decreases the need for near-term precautionary abatement. The long-term effect is the opposite; assuming that marginal costs of mitigation decrease with time while marginal climate change damages increase, then air capture increases long-run abatement. Air capture produces an environmental Kuznets curve, in which concentrations are returned to preindustrial levels.

Stolaroff, J. K., Keith, D. W., & Lowry, G. V. (2008). Carbon dioxide capture from atmospheric air using sodium hydroxide spray. *Environmental Science & Technology*, *42*(8), 2728–2735.

**Abstract:** In contrast to conventional carbon capture systems for power plants and other large point sources, the system described in this paper captures $CO_2$ directly from ambient air. This has the advantages that emissions from diffuse sources and past emissions may be captured. The objective of this research is to determine the feasibility of a NaOH spray-based contactor for use in an air capture system by estimating the cost and energy requirements per unit $CO_2$ captured. A prototype system is constructed and tested to measure $CO_2$ absorption, energy use, and evaporative water loss and compared with theoretical predictions. A numerical model of drop collision and coalescence is used to estimate operating parameters for a full-scale system, and the cost of operating the system per unit $CO_2$ captured is estimated. The analysis indicates that $CO_2$ capture from air for climate change mitigation is technically feasible using off-the-shelf technology. Drop coalescence significantly decreases the $CO_2$ absorption efficiency; however, fan and pump energy requirements are manageable. Water loss is significant (20 mol $H_2O$/mol $CO_2$ at 15 °C and 65% RH) but can be lowered by appropriately designing and operating the system. The cost of $CO_2$ capture using NaOH spray (excluding solution recovery and $CO_2$ sequestration, which may be comparable) in the full-scale system is 96 \$/ton-$CO_2$ in the base case, and ranges from 53 to 127 \$/ton-$CO_2$ under alternate operating parameters and assumptions regarding capital costs and mass transfer rate. The low end of the cost range is reached by a spray with 50 μm mean drop diameter, which is achievable with commercially available spray nozzles.

**Some notable writings by authors not affiliated with our Centers:**

Benson, S. M., & Surles, T. (2006). Carbon dioxide capture and storage: An overview with emphasis on capture and storage in deep geological formations. *Proceedings of the IEEE*, *94*(10), 1795–1805.

Benson, S. M., & Orr, F. M. (2008). Carbon dioxide capture and storage. *MRS Bulletin*, *33*(4), 303–305.

Bui, M. et al. (2018). Carbon capture and storage (CCS): The way forward. *Energy & Environmental Science*, *11*(5), 1062–1176.

de Coninck, H., & Benson, S. M. (2014). Carbon dioxide capture and storage: Issues and prospects. *Annual review of Environment and Resources*, *39*, 243–270.

Herzog, H. (2017). Financing CCS demonstration projects: Lessons learned from two decades of experience. *Energy Procedia*, *114*, 5691–5700.

House, K. Z., Baclig, A. C., Ranjan, M., van Nierop, E. A., Wilcox, J., & Herzog, H. J. (2011). Economic and energetic analysis of capturing $CO_2$ from ambient air. *Proceedings of the National Academy of Sciences*, *108*(51), 20428–20433.

National Academies of Sciences, Engineering, and Medicine. (2015). *Climate Intervention: Carbon Dioxide Removal and Reliable Sequestration*. National Academies Press.

Paltsev, S., Morris, J., Kheshgi, H., & Herzog, H. (2021). Hard-to-Abate Sectors: The role of industrial carbon capture and storage (CCS) in emission mitigation. *Applied Energy*, *300*, 117322.

Szulczewski, M. L., MacMinn, C. W., Herzog, H. J., & Juanes, R. (2012). Lifetime of carbon capture and storage as a climate-change mitigation technology. *Proceedings of the National Academy of Sciences*, *109*(14), 5185–5189.

Wilcox, J. (2012). *Carbon Capture*. Springer Science & Business Media.

## Notes

1 Except perhaps through biofuel, though biofuels were out of favor at the time because the life cycle effects of indirect land use change made them carbon-emitting.
2 A gigaton is a billion tons.
3 See for example Ranjan, M., & Herzog, H. J. (2011). Feasibility of air capture. *Energy Procedia*, *4*, 2869–2876 and House, K. Z., Baclig, A. C., Ranjan, M., Van Nierop, E. A., Wilcox, J., & Herzog, H. J. (2011). Economic and energetic analysis of capturing $CO_2$ from ambient air. *Proceedings of the National Academy of Sciences*, *108*(51), 20428–20433.
4 See https://carbonengineering.com

# 18

# A LAST RESORT – ENGINEERING THE PLANET

*Granger Morgan*

Earth's climate is roughly 60°F (33°C) warmer than it would be without any atmosphere because water vapor, carbon dioxide, and other greenhouse gases trap heat and make it harder for energy that comes in from the sun to be radiated back into space (lower left of Figure 18.1). There are basically two ways to cool the climate (lower right of Figure 18.1): either reduce the amount of carbon dioxide in the atmosphere (so that less heat is trapped) or increase the amount of sunlight that is reflected back into space (so that less energy is absorbed by the earth).

The fraction of sunlight reflected back into space is called the "planetary albedo." By reflecting just a few percent more sunlight (increasing the albedo a little bit), the warming caused by increased levels of $CO_2$ and other "greenhouse gases" in the atmosphere can be offset. Of course, the elevated level of $CO_2$ remains in the atmosphere, so other effects, such as ongoing acidification of the world's oceans and changes in the mix of plant species in natural ecosystems (because some grow better with higher concentrations of $CO_2$ than others), are not offset by changing the albedo.

There is really no uncertainty about whether increasing the earth's albedo will cool the planet. We have compelling evidence of this from past volcanic eruptions that have ejected great quantities of fine reflecting particles into the stratosphere. Figure 18.2 shows the cooling that followed the eruption of Mt. Pinatubo in the Philippines in 1991.

There is also powerful historical evidence from even larger past eruptions. For example, in 1815, fine particles ejected into the stratosphere from the eruption of the volcano Tambura in Indonesia gave rise to what is often termed "the year without a summer." The resulting cooling caused crops to fail (as well as disease to spread). Roughly 65,000 people died in Britain, France, and Ireland[1] and even more

DOI: 10.4324/9781003330226-18

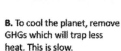

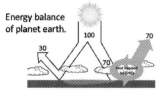

Energy balance
of planet earth.

**A.** To warm the planet,
add more GHGs which will
trap more heat.

**B.** To cool the planet, remove
GHGs which will trap less
heat. This is slow.

**C.** To cool the planet, increase
the albedo to reflect more
sunlight. This is fast.

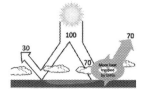

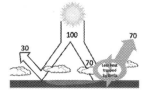

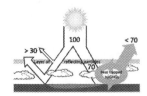

FIGURE 18.1 As shown at the top of this figure, about 30% of the energy from the sun that shines on the earth is immediately reflected back into space. This fraction is called the "planetary albedo." The remaining 70% is absorbed by the earth. Greenhouse gases like water vapor and carbon dioxide trap that energy and the system warms up until it is just warm enough so that the same amount of energy is radiated back to space from the top of the atmosphere in the form of infrared (heat) energy. If more $CO_2$ or other greenhouse gases are added to the atmosphere (**A**), the planet will warm up. If $CO_2$ or other greenhouse gases are removed from the atmosphere (**B**), less heat is trapped and the planet will cool. Since there is no way to get $CO_2$ out of the atmosphere quickly, this is slow. If a layer of reflecting particles is added to increase the albedo (**C**), more energy is reflected back into space and the planet will cool. Reflecting particles can be added very quickly.

Drawn by Granger Morgan.

died elsewhere around the world. Eruptions by volcanoes such as Krakatoa, also in Indonesia, have caused even larger cooling events.[2]

When we began our first NSF-supported center, we spent some time trying to identify all the various technologies and strategies that we believed should be considered as part of a comprehensive approach to managing the problem of global warming and climate change. In addition to identifying things, we should work on to reduce emissions; and to better understand the various impacts of a changing climate, we also noted there were possibilities to intentionally modify the earth's climate either by removing $CO_2$ from the atmosphere[3] or by increasing the planet's albedo. Together, these two strategies have often been referred to as "geoengineering." However, sometimes when people use this term, they only mean changing the albedo. Because the two activities are really very different and have very different implications in terms of risks, benefits, and public policy, we use several different terms today to keep things straight:

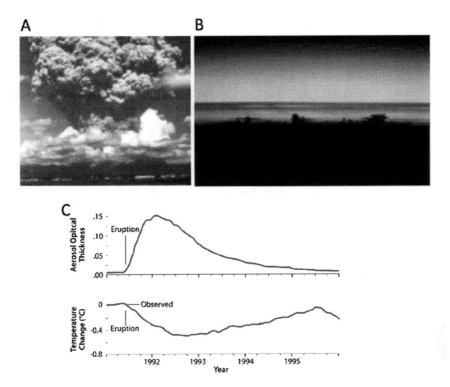

**FIGURE 18.2** The eruption of Mt. Pinatubo in the Philippines in 1991 (**A**), injected a layer of reflecting particles into the stratosphere (**B**). Over the course of the year that followed, the average temperature of the earth fell by just over 0.4°C (0.7°F) (**C**) before the particles finally washed and fell out of the stratosphere a couple of years later and the cooling came to an end. (Images from NASA, IPCC, and Novim.[13]).

U.S. Geological Survey/Dave Harlow.

U.S. Geological Survey.

This image is derived from data and images available at the website of the NASA Earth Observatory.

- SRM or "solar radiation management" or "solar radiation modification" is used to refer to activities that involve changing the albedo.
- CDR or "carbon dioxide removal" is used to refer to activities and processes that can remove $CO_2$ from the atmosphere.
- DAC or "direct air capture" refers to engineered systems that are designed to capture and remove $CO_2$ from the atmosphere (i.e., perform CDR with a machine).
- CCS or "carbon capture and sequestration" refers to engineered systems that capture $CO_2$ in the exhaust of systems like power plants that burn fossil fuels.[4]

We have already discussed the center's work on CCS in Chapter 11 and our work on CDR via DAC in Chapter 17. In this chapter, we focus on the center's work on SRM – modifying the earth's albedo to reflect more sunlight back into space.

In 1991, David Keith and Hadi Dowlatabadi prepared a review that they titled "A serious look at geoengineering," which appeared in *EOS*, a publication of the American Geophysical Union. That paper looked at both SRM and CDR. In introducing the topic, David and Hadi wrote:

> We do not advocate geoengineering, but we offer these justifications for a more systematic evaluation of geoengineering options.
>
> • Geoengineering may be needed if climate change is worse than we expect. That is, geoengineering could serve as fallback technology – one that puts an upper bound on the worst-case, thereby allowing more confidence in pursuing other policy options.
> • It seems very unlikely that world greenhouse gas (GHG) emissions can be kept below ~40% of 1990 levels – a prerequisite for averting climate change in the long-term…
>
> Doubt about the prospects for cooperative abatement of global GHG emissions is a pragmatic reason to consider geoengineering, whose implementation requires fewer cooperating actors than abatement. Though, geoengineering fills a unique niche because of its potential to mitigate catastrophic climate change.

After publishing that review, we set aside work on SRM to focus the bulk of our centers' research efforts on understanding strategies to reduce emissions of carbon dioxide and improving our understanding of the impacts of climate change.

However, over time it became increasingly clear that progress on abatement was proceeding *much* too slowly. It also became apparent that, in those days, virtually nobody in the community of experts in international relations was aware of the fact that SRM was a possibility and that it could be undertaken unilaterally by a single major nation. Accordingly, I recruited Kate Ricke, who had done an undergraduate degree in ocean engineering at MIT, to join EPP to pursue a PhD addressing a variety of issues related to SRM.

One of the first things Kate did was to work with us to organize a workshop at the Council on Foreign Relations in Washington in 2008 where, in collaboration with political scientists John Steinbruner and David Victor, we brought together a group of leading climate scientists and experts in international relations in order to begin to get the foreign policy community better informed about the possibility of SRM. After completing that workshop, we worked to further inform the community in international relations by publishing an article in the journal *Foreign Affairs* that we titled "The geoengineering option: a last resort against global warming?"

Participants in the Washington workshop were all drawn from across North America. To further expand the community of foreign policy experts who understood SRM, we arranged for the Government of Portugal[5] to invite us to conduct a second workshop at the Gulbenkian Foundation in Lisbon. For that workshop, we gathered climate and international security experts from North America, the EU, China, Russia, and India. Co-sponsors for the Lisbon workshop included the International Risk Governance Council and the University of Calgary.

While these workshop activities were going forward, Kate was working with me and others at Carnegie Mellon and with Miles Allen at Oxford University to perform a series of computer modeling studies of the likely consequences of using SRM to change the earth's albedo. To do that, Kate needed access to a global climate model that she could modify to simulate SRM. Models like this require a very large amount of computing power. Miles and his co-workers had developed a version of a British global climate model that could be chopped up into sub-elements that many hundreds of people could run on their personal computers at night when their computers were not being used. Kate went to Oxford to learn the ins and outs of using this system called climateprediction.net[6] and then arranged with the Oxford group to use this model to conduct a variety of experiments. The results of this research were published in a pair of papers in *Nature Geoscience* and in *Nature Climate Change* (see abstracts at the end of this chapter).

In her first paper, Kate showed that:

> Over time, simulated temperature and precipitation in large regions such as China and India vary significantly with different trajectories for solar-radiation management, and they diverge from historical baselines in different directions. Hence, it may not be possible to stabilize the climate in all regions simultaneously using solar-radiation management. Regional diversity in the response to different levels of solar-radiation management could make consensus about the optimal level of geoengineering difficult, if not impossible, to achieve.

As we discussed in Chapter 4, one of the key uncertainties about climate change is the value of "climate sensitivity," the amount of warming that would occur if the $CO_2$ concentration in the earth's atmosphere were doubled and then held constant. In her second paper, Kate found that:

> When SRM-S[7] is used to compensate for rising atmospheric concentrations of greenhouse gases, its effectiveness in stabilizing regional climates diminishes with increasing climate sensitivity. However, the potential of SRM-S to slow down unmitigated climate change, even regionally, increases with climate sensitivity. On average, in variants of the model with higher sensitivity, SRM-S reduces regional rates of temperature change by more than 90% and rates of precipitation change by more than 50%.

While these research activities were proceeding, several of us were participating in a flurry of national and international workshops, conferences, and other activities that had begun to address the issue of SRM (see Table 18.1). Kate and I also authored a policy brief on SRM that was published by the International Risk Governance

**TABLE 18.1** Chronology of some of the research and other activities by investigators in our centers on solar radiation management (SRM) between 1990 and 2015.

**1991:** David Keith and Hadi Dowlatabadi publish their results from a systematic review of the literature on geoengineering, focusing primarily on what today is termed solar radiation management. This was the last work researchers in our centers did on SRM until 2008.

**2008:** As society continued to fail to make significant progress on abating the emissions of $CO_2$ and other greenhouse gases, Granger Morgan grew concerned that the diplomatic community was completely unaware of the possibility of performing SRM and that it was likely that a single nation-state could undertake SRM unilaterally. Together with Jay Apt, David Victor, John Steinbruner, and newly arrived EPP PhD student Kate Rick, we organized a workshop at the Council on Foreign Relations in Washington, DC. Half the participants were senior climate scientists, and half were senior people from the U.S. foreign policy community. After the workshop, in 2009 the five of us published a paper in the journal *Foreign Affairs*.

**2009:** In June, the NRC conducted a workshop on *Geoengineering Options to Respond to Climate Change: Steps to Establish a Research Agenda.* David, Jay, John, and Granger all participated. Then in September, the Royal Society in Britain published a report titled *Geoengineering the Climate: Science, governance and uncertainty.* David was a member of the authoring team, and ideas we had developed in our paper in *Foreign Affairs* were used to shape the discussion of issues of governance. Later that same year, the NRC conducted another workshop.

**2010:** The participants in our workshop at the Council on Foreign Relations were all from North America. Because SRM is an issue that should be of international concern, we arranged for the Ministry of Science, Technology and Higher Education of the Government of Portugal to host a two-day workshop in Lisbon using the facilities of the Gulbenkian Foundation. This workshop's co-sponsors included: the International Risk Governance Council, CMU-CDMC, and the University of Calgary. Participants came from North America, the EU, China, Russia, and India.

Granger gave testimony at a hearing of the House Science Committee held jointly with the Science Committee of the UK House of Parliament. He, Kate, David, and others participated in discussions of risk governance at a meeting at Asilomar on geoengineering. Several members of the center participated in the first multi-university summer study program for graduate students on geoengineering held in Heidelberg, Germany. The second such study program was held in Calgary in 2011, and similar meetings have continued to be held in the years that followed.

The paper discussed in this chapter by Kate Ricke et al. was published in *Nature Geoscience*.

**2011:** Meetings in the UK of a workshop on the Solar Radiation Management Governance Initiative (SRMGI); at UCSD in La Jolla to explore possible ecosystem impacts of SRM.

Granger gave an overview and then, because several U.S. speakers could not attend because of flight cancellations he gave more detailed talks on geoengineering at an IPCC meeting in Lima, Peru.

**TABLE 18.1** Cont.

---

The second paper discussed in this chapter by Kate Ricke et al. was published in *Nature Climate Change.*

**2013:** The paper discussed in this chapter on research guidelines by Morgan, Nordhaus, and Gottlieb was published in *Issues in Science and Technology.*

**2014:** Workshops were held at Harvard and by EDF/CMU in San Francisco on SRM research and research governance.

**2015:** Two reports discussed in this chapter by the U.S. National Academies were published.

**After 2015:** Several investigators who had previously been involved with our centers, including David Keith, Kate Ricke, and Juan Moreno-Cruz, remained very active, pursuing research and organizing and participating in various meetings, workshops, and other activities. However, there was no more center involvement with issues related to SRM.

---

Counsel (IRGC). In that brief, which we wrote for a general audience, we began by noting:

> There is nothing new about the idea of modifying the climate by increasing albedo. Scientists have known for many years that this could be done… However, until very recently, there has been almost no serious research on how to do SRM, on what it might cost, on how well it might work, or what its undesirable side effects and risks might be. We believe that there are two reasons the climate research community has not devoted serious research attention to these issues:
>
> • Scientists have been reluctant to divert scarce research funds away from the urgent task of studying the climate system, climate change, and its impacts.
> • Scientists have been legitimately concerned that studying this topic might increase the likelihood that someone might actually do it. Humans have a dismaying track record of changing their intentions as their capabilities change.

In our view, today the world has passed a tipping point and there are two reasons why it is too dangerous *not* to study and understand SRM:

> 1. There is a growing chance that some part of the world will find itself pushed past a critical point where, for example, patterns of rainfall have shifted so much that agriculture in the region can no longer feed the people. Believing this shift is the result of rising global temperatures, such a region might be tempted to unilaterally start doing SRM to solve its problem. If this situation arises, and no research has been done on SRM, the rest of the world could not respond in an informed way.

2.  With luck, the major effects of climate change will continue to occur slowly, over periods of decades. However, if the world is unlucky and a serious change occurs very rapidly, the countries of the world might need to consider collectively doing SRM. If this situation arises, and no research has been done, SRM would involve a hopeful assumption that the uncertain benefits would outweigh the uncertain and perhaps unknown costs.

While there is great uncertainty about SRM, we are confident that it has "three essential characteristics: it is cheap, fast and imperfect."[8]

We'd always known that doing SRM was likely to be pretty cheap, but we got confirmation of this fact when David and Jay Apt arranged to have Aurora Flight Systems, a leading aerospace design group, do an analysis to determine the most efficient strategy to deploy aerosols to the stratosphere. When compared with estimates that the IPCC had made, results from the Aurora Flight Systems study suggested that the direct economic costs of SRM might be only about 1/100th of the cost of a serious program of reducing $CO_2$ emissions.

In our IRGC policy brief, we argued that the research questions on SRM should address:

*   What methods and strategies might work to implement SRM?
*   How well are these various proposed methods likely to work and how well could they be controlled?
*   How much would these different methods cost?
*   What undesired side effects might arise and what new risks might be associated with these various methods?
*   How will the direct effects of these various methods be distributed over time and across the world?
*   What uncertainties remain because of incomplete understanding of the complex climate system?

We laid out an argument for agreeing on an "allowed zone," which identified experiments that could safely be done in the atmosphere without any risk of causing changes, and we laid out a decision tree framework for how to think about research and what it might learn (Figure 18.3).

Then we used several modifications of this diagram to explore the implications for situations in which:

1.  Some nation or group chooses to engage unilaterally in SRM in order to address a local or regional climate problem and, as a consequence, imposes large externalities on all the rest of the world.
2.  The world finds itself facing a climate emergency (i.e., some serious global-scale outcome turns out to lie in the high tail of our present probability distribution).

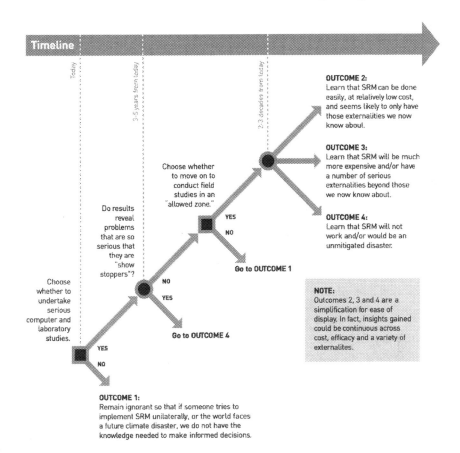

**Timeline**

Today

3-5 years from today

2-3 decades from today

**OUTCOME 2:**
Learn that SRM can be done easily, at relatively low cost, and seems likely to only have those externalities we now know about.

**OUTCOME 3:**
Learn that SRM will be much more expensive and/or have a number of serious externalities beyond those we now know about.

Choose whether to move on to conduct field studies in an "allowed zone."

Do results reveal problems that are so serious that they are "show stoppers"?

YES

NO

**OUTCOME 4:**
Learn that SRM will not work and/or would be an unmitigated disaster.

Go to OUTCOME 1

Choose whether to undertake serious computer and laboratory studies.

NO

YES

**NOTE:**
Outcomes 2, 3 and 4 are a simplification for ease of display. In fact, insights gained could be continuous across cost, efficacy and a variety of externalites.

Go to OUTCOME 4

YES

NO

**OUTCOME 1:**
Remain ignorant so that if someone tries to implement SRM unilaterally, or the world faces a future climate disaster, we do not have the knowledge needed to make informed decisions.

**FIGURE 18.3** Simple decision tree developed for IRGC by Morgan and Ricke to explore what might be learned through a program of SRM research and the actions those findings would imply.

International Risk Governance Council (IRGC), Ricke, K., Morgan, M. G., (2018). Cooling the Earth Through Solar Radiation Management: The need for research and an approach to its governance, International Risk Governance Council.

After spending time at NCAR and then in a research group at Harvard, David Keith moved to a position as Canadian Research Chair Professor at the University of Calgary. There he supervised the work of PhD student Juan Moreno-Cruz to pursue a PhD on the economics of geoengineering – the first study of this kind.

Working with David and Kate, Juan developed a theoretical framework to capture what they saw as the most relevant aspects of SRM. They combined economics and simulations from general circulation models to suggest that SRM could benefit most regions of the world, but only if used with the aim of reducing unequal actions and not simply to achieve efficiency gains. In a follow-up paper, Juan and David added uncertainty into the same framework to discuss the trade-off involved in introducing a novel technology to deal with the risks of climate change and showed that the flexibility of having SRM could be worth several trillion dollars in a future scenario where climate change turns out to be more dangerous than expected.

After exploring how SRM could change understanding of the design of an optimal climate policy, Juan turned to modeling how SRM would modify international politics. In a first paper, he examined the economic issues introduced when SRM becomes available in a world where strategic interaction leads to countries not doing enough mitigation due to what economists call free-riding. Free-riding is the tendency to let others pick up the costs of a mutually desirable outcome. This paper showed how the impact of introducing SRM depends quite delicately on the degree of similarity between countries. When countries differ in their underlying characteristics, the levels of mitigation can increase rather than decrease. This result ran counter to the prevailing theory at the time – that pursuing SRM would lessen efforts to reduce emissions. That finding has since changed the tone of the conversation and led to a more nuanced approach to the economics of SRM.

In a next paper, using results from Kate's general circulation model, Juan, Kate, and Ken Caldeira of the Carnegie Institution (where Kate had gone after graduation to do a post-doc) showed that regional differences in climate outcomes create incentives for countries to form international coalitions that are as small as possible while still powerful enough to deploy SRM. These incentives are very different from those that dominate the international politics of greenhouse-gas emissions reduction, where the central challenge is to compel free-riders to participate. A final contribution Juan made before moving on from working with the center involved a response to a widely held belief that the introduction of SRM would diminish the incentives to reduce emissions in future generations. In this paper, Juan and his coauthors Daniel Heyen and Timo Goeschl showed a situation where mitigation and geoengineering are strategic complements, not substitutes as previous literature had argued.

As it became apparent to more and more people that there was a need to pursue serious research on SRM, discussions began on how best to govern, and provide transparency and oversight for, that research. I teamed up with two experienced lawyers, Bob Nordhaus and Paul Gottlieb, to develop a proposal to show how the U.S. could lead by example. Bob had previously served as General Counsel for both the U.S. Department of Energy and for the Federal Energy Regulatory Commission; Paul had worked with Bob as the Assistant General Counsel for Technology Transfer and Intellectual Property at the U.S. Department of Energy.

In a paper in *Issues in Science & Technology*, we argued that while it was important to develop a research agenda for SRM, it was equally important to develop what we called "a code of best SRM research practices." We outlined a process by which this might be done and stressed: (1) the need to define which research would and would not count as research on SRM; (2) the types of activities that should be subject to requirements for prior notification; (3) the need to spell out policies with respect to public access to research results; (4) the need to define what kinds of research conducted in the atmosphere would have minimal actual impact and thus be allowed; and (5) a number of other policy issues. I felt it was especially important to establish a precedence of openness, given the possibility that some nations might conduct SRM research in a classified manner through their defense or intelligence communities.

We argued that "Federal research agreements should include provisions requiring delivery to the government of publicly releasable research results commensurate with the SRM research code of best practice." We noted that such action by the U.S. government "…would set a powerful precedent by a major player in the world economy and world research community, giving the nation better standing to advocate for international action in SRM research. Specific U.S. action, developed with input from stakeholders including public interest groups, would establish a model that ensures appropriate public availability of information without unnecessarily affecting commercial interests." We argued that once these norms had become developed and adopted by federal research programs, they should also be urged upon all privately funded research, "… after which it should be possible to persuade others across the international research community to adopt similar norms." We concluded:

> …as the prospect of large-scale field studies—or actual implementation—of SRM becomes more real, the need for and pressure to develop such regulation will grow. Because future regulations should be based on solid well-developed science, the creation of a serious program of SRM research, combined with procedures to ensure open access to SRM knowledge, is now urgent.

A few years later, I found myself serving as a member of consensus study by the U.S. National Academies (NASEM) that was charged with examining a range of issues related to geoengineering. Several of us successfully argued to break the study into two separate reports – one on CDR including CCS and DAC and the second on SRM.[9] I was able to help the committee build on much of our centers' work, including the work Bob, Paul, and I had done on SRM governance, as we prepared those reports.

While investigators associated with our climate centers were quite involved in a variety of early work on SRM, since my involvement with the 2015 NASEM reports, neither I nor anyone else supported by our climate center has worked on these issues. However, others who first worked on these issues through the center have continued to play leading roles in exploring both the technical and policy issues related to SRM.

In 2011, David moved from Calgary to the faculty at Harvard, where he has pursued an active program exploring both the policy and science aspects of SRM. Perhaps most important has been the work on developing designs for a set of careful, small-scale experimental studies in the stratosphere. David has also written extensively about SRM for both general and policy audiences. Citations to all his writings can be found on his website.[10]

After completing her post-doc at the Carnegie Institution, Kate joined the faculty of the University of California at San Diego.[11] Juan is now on the faculty at the University of Waterloo[12] in Canada. Along with pursuing research on other topics, both continue to address issues related to SRM.

## Technical Details for Chapter 18

Shortly after our first center began its work, we conducted a systematic look at all aspects of climate change including SRM and CDR. David Keith and Hadi Dowlatabadi published this systematic review:

Keith, D. W., & Dowlatabadi, H. (1992). A serious look at geoengineering. *Eos, Transactions American Geophysical Union*, 73(27), 289–293.

This review can be downloaded from: https://agupubs.onlinelibrary.wiley.com/doi/ pdf/10.1029/91EO00231?casa_token=QTar5j9c3YoAAAAA:DwLDJ7L9FSqG xlC-x_EbldFAFcPzQKcGLkCvzNroAWcq32b8Q4E-IITzYiwQZY1ihDAHz CGAlK9ipyyP

For the next several years after David and Hadi published their review, the work of our centers focused on strategies to reduce emissions and on developing a better understanding of the impacts of climate change. However, as it became apparent that most major countries were not getting serious about climate policy, and that the foreign policy community was largely unaware of the possibility of SRM that could be done quickly by just a single major player, we ran a workshop at the Council on Foreign Relations, after which we published this article in *Foreign Affair*.

Victor, D. G., Morgan, M. G., Apt, J., Steinbruner, J., & Ricke, K. (2009). The geoengineering option: A last resort against global warming?. *Foreign Affairs*, 64–76.

Here are the two papers on modeling SRM that made up the bulk of Kate Ricke's PhD thesis:

Ricke, K. L., Morgan, M. G., & Allen, M. R. (2010). Regional climate response to solar-radiation management. *Nature Geoscience*, 3(8), 537.

**Summary paragraph:** Concerns about the slow pace of climate mitigation have led to renewed dialogue about solar-radiation management, which could be achieved by adding reflecting aerosols to the stratosphere.... Modelling studies suggest that solar-radiation management could produce stabilized global temperatures and reduced global precipitation...Here we present an analysis of regional differences in a climate modified by solar radiation management, using a large-ensemble modelling experiment that examines the impacts of 54 scenarios for global temperature stabilization. Our results confirm that solar-radiation management would generally lead to less extreme temperature and precipitation anomalies, compared with unmitigated greenhouse gas emissions. However, they also illustrate that it is physically not feasible to stabilize global precipitation and temperature simultaneously as long as atmospheric greenhouse gas concentrations continue to rise. Over time, simulated temperature and precipitation in large regions such as China and India vary significantly with different trajectories for solar-radiation management, and they diverge from historical baselines in different directions. Hence, it may not be possible to stabilize the climate in all regions simultaneously using solar-radiation management. Regional diversity in the response to different levels of solar-radiation management could make consensus about the optimal level of geoengineering difficult, if not impossible, to achieve.

Ricke, K. L., Rowlands, D. J., Ingram, W. J., Keith, D. W., & Morgan, M. G. (2012). Effectiveness of stratospheric solar-radiation management as a function of climate sensitivity. *Nature Climate Change, 2*(2), 92–96.

**Summary paragraph:** If implementation of proposals to engineer the climate through solar-radiation management (SRM) ever occurs, it is likely to be contingent on climate sensitivity. However, modelling studies examining the effectiveness of SRM as a strategy to offset anthropogenic climate change have used only the standard parameterizations of atmosphere–ocean general circulation models that yield climate sensitivities close to the Coupled Model Intercomparison Project mean. Here, we use a perturbed-physics ensemble modelling experiment to examine how the response of the climate to SRM implemented in the stratosphere (SRM-S) varies under different greenhouse-gas climate sensitivities. When SRM-S is used to compensate for rising atmospheric concentrations of greenhouse gases, its effectiveness in stabilizing regional climates diminishes with increasing climate sensitivity. However, the potential of SRM-S to slow down unmitigated climate change, even regionally, increases with climate sensitivity. On average, in variants of the model with higher sensitivity, SRM-S reduces regional rates of temperature change by more than 90% and rates of precipitation change by more than 50%.

In order to assess alternative strategies that could be used to loft fine particles to the stratosphere as well as the cost of doing that, David Keith and Jay Apt worked with folks at a leading aerospace research firm called Aurora Flight Systems. Here is a summary of the study that resulted:

McClellan, J., Keith, D. W., & Apt, J. (2012). Cost analysis of stratospheric albedo modification delivery systems. *Environmental Research Letters, 7*(3), 034019.

**Abstract:** We perform engineering cost analyses of systems capable of delivering 1–5 million metric tonnes (Mt) of albedo modification material to altitudes of 18–30 km. The goal is to compare a range of delivery systems evaluated on a consistent cost basis. Cost estimates are developed with statistical cost estimating relationships based on historical costs of aerospace development programs and operations concepts using labor rates appropriate to the operations. We evaluate existing aircraft cost of acquisition and operations, perform in-depth new aircraft and airship design studies and cost analyses, and survey rockets, guns, and suspended gas and slurry pipes, comparing their costs to those of aircraft and airships. Annual costs for delivery systems based on new aircraft designs are estimated to be \$1–3B to deliver 1 Mt to 20–30 km or \$2–8B to deliver 5 Mt to the same altitude range. Costs for hybrid airships may be competitive, but their large surface area complicates operations in high altitude wind shear, and development costs are more uncertain than those for airplanes. Pipes suspended by floating platforms provide low recurring costs to pump a liquid or gas to altitudes as high as 20 km, but the research, development, testing and evaluation costs of these systems are high and carry a large uncertainty; the pipe system's high operating pressures and tensile strength requirements bring the feasibility of this system into question. The costs for rockets and guns are significantly higher than those for other systems. We conclude that (a) the basic technological capability to deliver material to the stratosphere at million tonne per year rates exists today, (b) based on prior literature, a few million tonnes per year would be sufficient to alter radiative forcing by an amount roughly equivalent to the growth of anticipated greenhouse gas forcing over the next half century, and that (c) several different methods could possibly deliver this quantity for less than \$8B per year. We do not address here the science of aerosols in the stratosphere, nor issues of risk, effectiveness or governance that will add to the costs of solar geoengineering.

Here are four papers that Juan, Kate, David, and others wrote on the economics of SRM:

Moreno-Cruz, J. B., Ricke, K. L., & Keith, D. W. (2012). A simple model to account for regional inequalities in the effectiveness of solar radiation management. *Climatic Change, 110*(3–4), 649–668.

**Abstract:** We present a simple model to account for the potential effectiveness of solar radiation management (SRM) in compensating for anthropogenic climate change. This method provides a parsimonious way to account for regional inequality in the assessment of SRM effectiveness and allows policy and decision makers to examine the linear climate response to different SRM configurations. To illustrate how the model works, we use data from an ensemble of modeling experiments conducted with a general circulation model (GCM). We find that an SRM scheme optimized to restore population-weighted temperature changes to their baseline compensates for 99% of these changes while an SRM scheme optimized for population-weighted precipitation changes compensates for 97% of these changes. Hence, while inequalities in the effectiveness of SRM are important, they may not be as severe as it is often assumed.

Ricke, K. L., Moreno-Cruz, J. B., & Caldeira, K. (2013). Strategic incentives for climate geoengineering coalitions to exclude broad participation. *Environmental Research Letters, 8*(1), 014021.

**Abstract:** Solar geoengineering is the deliberate reduction in the absorption of incoming solar radiation by the Earth's climate system with the aim of reducing impacts of anthropogenic climate change. Climate model simulations project a diversity of regional outcomes that vary with the amount of solar geoengineering deployed. It is unlikely that a single small actor could implement and sustain global-scale geoengineering that harms much of the world without intervention from harmed world powers. However, a sufficiently powerful international coalition might be able to deploy solar geoengineering. Here, we show that regional differences in climate outcomes create strategic incentives to form coalitions that are as small as possible, while still powerful enough to deploy solar geoengineering. The characteristics of coalitions to geoengineer climate are modeled using a 'global thermostat setting game' based on climate model results. Coalition members have incentives to exclude non-members that would prevent implementation of solar geoengineering at a level that is optimal for the existing coalition. These incentives differ markedly from those that dominate international politics of greenhouse-gas emissions reduction, where the central challenge is to compel free riders to participate.

Goeschl, T., Heyen, D., & Moreno-Cruz, J. (2013). The intergenerational transfer of solar radiation management capabilities and atmospheric carbon stocks. *Environmental and Resource Economics, 56*(1), 85–104.

**Abstract:** Solar radiation management (SRM) technologies are considered one of the likeliest forms of geoengineering. If developed, a future generation could deploy them to limit the damages caused by the atmospheric carbon stock inherited from the current generation, despite their negative side effects. Should the current generation develop these geoengineering capabilities for a future generation? And how would a decision to develop SRM impact on the current generation's abatement efforts? Natural scientists, ethicists, and other scholars argue that future generations could be more sanguine about the side effects of SRM deployment than the current generation. In this paper, we add economic rigor to this important debate on the intergenerational transfer of technological capabilities and pollution stocks. We identify three conjectures that constitute potentially rational courses of action for current society, including a ban on the development of SRM. However, the same premises that underpin these conjectures also allow for a novel possibility: If the development of SRM capabilities is sufficiently cheap, the current generation may for reasons of intergenerational strategy decide not just to develop SRM technologies, but also to abate more than in the absence of SRM.

Moreno-Cruz, J. B. (2015). Mitigation and the geoengineering threat. *Resource and Energy Economics, 41*, 248–263.

**Abstract:** Recent scientific advances have introduced the possibility of engineering the climate system to lower ambient temperatures without lowering greenhouse gas concentrations. This possibility has created an intense debate given the ethical, moral and scientific questions it raises. This paper examines the economic issues introduced when geoengineering becomes available in a standard model where strategic interaction leads to suboptimal mitigation. Geoengineering introduces the possibility of technical substitution away from mitigation, but it also affects the strategic interaction across countries: mitigation decisions directly affect geoengineering decisions. With similar countries, I find these strategic effects create greater incentives for free-riding on mitigation, but with asymmetric countries, the prospect of geoengineering can induce inefficiently high levels of mitigation.

In 2013, working with two experienced lawyers, we proposed a strategy for the U.S. to provide an example for the world of how research on SRM might best be governed:

Morgan, M. G., Nordhaus, R. R., & Gottlieb, P. (2013). Needed: Research guidelines for solar radiation management. *Issues in Science & Technology, 29*(3), 37–44.

This paper can be downloaded at: www.jstor.org/stable/pdf/43315739.pdf?casa_token= JAVqa1v2QyUAAAAA:tyt9AAy9GIl5SyC8hGK8ASdvdeqNmV1lQowBUXWTkHV DEkcKsgfvvzt1Bor4kAodQqLZzVxR2wWIu3p3q1kJQpTkEGosSfttApSwDe-iNk75r 4GtkTGo

Finally, here are a pair of opinion pieces that we published on the need to start research on SRM:

Keith, D. W., Parson, E., & Morgan, M. G. (2010). Research on global sun block needed now. *Nature, 463*(7280), 426.
Long, J. C., Loy, F., & Morgan, M. G. (2015). Start research on climate engineering. *Nature, 518*(7537), 29.

**Some notable writings by authors not affiliated with our Centers:**

Caldeira, K., Bala, G., & Cao, L. (2013). The science of geoengineering. *Annual Review of Earth and Planetary Sciences, 41*, 231–256.
Keith, D. W. (2000). Geoengineering the climate: History and prospect. *Annual Review of Energy and the Environment, 25*(1), 245–284.
Keith, D. W. (2021). Toward constructive disagreement about geoengineering. *Science, 374*(6569), 812–815.
National Academies of Sciences, Engineering, and Medicine (2021). *Reflecting sunlight: Recommendations for solar geoengineering research and research governance.* National Academies Press.
National Academies of Sciences, Engineering, and Medicine (2015). *Climate intervention: Reflecting sunlight to cool earth.* National Academies Press.

Schneider, S. H. (1996). Geoengineering: Could—or should—we do it?. *Climatic Change*, *33*(3), 291–302.

Victor, D. G. (2008). On the regulation of geoengineering. *Oxford Review of Economic Policy*, *24*(2), 322–336.

Weisenstein, D. K., Visioni, D., Franke, H., Niemeier, U., Vattioni, S., Chiodo, G., Peter, T. & Keith, D. W. (2022). An interactive stratospheric aerosol model intercomparison of solar geoengineering by stratospheric injection of $SO_2$ or accumulation-mode sulfuric acid aerosols. *Atmospheric Chemistry and Physics*, *22*, 2955–2973.

# Notes

1 For details, see: https://en.wikipedia.org/wiki/List_of_disasters_in_Great_Britain_and_Ireland_by_death_toll#Over_200_fatalities
2 For details, see: https://en.wikipedia.org/wiki/Krakatoa
3 See Chapter 17 for a description of the center's work on removing $CO_2$ from the atmosphere.
4 The S in CCS is sometimes used to refer to "storage," but as explained in Chapter 11, I do not use that term because I think it is intentionally misleading.
5 Carnegie Mellon and Portugal have long collaborated on a variety of educational and other initiatives. See: www.cmuportugal.org
6 For details, see: www.climateprediction.net
7 The final S in SRM-S refers to the fact that the SRM has been implemented in the stratosphere.
8 The quote is from Keith, D. W., Parson, E., & Morgan, M. G. (2010). Research on global sun block needed now. *Nature*, *463*(7280), 426427.
9 As with all reports of the National Academies, these two reports published in 2015 are available online for free download. One is called:

   *Climate Intervention: Carbon dioxide removal and reliable sequestration.*

   The other is called:

   *Climate Intervention: Reflecting sunlight to cool the earth.*
10 See: https://keith.seas.harvard.edu/people/david-keith
11 See: https://gps.ucsd.edu/faculty-directory/kate-ricke.html AND https://kricke.scrippsprofiles.ucsd.edu
12 See:https://uwaterloo.ca/school-environment-enterprise-development/people-profiles/juan-moreno-cruz
13 Blackstock, J. et al. (2009). *Climate Engineering Responses to Climate Emergencies*, Novim. Available online at: https://arxiv.org/pdf/0907.5140.pdf

# 19

# WHAT WE HAVE LEARNED[1]

*Granger Morgan (informed by many others)*

After working together for over 30 years, we've learned a lot about the problems posed by global change. We've also learned a lot about how to build and sustain a strong interdisciplinary, research collaboration, even when members of the team are working at many different institutions. In this final chapter, we summarize a few general insights about both.

## What We Learned about Deep Uncertainties, Integrated Assessment, and Climate Policy

As we explained in Chapter 2, after successfully building an integrated model of the problem of acid rain that was able to provide clear policy prescriptions in the 1980s, in 1991 we set out (with some hubris) to do the same thing for the problem of climate change. After a few years of work on progressively more sophisticated versions of the ICAM model, we reached the conclusion that, even in the face of all the uncertainties about climate and other things, there are strategies for mitigation and adaptation that lead to dominant outcomes in the long run. However, that did not provide a sufficiently strong narrative to offset short-term interests and motivate substantive policies.

In ICAM-3, Hadi hypothesized that extreme climate events that occurred in a number of regions, and within a short time period, could finally lead to substantive and multi-lateral action in climate change mitigation. He simulated this using autonomous adaptive agents sensitive to extreme events attributed to climate change as well as intense pressure from special interests in the economy resisting high carbon taxes. In these model simulations, we saw the spontaneous development of multi-lateral global mitigation in a high fraction of our simulation runs due to the increased frequency of extreme events across different regions. However, we also saw that unless there were policies designed to prevent backsliding, and the

DOI: 10.4324/9781003330226-19

fossil industry did not aggressively offer competitive alternatives, the high energy costs could lead to policy failures that would cascade through regional coalitions. Maintaining decades-long pressure to transition to a low-carbon future succeeded when alternatives to the fossil fuels were widely available and affordable. However, even today, with wind and solar prices at or below those of fossil fuels, deployment has been relatively slow (~12% in the U.S.), and we have yet to solve the problem of long duration and inter-seasonal energy storage.

Given the uncertainties in system response, our best policy advice from Hadi Dowlatabadi's work with ICAM was to adopt a dual control system where, by implementing moderate mitigation policy we could improve our understanding of how the system responds and establish more informed quantitative policy prescriptions. By the end of the 1990s, with the continued unwillingness of governments to try an adaptive management approach to climate change mitigation, we concluded that there was little reason to continue developing the ICAM family of models. We concluded that in our later Centers our time would be better spent on understanding impacts and focusing on specific technologies and strategies to decarbonize the energy system and the rest of the economy.

Our work on the ICAM family of models made it clear that there are different time-scales over which it is sensible to try to build detailed predictive models of different components of the relevant social, economic, demographic, and geophysical systems. We discussed this issue in a paper we published in *Risk Analysis* in 1999 titled "Mixed Levels of Uncertainty in Complex Policy Models." In that paper, we noted that physical models of the climate system are probably more reliable many decades into the future than models of many social processes and the economy. We suggested ways to indicate how much confidence one should place in a model over time and how the probability of a "surprise" might evolve over time. We also illustrated a method that would allow an analyst to start with a very detailed model, transition to a much simpler model, and then finally transition to a simple bounding analysis as one moves further and further into a realm in which detailed models are simply not reliable (see the discussion of energy system bounding analysis in Chapter 15).

## What We Learned about Conventional Tools for Analysis and Policy Advice

Many of the issues that our work has addressed fall comfortably within the framework of conventional tools of policy research and analysis of the sort discussed in the opening chapters of my book *Theory and Practice in Policy Analysis*.[2] However, in the course of our work, we have encountered a number of problems for which the underlying assumptions on which such conventional tools are based are simply not appropriate. I will first illustrate this point with two examples and then generalize and offer some broader guidance on the issues that analysts should keep in mind.

In a first example, when charged by the Obama Administration to come up with a strategy to put a dollar value on the "social cost of carbon" (SCC), the Interagency Working Group that was created to do this assumed that the changes induced by climate change could be assessed using *marginal* analysis[3] in the same way that such analysis had been performed in the past for conventional pollutants. Figure 19.1 shows their basic approach. However, it is clear that in order to keep climate change impacts below catastrophic harms, global greenhouse gas mitigation needs to be cut by more than 80% from current emissions. The change that is needed is by no means "marginal" and we do not have real world experience of achieving such non-marginal changes in the foundations of our economies to calibrate and validate appropriate models for their simulation in policy analysis.

Furthermore, as readers can infer from the discussion in Chapter 3 about the very limited capabilities of integrated assessment models, analysts are ill equipped to characterize non-market damages as experienced by different sectors and regions of the world. We do not have a sensible and equitable method to determine what might be equivalent losses across groups and generations. Therefore, we do not believe that there is any way to sensibly place a dollar value on all the present and likely future consequences of climate change or to compute their net present value.

However, our greatest concern with this approach is not with the analytical limitations of the method. There is a very big difference between carbon dioxide and

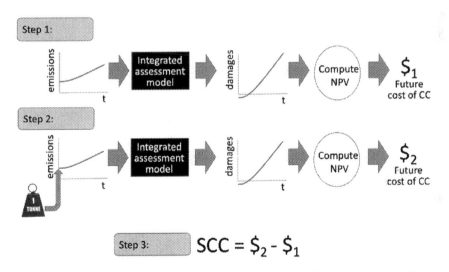

FIGURE 19.1 Basic marginal analytical approach adopted by the Interagency Working Group on the social cost of carbon (SCC). This approach requires four things: a projection of how future global emissions of GHGs are likely to evolve; a model that estimates how those future emissions of GHGs will change the climate; a model of all the consequences of that change in the climate; and a way to assign monetary values to all those consequences so that qualitatively disparate damages may be combined. We do not believe that *any* of the steps in this process can be done in a way that provides reliable quantitative results. However, our concern is more fundamental: many of the results of climate change will involve very fundamental, not marginal (i.e., small), changes.

Drawn by Granger Morgan.

conventional air pollutants (see Chapter 5). Many of the health and ecological effects of conventional pollutants become apparent in days or a few years. Once emissions cease, conventional pollutants disappear from the atmosphere in just hours or days. Health effects show up in days to decades. Hence, it is reasonable to base regulatory policy on an estimate of the damage caused by the emission of an incremental amount of conventional air pollution – that is, on the "marginal damage."

The same is *not* true for carbon dioxide. A substantial fraction of the carbon dioxide that enters the atmosphere remains there for centuries. Many of its effects become apparent only over decades to millennia, at which point they cannot be reversed by stopping emissions. For this reason, using conventional assessments of marginal damage in benefit-cost analysis to support climate policy fails to consider how little we know about long-term effects of climate change and how these effects should be valued by today's decision makers.

We know that the response of the climate system to changes in radiative forcing is nonlinear. Geologic evidence indicates that the earth has several quasi-equilibrium climate states. The feedbacks in the climate system that have blessed the planet with a stable "climate optimum" for the past ten thousand years are uncertain in magnitude and operate over a limited range of disruptions or perturbations. Beyond that range of perturbations, climate system dynamics may tip to a very different climate state. Nobody can adequately assess the probability and consequences of such climate transitions. If and when such transitions occur, many resulting changes will *not* be marginal.

Even in the simple case of inundation caused by sea level rise, experience with displaced populations from places such as the Bikini Atoll and Diego Garcia suggests that compensation of the "value of lost real estate" does not begin to make up for the loss experienced by the affected peoples. The inhabitants of these communities were moved during the Cold War. Their resultant high suicide rates, short life expectancy, and broken social structures make it clear that they have failed to "adapt" to their new locations, even after half a century. The problem grows only more complex when other damages are considered for valuation.

Climate change and its effects will vary by location, ecosystem, and socio-economic context. The responses of social, economic, and ecological systems are also likely to be nonlinear, with some entering protracted periods of unstable chaos while others undergo rapid transition to conditions fundamentally, not marginally, different from today. We neither know how to characterize such effects or how they will be valued across different cultures, societies, and future generations. Indeed, monetizing, combining, and discounting these heterogeneous and contextual effects as a single global monetary metric displays a hubris that has been roundly condemned by ethicists and decision analysts.

In a paper we wrote in 2017 for *Issues in Science & Technology* (see technical details at the end of this chapter), we argued that there is no way that conventional modeling methods can appropriately address multi-jurisdictional and multi-generational problems. We elaborated these and other concerns and proposed a very different approach that was much more explicitly normative – choose a socially acceptable level of warming and work backwards from emission reduction supply curves to choose an investment rate for abatement (that may vary over time and across nations). Our advice was to use those values in place of the SCC. This is, in fact, the approach that had been adopted by the UK government before our paper appeared.

A second example of where the uncritical application of conventional analytical approaches can lead to serious problems is provided by the way in which analysts in Working Group III of the IPCC's 2nd assessment used different values of statistical life (VSL) in different countries in order to estimate an overall value of the impacts of climate change.

The value of a statistical life depends critically on an individual's and a country's level of income. If one is addressing an entirely domestic question, such as "Should automatic pedestrian avoidance systems be installed on buses," it is appropriate to use different levels of VSL for decisions that are to be made entirely within different societies. For example, in an assessment of such technology in Bogotá, Colombia (VSL=$160,000) and New York (VSL=$9-million), Sonia Cecilia Mangones Matos[4] showed that investing in such technology could be justified in New York society but not in Bogotá society.

However, it makes no sense to add different VSLs together from different countries to assess a global cost. Quite appropriately, when that was done by folks working for the IPCC, enormous controversy ensued.[5] Given that the developed world (with high VSLs) had produced most of the emissions that cause climate change, using low VSLs to assess damages in the heavily impacted developing world obviously raised serious questions of inter-nation equity![6]

The first of these two examples illustrated a situation in which some of the changes being considered involve time-scales of centuries, with impacts spanning across multiple societies and cultures. The second example also involves impacts across multiple societies and cultures.

In an editorial comment that several of us published in *Climatic Change*, we argued that conventional policy analysis typically makes one or more of the six assumptions listed in Table 19.1.

**TABLE 19.1** The conventional tools of policy analysis typically make some or all of these six assumptions, which are often not appropriate when analyzing issues related to climate change.

| | |
|---|---|
| 1) | There is a single public-sector decision maker who faces a single problem in the context of a single polity; |
| 2) | Values are known (or knowable) and static; |
| 3) | The decision maker should select a policy by maximizing expected utility; |
| 4) | The impacts involved are of manageable size and can be valued at the margin; |
| 5) | Time preference is accurately described by conventional exponential discounting of future costs and benefits; |
| 6) | The system under study can reasonably be treated as linear; |
| 7) | Uncertainty is modest and manageable. |

Of course, not all conventional policy analysis makes all of these assumptions. Even when it does, good analysts are often aware of the limitations this imposes and take steps to address them, or at least discuss their implications for the results they have obtained. But the difficulties that these assumptions pose appear to us to be greater in the context of global change than they are for most other domains of policy analysis. In our editorial comment, we examined each in turn and noted that while we have tried to be aware of these issues as we have addressed problems in global change, we readily admit that some of the criticisms we discuss can be leveled at some of our own work as well as that of others.

A key source of the difficulty is illustrated in Figure 19.2. Most tools of modern quantitative policy analysis were developed to address problems that lie near the origin in this space. As one moves outward from the origin, more and more of the underlying assumptions upon which conventional tools are based begin to break down. Because many problems in global change lie far from the origin on all three dimensions, one can expect that the straightforward application of standard ideas and methods will often fail.

## What We Learned about the Need to Treat Climate as Just One Element of Global Change

In thinking about climate change, it is easy to forget that climate is not the only thing that is changing. The early work that Liz Casman, Richard Tol, and Hadi Dowlatabadi did on the environmental determinants of malaria (described in Chapter 7) provides an excellent illustration of the risk of adopting an overly climate-centric perspective. Climate is certainly a factor that influences the prevalence of disease. However, arguments about malaria that were advanced by the office of the Vice President to raise public concern near the end of the Clinton Administration, neglected to note that in the late 1800s malaria occurred in the U.S. along much of the East Coast, throughout the South, the Midwest, and in the Central Valley of California. Beginning in 1947, public health interventions in the U.S. have eliminated many vector-borne diseases that still wreak havoc in less developed countries. Thousands of infected travelers land in the U.S. every year; however, there are almost no cases of community transmission of malaria thanks to aggressive programs of surveillance, treatment, and isolation of infected travelers.[7] While climate does play a role in the presence or absence of potential vectors[8] for diseases like malaria and dengue, the presence or absence of public health measures is a far more important factor.

In 2000, the first U.S. National Assessment made a very similar point with respect to the prevalence of dengue in Texas and the adjacent border states of northern Mexico. In the years between 1980 and 1996, while there were over 50,000 cases of dengue in northern Mexico, there were only 43 cases in the adjacent regions of Texas that experience an almost identical climate. Back in 1922, there were an estimated 500,000 cases in Texas. The precipitous drop illustrates

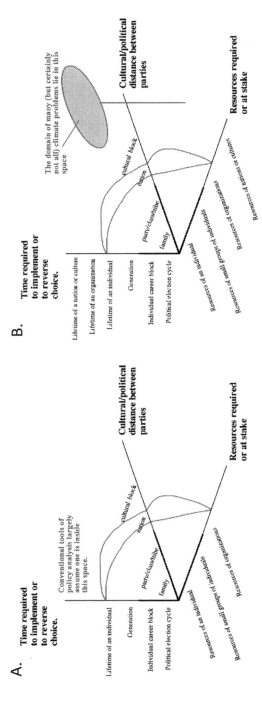

**FIGURE 19.2** While it is appropriate to apply the conventional tools of policy analysis to most problems that lie in the inner part of the space shown on the left (**A**), for problems such as some aspects of climate change that involve issues that lie well outside that space, analysts should not make use of those tools without first carefully examining the appropriateness of the assumptions on which they are based, as shown on the right (**B**). Figure is from Morgan, Kandlikar, Risbey, & Dowlatabadi, 1999.

Morgan, M. G., Kandlikar, M., Risbey, J., & Dowlatabadi, H. (1999). Why conventional tools for policy analysis are often inadequate for problems of global change. *Climatic Change, 41*(3–4), 271.

the importance of factors beyond public health and climate such as how the availability of air-conditioning and window screens has dramatically altered how people socialize while they are isolated from mosquitoes during their peak blood-questing periods.[9]

Just as our work has illustrated the importance of remembering that many things are changing at the same time that climate is changing, it has also illustrated that in many cases the early damage from climate change comes from extreme events and not from the gradual increase in things like temperature and sea level. In early studies of impacts from sea level rise many analysts adopted a "coloring book" approach – color in the region of coastal property that will be inundated as sea levels gradually increase and value the resulting damage in terms of the value of property that lies within the colored region. In Chapter 16, we described work done by Hadi Dowlatabadi, Jason West, and Mitch Small that showed that most of the early damage from sea level rise will arise from wave action during major storm events. They showed that absent local development regulations, whether coastal residents concluded that they should retreat or rebuild is likely to depend strongly on the recurrence frequency of storms. A future timeline of small storms could generate the greatest damage (because homeowners kept rebuilding and not retreating) while one early large storm that caused people to retreat could lead to lower long-term damages.

Jason, Mitch, and Hadi also found that, because storm events are infrequent and the memories of most coastal residents and other decision makers are short, significant damage from storms can result even in the absence of sea level rise. Looking at long-term processes using a short time-horizon is also a challenge for effective signaling and coverage by the insurance industry. When we ran an early workshop on climate change and the insurance industry, we found that unless we could tell insurers what damages were likely to be next year, they weren't particularly interested because they often set their rates on a yearly basis. The reinsurers did adopt a somewhat longer-time perspective, but short timelines, combined with high market discount rates, have been a serious problem in persuading decision makers to pay serious attention to the long-term consequences of climate change. Two members of the insurance industry provided a striking illustration of this problem when, in a later workshop at the Wharton School, they described hurricane Katrina, the category 5 event which killed 1800 people and caused roughly $125 billion in damages, as a "black swan event."[10] Anybody who has paid attention to the history of hurricanes along the U.S. Gulf Coast over the past couple of centuries, to say nothing about the longer term evidence in coastal geology, should be expecting storms like Katrina.[11] This failure to pay attention to historical data is not limited to coastal storms. As Luke Lavin and Jay Apt pointed out, the three-day 2021 Texas winter blackout was caused by failure to winterize energy systems against cold weather that appears roughly once a decade.[12] All around the country, power grid operators assume that each generator fails at a random time independent of others. Our research showed that, historically, many generators fail simultaneously in cold or hot weather and that must be taken into account when computing the reserves required to keep grids operating in the face of failures.[13]

Closely related to this problem of not paying attention to the frequency of past events is the broadly held misconception that past frequencies will remain unchanged in the future. Local TV weather forecasters routinely talk about how very hot or otherwise extreme forecasts are "unusual" or "record breaking," comparing them to occurrences over past decades, while ignoring the fact that, due to climate change, the statistics for weather events are now changing (i.e., are "not stationary").

## What We Learned about Building and Sustaining a Distributed Interdisciplinary Research Collaboration

Five lines of advice summarize much of what we've learned about running a large, distributed interdisciplinary research program over a period of many years:

1.  Work with colleagues who like and respect each other, enjoy collaborative work, and share a basic perspective on problem-solving but bring to bear a variety of different perspectives on what is important.
2.  Involve first-rate students in an institutional environment that places value on interdisciplinary collaboration and supports the idea that the nature of the problem should drive the choice of methods.
3.  Adopt a light touch in coordination that uses a few simple strategies to choose the focus of the work and promotes periodic interactions across all project participants.
4.  Assemble a diverse group of outside advisors who develop an ongoing understanding of what the program is doing and can offer advice and open doors for discussing findings with key decision-makers.
5.  Secure sustained financial support that imposes minimal reporting requirements but expects substantial refereed and other publication and various forms of outreach.

When an opportunity is announced to fund some interdisciplinary research programs, universities and other research organizations are remarkably good at coming up with groups of experts who have all the right disciplinary labels. However, too often the program officers and reviewers for these initiatives don't ask the simple question: "How many of these people have previously published together?" It is a relatively small subset of first-rate scholars who really enjoy mixing it up with others who have very different backgrounds and who have done that successfully over an extended period.

In the case of our three Centers, we have had the advantage that most of our researchers have been acculturated to such collaborative work through involvement with Carnegie Mellon's unique Department of Engineering and Public Policy

(EPP).[14] EPP is a department in the College of Engineering that works on policy problems in which the technical details are of central importance. While a number of us worked hard to build the Department of Engineering and Public Policy, the fact that the department exists at Carnegie Mellon is in large part the result of Carnegie Mellon's long-standing commitment to addressing applied problems that lie at the boundary between traditional disciplines.[15]

In some cases, this acculturation has occurred through years of involvement as members of the faculty in EPP. In others, it has occurred through doing PhDs or post-docs with the department or in other parts of CMU, such as the Heinz College of Information Systems and Public Policy or the Department of Social and Decision Sciences, and then continuing to work with us. While our collaborations have also involved folks from other institutions who have never spent time in EPP, almost all our core group (and all the co-authors of this book) have had extended affiliations with EPP, at least during some stage in their career.

For most of our work, rather than starting with specific analytical tools and looking for problems to address, we've chosen problems we thought were important and then let the problems drive our choice of methods. In the process, we've often ended up advancing new tools and methods or using existing tools in new ways. This has certainly been the case with the many expert elicitations that we have conducted (Chapter 4). It's also been true for things like Hadi Dowlatabadi's work on integrated assessment (Chapter 3), the analysis of the marginal impacts of various electricity-generating technologies pioneered by Inês Azevedo and Kyle Siler-Evans (Chapter 10), the use of power spectral analysis of the output of wind farms pioneered by Jay Apt (also Chapter 10), and the use of huge data sets to examine temperature-dependent power generator failures by Sinnott Murphy and Jay Apt (Chapter 13).

We've used a variety of strategies to keep the group together and informed about each other's work. Beyond formal structures like regular meetings of the Center's Executive Committee, every year except two during COVID, we've run an annual face-to-face meeting in the spring at which faculty and graduate students have presented completed or "half-baked" work. All the participants in these meetings then offer comments and advice. At these meetings, we've also run periodic tutorial sessions with invited experts who have exposed our investigators to new concepts and methods and conducted strategic planning sessions in which we have explored issues we think we should be working on in the future. In addition to faculty and graduate students from the various institutions that have been part of our Centers, the annual meetings have always also been attended by members of our outside Advisory Board who, in addition to providing advice and feedback on our work, have also been active participants in much of the discourse over the course of the meetings. Figure 19.3 shows views from three different years of our annual meetings.

In addition to annual meetings, the Centers have conducted over a dozen invitational expert workshops focused on specific issues. Some, like the workshop we

**FIGURE 19.3** Views of several of the annual meetings of our Centers, which have been held every May. The image for 2018 shows the Center's Advisory Board around the table in the foreground.

Photos taken by Granger Morgan.

ran in Lausanne, Switzerland in 2000 on the environmental determinants of malaria (see Chapter 7), the one we ran in Washington, DC in 2011 on the rebound effect (see Chapter 9), or the one on decarbonizing heavy freight that we held in Pittsburgh in 2017 (see Chapter 14), brought experts from around the world to address topics that had not yet been sufficiently well defined or explained. Others, like the ones we ran in Washington, DC on strategies for energy forecasting in 2013 and on identifying and avoiding dead ends in climate mitigation policy in 2016, were focused on developing new theory and methods. Figure 19.4 shows views from just three of these gatherings.

A key element of our successful collaborations has been the fun we've had working together. For example, for several years our colleague Baruch Fischhoff ran a series of Sackler Forums at the National Academies of Science on "The Science of Science Communication." In December of 2017, he asked psychologist Wändi Bruine de Bruin and me (an engineer) to give a joint talk about our 15 years of collaborative research.[16] We talked about joint work we had done on public perceptions of carbon capture and deep geological sequestration and on public understanding of the atmospheric residence time of carbon dioxide (see Chapter 5). We subsequently published a paper describing this joint work in the *Proceedings of the National Academies of Sciences*. Table 19.2 summarizes some of our conclusions about the factors that made our collaborations successful.

**FIGURE 19.4** Views of three of the many expert workshops the Centers have run. All three of these were held in the AAAS building in Washington, where for years EPP maintained a small office.

Photos taken by Granger Morgan.

**TABLE 19.2** Summary of attributes that Wändi Bruine de Bruin and Granger Morgan listed as key to their successful collaboration. Elaborated discussions of each can be found in their joint paper.

**Shared Research Goal.** Our projects are motivated by a common goal that is agreed upon beforehand.

**Shared Methodology.** We agree on a shared methodology that combines those in which we are trained.

**Shared Effort.** On each project, we commit to collaborate from the start. We consult each other on every step and take care to understand and address each other's concerns.

**Shared Benefits.** We both believe that our collaborations make our projects stronger than they would have been if we were working solely within our own discipline. Together, we are gaining more insight into understanding complex societal problems, which we would not have by relying solely on our own discipline.

**Interpersonal Connection.** Interdisciplinary research is not for everyone, but whether a collaboration is within or across disciplines, we have found that projects progress more smoothly if people get along well. Even seemingly small gestures may promote team cohesion....

**Excellent Students.** We have been able to attract and co-supervise excellent graduate students.

**Adequate Funding.** In recent years, the National Science Foundation has been supporting interdisciplinary research, and we have been fortunate to write successful proposals.

**Supportive Environment.** These days, while most universities talk a good line about interdisciplinary research and education, the number in which that rhetoric has been matched by supportive reality is relatively low.

FIGURE 19.5 Wändi and Granger having fun together with color-coordinated slides and outfits as they jointly talk about their 15 years of collaboration on the main stage at the National Academy of Science.

"Photo and 'National Academy of Sciences' are used with permission from the National Academy of Sciences."

In our paper, we explained that even seemingly small gestures may promote team cohesion, as illustrated by the following example. Wändi really likes purple. When she was at Carnegie Mellon, she even had a purple desk. For the talk that we gave on this paper at the National Academy of Sciences' Colloquium on The Science of Science Communication, Granger proposed that when Wändi talked, the headings on the slides should be purple, and when Granger talked, his headings should be green because "green goes well with purple." For a bit of fun, Wändi selected her favorite purple outfit and Granger bought a green shirt so we could each dress in line with the color scheme on our slides.

Over the years, our Centers have engaged in a wide variety of outreach activities to both professional and lay audiences. One of the more interesting activities has been a program for junior high students and teachers called SUCCEED (Summer Center for Climate, Energy, and Environmental Decision Making). This program began in 2011. It has been organized and coordinated by EPP PhD students who have drawn upon speakers from Center investigators and others. The programs for students have often involved hands-on projects and field trips to places like power plants. In addition to overview talks across a wide range of issues, the program for teachers has provided participants with a great deal of resource material for classroom use.

## Some Final Thoughts

After roughly 30 years and several tens of millions of dollars of NSF, EPRI, and other support, what did we accomplish?

### In the small, a great deal

The appendices in the back of this book list many of the PhD students we educated and the hundreds of research papers that Center investigators and students have published in the refereed literature.

We believe, and members of our Center's Advisory Board tell us, that this body of research, and the many presentations and formal and informal briefings that we have done, have helped to reframe public and private discourse about climate change. We have briefed legislators and corporate decision makers and engaged with groups like the National Association of Regulatory Utility Commissioners (NARUC) and the Federal Energy Regulatory Commission (FERC).

Through presentations at professional conferences, regular topical meetings in places like Snowmass, CO, and private extensive discussions, we have helped shape programs such as those of the Electric Power Research Institute (EPRI) and the U.S. National Academies of Science, Engineering, and Medicine (NASEM). Through communication materials we have developed, we have helped to improve understanding by both schoolteachers and the general public.[17] Through interviews with popular media, and many published opinion pieces, we have also worked to help members of the general public become more informed participants in public discourse about climate change and climate policy.

### In the large, not as much as we would have liked

Even before I got acquainted with climatologist Steve Schneider when, as graduate students, we were attending sessions on climate science at annual meetings of the American Geophysical Union (AGU) in the late 1960s, it has been clear to experts working in the field that the world has a climate problem. Since Steve and I first met, the average temperature of the planet has risen by almost a full degree centigrade (just under 2°F). If things keep going as they now are, the average warming will have reached something between 3° and 4°C (roughly 5.5° to 7°F) by 2100, and the results will be catastrophic.

Well before we started our first Center roughly 30 years ago, it was pretty clear what the world needed to do: protect the planet's lands, air, and waters; make the transition to a more sustainable economy that imposes a far lighter footprint on the world's ecosystems; and reduce dramatically the emissions of greenhouse gases. Over the course of those 30 years, humans have burned many millions of tons of coal, cleared vast regions of tropical rain forest, added millions of tons of plastic to the world's oceans, and increased the concentration of $CO_2$ in the earth's atmosphere by roughly 18% – from an average of about 355 ppm to an average of almost 420 ppm.

The world has made only the feeblest of efforts to reduce emissions. If it had met the goals of the Kyoto Protocol, or managed to honor the terms of the 2015 Paris agreement, that would have only slowed the warming. U.S. international commitments to reduce emissions have been an on-again, off-again thing. Some U.S. states and cities have tried valiantly to decarbonize their economies, but progress has been *much* too slow. U.S. emissions have now stabilized and began to gradually decline, but given that most of the $CO_2$ that enters the atmosphere stays there for centuries (see Chapter 5), even those lower U.S. emissions are steadily making the problem worse. $CO_2$ emissions from major developing regions such as China and India continue to grow.

We know what needs to be done, but the world has not figured out how to overcome myopic and selfish short-term political and economic interests to make the major changes that are needed.

Is the world we will leave our children and their children toast...or even burnt toast?

Things don't look great, but we do see two modest rays of hope that might allow humankind to avoid at least the worst consequences of climate change: (1) dramatic technology innovation and rapid adoption and (2) carbon management from the bottom up.

Recent years have seen dramatic reductions in the costs of commercial-scale wind and solar. The result is that, while they still constitute much too small a proportion of total energy consumption (2%), the use of wind and solar has grown rapidly. If similar progress can be made on several other technologies, together they might be used to head off the most dire warming and associated changes, although given the inertia in the climate and economic systems, earth and its ecosystems will still be in for at least half a century of continued warming with ever more dire consequences. Technologies and strategies that should get special attention include:

- The development of a wide range of much lower cost zero-emission sources of energy;
- Cost-effective strategies to capture $CO_2$, both after combustion of fossil fuels and directly from the atmosphere;
- Cost-effective strategies to convert captured $CO_2$ into products of value (e.g., construction aggregate and building materials);
- The development of low-cost technologies for long-duration energy storage;
- The development of net-zero or net-negative liquid and gaseous fuels;
- And, while we are reluctant to say it, at this point we also need to work on developing strategies that could be used to reduce planetary albedo (see Chapter 17) for at least a few decades during peak warming.

None of these things will happen fast enough if investors and innovators are left to their own devices in the present marketplace. That leads us to our second modest ray of hope. In a Policy Forum piece in the journal *Science* titled "Managing carbon from the bottom up" that I published in 2000,[18] I lamented the lack of progress on international negotiations and argued that "diplomats will put a good face on things, but for at least the next decade, it is unlikely that all the world's major states will simultaneously agree to a serious program to curtail emissions of $CO_2$ and other greenhouse gases." Twenty years later, international negotiations are still basically stuck in the same place.

I argued that a more likely route to serious emission abatement lies in encouraging the development of local and regional programs that could gradually coalesce into successful global policy. In a subsequent book, UCSD's David Victor provided a thoughtful elaboration of this basic idea,[19] and Bill Nordhaus has advanced similar ideas in proposals for "climate clubs."[20] Such strategies probably still offer the best hope for progress in the next few decades.

At the federal level, the U.S. has wasted most of the last 30 years, at best making modest and intermittent efforts. Several U.S. states, and a number of countries in Europe, have taken more aggressive measures. If these continue, and if major regions like the U.S., China, and the EU adopt aggressive policies to accelerate technical innovation, promote adoption, and keep their eye on the ultimate objective of almost completely eliminating the emissions of greenhouse gases, by the end of the century it might be possible to coalesce these efforts into a significant planet-wide effort to address the problem. Let's hope so, because we have only one planet.

## Technical and Other Details for Chapter 19

This is a paper that discusses several years of collaboration between Wändi Bruine de Bruin and Granger Morgan:

Bruine de Bruin, W., & Morgan, M. G. (2019). Reflections on an interdisciplinary collaboration to inform public understanding of climate change, mitigation, and impacts. *Proceedings of the National Academy of Sciences, 116*(16), 7676–7683.

**Abstract:** We describe two interdisciplinary projects in which natural scientists and engineers, as well as psychologists and other behavioral scientists, worked together to better communicate about climate change, including mitigation and impacts. One project focused on understanding and informing public perceptions of an emerging technology to capture and sequester carbon dioxide from coal-fired power plants, as well as other low-carbon electricity-generation technologies. A second project focused on public understanding about carbon dioxide's residence time in the atmosphere. In both projects, we applied the mental-models approach, which aims to design effective communications by using insights from interdisciplinary teams of experts and mental models elicited from intended audience members. In addition to summarizing our findings, we discuss the process of interdisciplinary collaboration that we pursued in framing and completing both projects. We conclude by describing what we think we have learned about the conditions that supported our ongoing interdisciplinary collaborations. That paper is only about collaboration in work on public understanding but many of the same arguments extend to the other issue we have addressed.

This paper suggests several strategies for doing analysis on problems in which different parts of the system being analyzed involve very different time-scales with different levels of uncertainty:

Casman, E. A., Morgan, M. G., & Dowlatabadi, H. (1999). Mixed levels of uncertainty in complex policy models. *Risk Analysis, 19*(1), 33–42.

**Abstract:** The characterization and treatment of uncertainty poses special challenges when modeling indeterminate or complex coupled systems such as those involved in the interactions between human activity, climate and the ecosystem. Uncertainty about model structure may become as, or more important than, uncertainty about parameter values. When uncertainty grows so large that prediction or optimization no longer makes sense, it may still be possible to use the model as a "behavioral test bed" to examine the relative robustness of alternative observational and behavioral strategies. When models must be run into portions of their phase space that are not well understood, different submodels may become unreliable at different rates. A common example involves running a time stepped model far into the future. Several strategies can be used to deal with such situations. The probability of model failure can be reported as a function of time. Possible alternative "surprises" can be assigned probabilities, modeled separately, and combined. Finally, through the use of subjective judgments, one may be able to combine, and over time shift between models, moving from more detailed to progressively simpler order-of-magnitude models, and perhaps ultimately, on to simple bounding analysis.

Here is the opinion piece we wrote about the limits to conventional tools for policy analysis:

Morgan, M. G., Kandlikar, M., Risbey, J., & Dowlatabadi, H. (1999). Why conventional tools for policy analysis are often inadequate for problems of global change. *Climatic Change, 41*(3–4), 271.

As explained in the chapter, this paper outlines a set of standard assumptions that are often made by policy analysis. It points out that in many cases these assumptions are not appropriate in the case of work that addresses issues of climate change and argues that in that case analysts should think carefully about adopting alternative strategies.

This is our paper on an alternative approach for putting a social cost on carbon dioxide emissions that does not depend upon the use of integrated assessment models:

Morgan, M. G., Vaishnav, P., Dowlatabadi, H., & Azevedo, I. L. (2017). Rethinking the social cost of carbon dioxide. *Issues in Science & Technology*, 43–50.

This paper argues that for a variety of reasons, attaching a marginal price to one additional ton of carbon dioxide emissions is really not a very sensible idea.

Here is the book that Granger wrote for the graduate core course in the Department of Engineering and Public Policy at Carnegie Mellon. Many of the ideas discussed in this book were first developed through research we conducted in our Centers:

Morgan, M. G. (2017). *Theory and Practice in Policy Analysis: Including applications in science and technology*. Cambridge University Press, 590pp.

**Summary:** Many books instruct readers on how to use the tools of policy analysis. This book is different. Its primary focus is on helping readers to look critically at the strengths, limitations, and the underlying assumptions analysts make when they use standard tools or problem framings. Using examples, many of which involve issues in science and technology, the book exposes readers to some of the critical issues of taste, professional responsibility, ethics, and values that are associated with policy analysis and research. Topics covered include policy problems formulated in terms of utility maximization such as benefit-cost, decision, and multi-attribute analysis, issues in the valuation of intangibles, uncertainty in policy analysis, selected topics in risk analysis and communication, limitations and alternatives to the paradigm of utility maximization, issues in behavioral decision theory, issues related to organizations and multiple agents, and selected topics in policy advice and policy analysis for government.

This paper describes the Department of Engineering and Public Policy at Carnegie Mellon (www.cmu.edu/epp):

Morgan, M. G., & Sicker, D. (2019). Technically based programs in science, technology and public policy (pp. 1–27). In *Science, Technology, and Society: New perspectives and directions,* Cambridge University Press, 270pp.

**Abstract:** In this chapter, we review and discuss academic programs in technology and public policy, focusing on those that are either located in an engineering college or have a strong engineering focus. We consider what constitutes technically focused research in programs melding engineering and policy, where and how this work is done, the focus of these programs at the undergraduate and graduate levels, and the challenges of building and sustaining such programs. Many academic programs in the United States and elsewhere focus on the social studies aspects of science, technology, and public policy. Indeed, most programs listed in the original American Association for the Advancement of Science guide to graduate education in science, engineering, and public policy were in this category (Levey, 1995). Few programs combine deep technical education and understanding with modern social science and policy-analytical skills and knowledge.

**Notable writings by authors unaffiliated with our Centers:**

**On climate and climate policy:**

**Assessment Reports**

From the Intergovernmental Panel on Climate Change are at: https://www.ipcc.ch
From the U.S. National Assessment are at: www.globalchange.gov/browse
National Academies of Sciences, Engineering, and Medicine. (2021). *Accelerating decarbonization of the US energy system.* National Academies Press.
Nordhaus, W. (2015). Climate clubs: Overcoming free-riding in international climate policy. *American Economic Review, 105*(4), 1339–1370.
Sabel, C. F., & Victor, D. G. (2022). *Fixing the Climate: Strategies for an Uncertain World.* Princeton University Press.
The Royal Society and the U.S. National Academy of Sciences. (2020). *Climate Change: Evidence and Causes: Update 2020.* National Academies Press, 34pp.
Victor, D. (2012). *Global Warming Grid Lock: Creating more effective strategies for protecting the planet.* Cambridge University Press.
**On interdisciplinary research:**
Derrick, E. G., Falk-Krzesinski, H. J., Roberts, M. R., & Olson, S. (2011). Facilitating interdisciplinary research and education: A practical guide. In report from the "Science on FIRE: Facilitating Interdisciplinary Research and Education" workshop of the American Association for the Advancement of Science.

Donovan, S. (2020). The landscape of challenges for cross-disciplinary activity. In *The Toolbox Dialogue Initiative* (pp. 48–57). CRC Press.

Lyall, C., & Meagher, L. (2007). A short guide to building and managing interdisciplinary research teams. *ISSTI Briefing Note*, (3), 1–4.

National Academy of Sciences, Engineering, and Medicine. (2005). *Facilitating Interdisciplinary Research*. The National Academies Press.

## Notes

1 Portions of the text in the first half of this chapter have been reworked from Morgan, 2017; Morgan et al., 1999; and Casman et al., 1999. See the end of the chapter for details.

2 Morgan, M. G. (2017). *Theory and Practice in Policy Analysis: Including applications in science and technology*, Cambridge University Press, 590pp. I wrote this book for the core introductory PhD course in our Department of Engineering and Public Policy at Carnegie Mellon. Many of the ideas and issues discussed in the textbook drew directly upon work we did in our several NSF-supported Centers.

3 A marginal change would be one that is small enough so that it is safe to assume that everything else remains pretty much the same.

4 Mangones, S. C., Fischbeck, P., & Jaramillo, P. (2017). Safety-related risk and benefit-cost analysis of crash avoidance systems applied to transit buses: Comparing New York City vs. Bogota, Colombia. *Safety Science, 91*, 122–131.

5 See Pearce, F. (1995). Price of life sends temperatures soaring. *New Scientist, 146*, 5.

6 For some taste of the controversy that arose, see: https://enthusiasmscepticismscience. wordpress.com/2013/01/15/enter-the-economists-the-price-of-life-and-how-the-ipcc-only-just-survived-the-other-chapter-controversy/

And: www.bishop-hill.net/blog/2013/2/26/the-price-of-life-the-ipccs-first-and-forgotten-controversy.html

7 For details on the history of malaria in the U.S. see: www.cdc.gov/malaria/about/hist ory/elimination_us.html

8 A disease vector is something like mosquitos or fleas that can carry disease to people.

9 Overview by the National Assessment Synthesis Team (2000). *Climate Change Impacts on the United States: The potential consequences of climate variability and change*. Cambridge University Press 154pp. Granger Morgan served as a member of the synthesis team.

10 Black swan events are major events that are entirely unexpected.

11 Note, however, that as more and more developers built vulnerable properties in Florida, the insurance industry did finally become unwilling to provide coverage, with the result that the State has stepped in. Unfortunately, one of the consequences of this has been that residences who are at lower risk in central Florida now subsidize the insurance premiums of residences along the coast.

12 Apt, J., & Lavin, L. (2021). What is happening in Texas will keep happening until we take action (Op-Ed). *The Washington Post*, February 18.

13 Murphy, S., Lavin, L., & Apt, J. (2020). Resource adequacy implications of temperature-dependent electric generator availability. *Applied Energy, 262*, 114424.

14 For details on EPP, see www.cmu.edu/epp

15 For some details on this environment, see Morgan and Sicker, 2019 referenced at the end of the chapter.

16 A video recording of the talk can be found at www.youtube.com/watch?v=OHi44_VT DWI. The paper that resulted is described at the end of the chapter.

17 For details on work on communication to the general public see Morgan, M. G., Fischhoff, B., Bostrom, A., & Atman, C. J. (2002). *Risk Communication: A mental models approach.* Cambridge University Press, 351pp.

18 See Morgan, G. (2000). Managing carbon from the bottom up. *Science, 289,* 2285.

19 Victor, D. G. (2011). *Global Warming Gridlock: Creating more effective strategies for protecting the planet.* Cambridge University Press, 2011.

20 Nordhaus, W. (2015). Climate clubs: overcoming free-riding in international climate policy. *American Economic Review, 105*(4), 1339–1370.

# APPENDIX 1

# SHORT BIOGRAPHICAL SKETCHES OF THE CO-AUTHORS

**Ahmed Abdulla** – Abdulla is an Assistant Professor in the Department of Mechanical and Aerospace Engineering at Carleton University in Ottawa, Canada. Prior to joining Carleton he held positions at UCSD and in the Department of Engineering and Public Policy at Carnegie Mellon where he was an Assistant Research Professor. He holds a PhD in Engineering and Public Policy from Carnegie Mellon University. Additional biographical details can be found at: https://carleton.ca/mae/profile/ahmed-abdulla/

**Jay Apt** – Apt is a Professor in the Tepper School of Business at Carnegie Mellon University where he holds a joint appointment with the Department of Engineering and Public Policy. He co-directs (with Granger Morgan) Carnegie Mellon's Electricity Industry Center. Prior to joining Carnegie Mellon he was Managing Director and Chief Technology Officer, iNetworks LLC Venture Capital (2000–2003); and Director, Carnegie Museum of Natural History (1997–2000). He held various positions with NASA, became an Astronaut and flew four missions on the space shuttle (1982–1997); before NASA, he was at Harvard and then JPL doing planetary astronomy (1978–1982). He holds a PhD in Physics from MIT. Additional biographical details can be found at: www.cmu.edu/epp/people/faculty/jay-apt.html

**Inês Azevedo** – Azevedo is an Associate Professor of Energy Resources Engineering at Stanford University where she is also Senior Fellow at the Woods Institute for the Environment and the Precourt Institute for Energy. Prior to joining Stanford, she was a Professor in the Department of Engineering and Public Policy at Carnegie Mellon. She holds a PhD in Engineering and Public Policy from Carnegie Mellon. Additional biographical details can be found at: https://profiles.stanford.edu/ines-azevedo

**Ann Bostrom** – Bostrom is the Weyerhaeuser Endowed Professor in Environmental Policy in the Evans School of Public Policy and Governance at the University of Washington. From 1999 to 2001 she was on leave from the university to co-direct the program in decision risk and management science at the National Science Foundation. She holds a PhD in Public Policy Analysis from the School of Urban and Public Affairs at Carnegie Mellon University. Additional biographical details can be found at: https://evans.uw.edu/profile/ann-bostrom/

**Wändi Bruine de Bruin** – Bruine de Bruin is Provost Professor of Public Policy, Psychology, and Behavioral Science in the Sol Price School of Public Policy and the Department of Psychology at the University of Southern California where she is also Senior Fellow and Director of Behavioral Science, Schaeffer Center of Health Policy and Economics. Prior to joining USC she was Professor of Behavioral Decision Making with University Leadership Chair at the University of Leeds in the UK (2012–2019); Associate Professor, Department of Engineering and Public Policy (EPP), as well as various other positions in EPP and in the Department of Social & Decision Sciences at Carnegie Mellon (2002–2014). She holds a PhD in Behavioral Decision Research and Psychology from Carnegie Mellon. Additional biographical details can be found at: https://priceschool.usc.edu/people/wandi-bruine-de-bruin/

**Elizabeth Casman** – Casman is an Associate Research Professor Emerita, of the Department of Engineering and Public Policy, Carnegie Mellon University. Prior to joining Carnegie Mellon, she was Chief of the Water Resources Mathematical Modeling Division at the Maryland Department of the Environment. Before that she worked as an Environmental Systems Engineer with the Interstate Commission on the Potomac River Basin. She holds a PhD from the Department of Geography and Environmental Engineering from The Johns Hopkins University and an MS in Microbiology from Northern Arizona University. Additional biographical details can be found at: www.cmu.edu/epp/people/emiriti-faculty/elizabeth-casman.html

**Hadi Dowlatabadi** – Dowlatabadi is a Professor in the Institute for Resources, Environment and Sustainability where he holds Canada Research Chair (T1, Applied Mathematics of Global Change) at the University of British Columbia. He has served as co-founder of half a dozen companies attempting to refine and market technologies that pave the way to a zero-carbon economy. He is a University Fellow at Resources for the Future. He holds a PhD in Physics from the University of Cambridge, UK. Additional biographical details can be found at: http://ires.ubc.ca/person/hadi-dowlatabadi/

**Mike Griffin** – Griffin is an Emeritus Research Professor in the Department of Engineering and Public Policy (EPP) at Carnegie Mellon University. He was previously Executive Director, Center for Climate and Energy Decision

Making, and Executive Director and Co-Director, Green Design Institute. Before joining Carnegie Mellon he was President, GMS Technologies, Inc. (1994–2001); Vice President, Research & Development, Applied CarboChemicals, Inc. (1996–1999); Director, Research and Development Services, CHMR and Director, Bioremediation Product Evaluation Center (BPEC) NETAC (1992–1994); Manager, Research and Development, Sybron Chemicals Inc. (1990–1992); Research Scientist, Biotechnology (and other positions), BP America Research and Development (previously Standard Oil of Ohio), Cleveland (1980–1990). He holds a PhD in Microbiology from the University of Rhode Island. Additional biographical details can be found at: www.cmu.edu/epp/people/emiriti-faculty/mich ael-griffin.htm

**Tim McDaniels** – McDaniels is an Emeritus Professor formerly appointed in two graduate interdisciplinary programs at the University of British Columbia. He is a specialist in decision sciences and policy analysis, particularly in managing environmental and technology-related societal risks. As a member of the series of EPP Centers on climate decision-making, his research focused on decision-making for adaptation of natural resource issues in linked human/ecological systems. He also conducted research projects concerned with building regional resilience in infrastructure systems. Tim was also an adjunct professor in Engineering and Public Policy at Carnegie Mellon University. He received his PhD from Carnegie Mellon in 1990. Additional biographical details can be found at: https://scarp.ubc.ca/peo ple/tim-mcdaniels

**M. Granger Morgan** – Morgan is the Hamerschlag University Professor of Engineering in the Department of Engineering and Public Policy (EPP) at Carnegie Mellon University. He also holds appointments in the Department of Electrical and Computer Engineering and in the Heinz College of Information Systems and Public Policy. He holds a PhD in Applied Physics from the University of California at San Diego. He was the founding Department Head in EPP, a job he held for 38 years. He is a member of the U.S. National Academy of Sciences and of the Academy of Arts and Sciences. Additional biographical details can be found at: www. cmu.edu/epp/people/faculty/m-granger-morgan.html

**Dalia Patiño-Echeverri** – Patiño-Echeverri is the Gendell Family Associate Professor of Energy Systems and Public Policy at the Nicholas School of the Environment at Duke University. She explores, assesses, and proposes technological, policy, and market approaches to pursue environmental sustainability, affordability, reliability, and resiliency in the energy sector. Most of her work focuses on the decisions that regulators and private actors must make regarding capital investment and operations in the electricity industry and quantifying the value of flexibility in multiple dimensions. Her work uses operations research tools to account for path dependencies, uncertainty, and risk's tradeoffs, ubiquitous in the energy system. Additional biographical details can be found at: https://nicholas.duke.edu/people/faculty/patino-echeverri

**Joshuah Stolaroff** – Stolaroff is the Chief Technology Officer at Mote Hydrogen which is developing systems that take $CO_2$ out of the air and make clean hydrogen from waste biomass. From 2009 to 2021 he was at Lawrence Livermore National Laboratory where he led policy-focused research related to climate change and sustainability. Prior to joining LLNL he served as AAAS Science and Technology Policy Fellow, U.S. Environmental Protection Agency (2007–2009). He holds a PhD in Engineering and Public Policy, and Civil & Environmental Engineering at Carnegie Mellon University. Additional details are at: www.motehydrogen.com/about

**Brinda Thomas** – Thomas is a Principal Data Scientist at Pacific Gas and Electric Company where she leads statistical and machine learning projects for fault localization and predictive maintenance for the Northern and Central California electric distribution grid. She has a B.S. in Physics from Stanford University, a PhD in Engineering and Public Policy from Carnegie Mellon, and was a visiting post-doc at Stanford University at the Precourt Energy Efficiency Center on data analytics for the smart grid. She has worked for a number of years at large energy companies including Tesla and GE and energy efficiency start-ups including Whisker Labs and Bidgley. She also served a two-year appointment as a member of the community advisory committee for East Bay Community Energy, a community choice aggregator in Alameda County, CA. Additional biographical details can be found at: https://medium.com/@mc2maven

**Parth Vaishnav** – Vaishnav is an Assistant Professor in the School of Environment and Sustainability at the University of Michigan. Prior to joining Michigan, he was an Assistant Research Professor in the Department of Engineering and Public Policy at Carnegie Mellon. Before pursuing his PhD he worked for a number of years at Royal Dutch Shell in the Netherlands. He holds a PhD in Engineering and Public Policy from Carnegie Mellon. Additional biographical details can be found at: https://seas.umich.edu/research/facu lty/parth-vaishnav

# APPENDIX 2

# PUBLICATIONS RESULTING FROM WORK SUPPORTED BY OUR THREE CENTERS

This appendix provides citations for most of the roughly 675 refereed publications and book chapters that have resulted from the work of our three NSF-supported Centers.

## CEDM

The following is an almost complete chronological list of publications, book chapters, and theses that were produced with support from the CEDM Center. Not included are roughly 90 reports and papers in conference proceedings which can be found online in a more complete list at www.cmu.edu/epp/climate/.

## *Journals*

### *2010*

1.   Fleishman, L. A., De Bruin, W. B., & Morgan, M. G. (2010). Informed public preferences for electricity portfolios with CCS and other low-carbon technologies. *Risk Analysis: An International Journal, 30*(9), 1399–1410.
2.   Gilmore, E. A., Apt, J., Walawalkar, R., Adams, P. J., & Lave, L. B. (2010). The air quality and human health effects of integrating utility-scale batteries into the New York State electricity grid. *Journal of Power Sources, 195*(8), 2405–2413.
3.   Gouge, B., Ries, F. J., & Dowlatabadi, H. (2010). Spatial distribution of diesel transit bus emissions and urban populations: Implications of coincidence and scale on exposure. *Environmental Science & Technology, 44*(18), 7163–7168.
4.   Gresham, R. L., McCoy, S. T., Apt, J., & Morgan, M. G. (2010). Implications of compensating property owners for geologic sequestration of $CO_2$. *Environmental Science & Technology, 44*(8), 2897–2903.

5. Grossmann, W. D., Grossmann, I., & Steininger, K. (2010). Indicators to determine winning renewable energy technologies with an application to photovoltaics. *Environmental Science & Technology, 44*(13), 4849–4855.

6. Hagerman, S., Dowlatabadi, H., Satterfield, T., & McDaniels, T. (2010). Expert views on biodiversity conservation in an era of climate change. *Global Environmental Change, 20*(1), 192–207.

7. Hagerman, S., Satterfield, T., & Dowlatabadi, H. (2010). Climate change impacts, conservation and protected values: understanding promotion, ambivalence and resistance to policy change at the World Conservation Congress. *Conservation and Society, 8*(4), 298–311.

8. Hittinger, E., Whitacre, J. F., & Apt, J. (2010). Compensating for wind variability using co-located natural gas generation and energy storage. *Energy Systems, 1*(4), 417–439.

9. Hoppock, D. C., & Patiño-Echeverri, D. (2010). Cost of wind energy: Comparing distant wind resources to local resources in the Midwestern United States. *Environmental Science & Technology, 44*(22), 8758–8765.

10. Jacob, C., McDaniels, T., & Hinch, S. (2010). Indigenous culture and adaptation to climate change: Sockeye salmon and the St'át'imc people. *Mitigation and Adaptation Strategies for Global Change, 15*(8), 859–876.

11. Jaffee, D., Kunreuther, H., & Michel-Kerjan, E. (2010). Long-term property insurance. *Journal of Insurance Regulation, 29*(7), 167–187.

12. Katzenstein, W., Fertig, E., & Apt, J. (2010). The variability of interconnected wind plants. *Energy Policy, 38*(8), 4400–4410.

13. Keith, D. W., Parson, E., & Morgan, M. G. (2010). Research on global sun block needed now. *Nature, 463*(7280), 426–427.

14. Lee, J., Veloso, F. M., Hounshell, D. A., & Rubin, E. S. (2010). Forcing technological change: A case of automobile emissions control technology development in the US. *Technovation, 30*(4), 249–264.

15. Michel-Kerjan, E. O., & Kousky, C. (2010). Come rain or shine: Evidence on flood insurance purchases in Florida. *Journal of Risk and Insurance, 77*(2), 369–397.

16. Moreno-Cruz, J. B. (2010). Mitigation and the geoengineering threat. *Resource and Energy Economics, 41*, 248–263.

17. Morgan, M. G. (2010). Why Geoengineering? *Technology Review*, Jan/Feb 2010, 14–15.

18. Peterson, S. B., Apt, J., & Whitacre, J. F. (2010). Lithium-ion battery cell degradation resulting from realistic vehicle and vehicle-to-grid utilization. *Journal of Power Sources, 195*(8), 2385–2392.

19. Peterson, S. B., Whitacre, J. F., & Apt, J. (2010). The economics of using plug-in hybrid electric vehicle battery packs for grid storage. *Journal of Power Sources, 195*(8), 2377–2384.

20. Posen, I. D., Jaramillo, P., Landis, A. E., & Griffin, W. M. (2010). Greenhouse gas mitigation for US plastics production: Energy first, feedstocks later. *Environmental Research Letters, 12*(3), 034024.

21. Reynolds, T. W., Bostrom, A., Read, D., & Morgan, M. G. (2010). Now what do people know about global climate change? Survey studies of educated laypeople. *Risk Analysis: An International Journal*, *30*(10), 1520–1538.

22. Ricke, K. L., Morgan, M. G., & Allen, M. R. (2010). Regional climate response to solar-radiation management. *Nature Geoscience*, *3*(8), 537–541.

23. Shiau, C. S. N., Kaushal, N., Hendrickson, C. T., Peterson, S. B., Whitacre, J. F., & Michalek, J. J. (2010). Optimal plug-in hybrid electric vehicle design and allocation for minimum life cycle cost, petroleum consumption, and greenhouse gas emissions. *Journal of Mechanical Design*, *132*(9).

24. Vaccaro, A., & Echeverri, D. P. (2010). Corporate transparency and green management. *Journal of Business Ethics*, *95*(3), 487–506.

25. Vignola, R., Koellner, T., Scholz, R. W., & McDaniels, T. L. (2010). Decision-making by farmers regarding ecosystem services: Factors affecting soil conservation efforts in Costa Rica. *Land Use Policy*, *27*(4), 1132–1142.

26. Weber, C. L., Jaramillo, P., Marriott, J., & Samaras, C. (2010). Life Cycle Assessment and Grid Electricity: What do we know and what can we know? *Environmental Science & Technology*, *44*(6), 1895–1901.

27. Zhai, H., & Rubin, E. S. (2010). Performance and cost of wet and dry cooling systems for pulverized coal power plants with and without carbon capture and storage. *Energy Policy*, *38*(10), 5653–5660.

## *2011*

28. Apt, J., Hendrickson, C. T., & Morgan, M. G. (2011). Lester Lave, visionary economist. *Environmental Science & Technology*, Vol. 45, No. 13, 2011, 5457–5458.

29. Apt, J., Peterson, S., & Whitacre, J. F. (2011). Battery vehicles reduce CO2 emissions. *Science*, Vol. 333, No. 6044, August 2011, 823 (DOI: 10.1126/science.333.6044.823-a).

30. Azevedo, I. M. L., Morgan, M. G., & Lave, L. (2011). Residential and regional electricity consumption in the US and EU: How much will higher prices reduce CO2 emissions?. *The Electricity Journal*, *24*(1), 21–29.

31. Blackhurst, M., Azevedo, I. L., Matthews, H. S., & Hendrickson, C. T. (2011). Designing building energy efficiency programs for greenhouse gas reductions. *Energy Policy*, *39*(9), 5269–5279.

32. Blackhurst, M., Matthews, H. S., Sharrard, A. L., Hendrickson, C. T., & Azevedo, I. L. (2011). Preparing US community greenhouse gas inventories for climate action plans. *Environmental Research Letters*, *6*(3), 034003.

33. Chung, T. S., Patiño-Echeverri, D., & Johnson, T. L. (2011). Expert assessments of retrofitting coal-fired power plants with carbon dioxide capture technologies. *Energy Policy*, *39*(9), 5609–5620.

34. Dowlatabadi, H. (2011). Lifecycle analysis of GHG intensity in BC's energy sources. *Pacific Institute for Climate Solutions, 32*, 1–6.

35. Espinosa-Romero, M. J., Chan, K. M., McDaniels, T., & Dalmer, D. M. (2011). Structuring decision-making for ecosystem-based management. *Marine Policy, 35*(5), 575–583.

36. Fertig, E., & Apt, J. (2011). Economics of compressed air energy storage to integrate wind power: A case study in ERCOT. *Energy Policy, 39*(5), 2330–2342.

37. Fischhoff, B. (2011). Applying the science of communication to the communication of science. *Climatic Change, 108*(4), 701–705.

38. Grossmann, I. (2011). Review of 'Informing Decisions in a Changing Climate' by the National Research Council. *The Quarterly Review of Biology*, Vol. 86, No. 4, 2011, 341–342.

39. Grossmann, I., & Morgan, M. G. (2011). Tropical cyclones, climate change, and scientific uncertainty: What do we know, what does it mean, and what should be done?. *Climatic Change, 108*(3), 543–579.

40. Hashambhoy, A. M., & Whitacre, J. F. (2011). Li diffusivity and phase change in $LiFe_{0.5}Mn_{0.5}PO_4$: A comparative study using galvanostatic intermittent titration and cyclic voltammetry. *Journal of the Electrochemical Society, 158*(4), A390.

41. Klima, K., Morgan, M. G., Grossmann, I., & Emanuel, K. (2011). Does it make sense to modify tropical cyclones? A decision-analytic assessment. *Environmental Science & Technology, 45*(10), 4242–4248.

42. Mantripragada, H. C., & Rubin, E. S. (2011). Techno-economic evaluation of coal-to-liquids (CTL) plants with carbon capture and sequestration. *Energy Policy, 39*(5), 2808–2816.

43. Michalek, J. J., Chester, M., Jaramillo, P., Samaras, C., Shiau, C. S. N., & Lave, L. B. (2011). Valuation of plug-in vehicle life-cycle air emissions and oil displacement benefits. *Proceedings of the National Academy of Sciences, 108*(40), 16554–16558.

44. Mitchell, A. L., & Casman, E. A. (2011). Economic incentives and regulatory framework for shale gas well site reclamation in Pennsylvania. *Environmental Science & Technology, 45*(22), 9506–9514.

45. Norman Shiau, C. S., & Michalek, J. J. (2011). Global optimization of plug-in hybrid vehicle design and allocation to minimize life cycle greenhouse gas emissions. *Journal of Mechanical Design, 133*(8).

46. Peterson, S. B., Whitacre, J. F., & Apt, J. (2011). Net air emissions from electric vehicles: The effect of carbon price and charging strategies. *Environmental Science & Technology, 45*(5), 1792–1797.

47. Schwietzke, S., Griffin, W. M., & Matthews, H. S. (2011). Relevance of emissions timing in biofuel greenhouse gases and climate impacts. *Environmental Science & Technology, 45*(19), 8197–8203.

48. Tokdar, S. T., Grossmann, I., Kadane, J. B., Charest, A. S., & Small, M. J. (2011). Impact of beliefs about Atlantic tropical cyclone detection on conclusions about trends in tropical cyclone numbers. *Bayesian Analysis, 6*(4), 547–572.

49.   Versteeg, P., & Rubin, E. S. (2011). A technical and economic assessment of ammonia-based post-combustion CO2 capture at coal-fired power plants. *International Journal of Greenhouse Gas Control, 5*(6), 1596–1605.

50.   Wiesmann, D., Azevedo, I. L., Ferrão, P., & Fernández, J. E. (2011). Residential electricity consumption in Portugal: Findings from top-down and bottom-up models. *Energy Policy, 39*(5), 2772–2779.

51.   Wong-Parodi, G., Dowlatabadi, H., McDaniels, T., & Ray, I. (2011). Influencing attitudes toward carbon capture and sequestration: A social marketing approach. *Environmental Science & Technology, 45*(16), 6743–6751.

52.   Zhai, H., Rubin, E. S., & Versteeg, P. L. (2011). Water use at pulverized coal power plants with postcombustion carbon capture and storage. *Environmental Science & Technology, 45*(6), 2479–2485.

## *2012*

53.   Azevedo, I. L., Sonnberger, M., Thomas, B. A., Morgan, G., & Renn, O. (2012). Developing robust energy efficiency policies while accounting for consumer behavior. *International Risk Governance Council (IRGC) report.*

54.   Baptista, P. C., Azevedo, I. L., & Farias, T. L. (2012). ICT solutions in transportation systems: Estimating the benefits and environmental impacts in the Lisbon. *Procedia-Social and Behavioral Sciences, 54,* 716–725.

55.   Changala, D., Dworkin, M., Apt, J., & Jaramillo, P. (2012). Comparative analysis of conventional oil and gas and wind project decommissioning regulations on federal, state, and county lands. *The Electricity Journal, 25*(1), 29–45.

56.   Cooley, S. R. (2012). How human communities could 'feel' changing ocean biogeochemistry. *Current Opinion in Environmental Sustainability, 4*(3), 258–263.

57.   Cooley, S. R., Lucey, N., Kite-Powell, H., & Doney, S. C. (2012). Nutrition and income from molluscs today imply vulnerability to ocean acidification tomorrow. *Fish and Fisheries, 13*(2), 182–215.

58.   Cooley, S., Doney, S. (2012). Ocean acidification. *Encyclopedia of Environmetrics, Second Edition,* 2012, 1828–1834. (DOI: 10:1002/9780470057339.vnn124).

59.   Doney, S. C., Ruckelshaus, M., Emmett Duffy, J., Barry, J. P., Chan, F., English, C. A., … & Talley, L. D. (2012). Climate change impacts on marine ecosystems. *Annual Review of Marine Science, 4,* 11–37.

60.   Fertig, E., Apt, J., Jaramillo, P., & Katzenstein, W. (2012). The effect of long-distance interconnection on wind power variability. *Environmental Research Letters, 7*(3), 034017.

61.   Fleishman, L. A., Bruine de Bruin, W., Morgan, M. G. (2012). The value of CCS public opinion research: A letter in response to Malone, Dooley, and Bradbury. *International Journal of Greenhouse Gas Control,* Vol. 7, 2012, 265–266.

62.   Grossmann, W., Steininger, K. W., Schmid, C., & Grossmann, I. (2012). Investment and employment from large-scale photovoltaics up to 2050. *Empirica, 39*(2), 165–189.

63. Halpern, B. S., Longo, C., Hardy, D., McLeod, K. L., Samhouri, J. F., Katona, S. K., … & Zeller, D. (2012). An index to assess the health and benefits of the global ocean. *Nature, 488*(7413), 615–620.

64. Hittinger, E., Mullins, K. A., & Azevedo, I. L. (2012). Electricity consumption and energy savings potential of video game consoles in the United States. *Energy Efficiency, 5*(4), 531–545.

65. Hittinger, E., Whitacre, J. F., & Apt, J. (2012). What properties of grid energy storage are most valuable?. *Journal of Power Sources, 206*, 436–449.

66. Irvine, P. J., Sriver, R. L., & Keller, K. (2012). Tension between reducing sea-level rise and global warming through solar-radiation management. *Nature Climate Change, 2*(2), 97–100.

67. Katzenstein, W., & Apt, J. (2012). The cost of wind power variability. *Energy Policy, 51*, 233–243.

68. Klima, K., & Morgan, M. G. (2012). Thoughts on whether government should steer a tropical cyclone if it could. *Journal of Risk Research, 15*(8), 1013–1020.

69. Klima, K., Bruine de Bruin, W., Morgan, M. G., & Grossmann, I. (2012). Public perceptions of hurricane modification. *Risk Analysis: An International Journal, 32*(7), 1194–1206.

70. Klima, K., Lin, N., Emanuel, K., Morgan, M. G., & Grossmann, I. (2012). Hurricane modification and adaptation in Miami-Dade County, Florida. *Environmental Science & Technology, 46*(2), 636–642.

71. Klinsky, S., Dowlatabadi, H., & McDaniels, T. (2012). Comparing public rationales for justice trade-offs in mitigation and adaptation climate policy dilemmas. *Global Environmental Change, 22*(4), 862–876.

72. Krishnamurti, T., Schwartz, D., Davis, A., Fischhoff, B., de Bruin, W. B., Lave, L., & Wang, J. (2012). Preparing for smart grid technologies: A behavioral decision research approach to understanding consumer expectations about smart meters. *Energy Policy, 41*, 790–797.

73. Lueken, C., Carvalho, P. M., & Apt, J. (2012). Distribution grid reconfiguration reduces power losses and helps integrate renewables. *Energy Policy, 48*, 260–273.

74. Lueken, C., Cohen, G. E., & Apt, J. (2012). Costs of solar and wind power variability for reducing CO2 emissions. *Environmental Science & Technology, 46*(17), 9761–9767.

75. Mashayekh, Y., Jaramillo, P., Samaras, C., Hendrickson, C. T., Blackhurst, M., MacLean, H. L., & Matthews, H. S. (2012). Potentials for sustainable transportation in cities to alleviate climate change impacts. *Environmental Science & Technology, 46*(5), 2529–2537.

76. Mauch, B., Carvalho, P. M., & Apt, J. (2012). Can a wind farm with CAES survive in the day-ahead market?. *Energy Policy, 48*, 584–593.

77. McClellan, J., Keith, D. W., & Apt, J. (2012). Cost analysis of stratospheric albedo modification delivery systems. *Environmental Research Letters, 7*(3), 034019.

78.  McDaniels, T., Mills, T., Gregory, R., & Ohlson, D. (2012). Using expert judgments to explore robust alternatives for forest management under climate change. *Risk Analysis: An International Journal*, *32*(12), 2098–2112.

79.  McDaniels, T., Mills, T., Gregory, R., & Ohlson, D. (2012). Using expert judgments to explore robust alternatives for forest management under climate change. *Risk Analysis: An International Journal*, *32*(12), 2098–2112.

80.  Michalek, J. J., Chester, M., & Samaras, C. (2012). Getting the most out of electric vehicle subsidies. *Issues In Science and Technology*, *28*(4), 25.

81.  Michel-Kerjan, E., Lemoyne de Forges, S., & Kunreuther, H. (2012). Policy tenure under the US national flood insurance program (NFIP). *Risk Analysis: An International Journal*, *32*(4), 644–658.

82.  Moreno-Cruz, J. B., Ricke, K. L., & Keith, D. W. (2012). A simple model to account for regional inequalities in the effectiveness of solar radiation management. *Climatic Change*, *110*(3), 649–668.

83.  Moreno-Cruz, J., Keith, D. W., (2012). Climate policy under uncertainty: A case for geoengineering. *Climatic Change*. (DOI 10.1007/s10584–012–0487–4).

84.  Patino-Echeverri, D., & Hoppock, D. C. (2012). Reducing the average cost of $CO_2$ capture by shutting-down the capture plant at times of high electricity prices. *International Journal of Greenhouse Gas Control*, *9*, 410–418.

85.  Patiño-Echeverri, D., & Hoppock, D. C. (2012). Reducing the energy penalty costs of postcombustion CCS systems with amine-storage. *Environmental Science & Technology*, *46*(2), 1243–1252.

86.  Patwardhan, A., Azevedo, I. Foran, T., Patankar, M. Rao, A. Raven, R. Samaras, C. Smith, A. Verbong, G., & Walawalkar, R. (2012). Transitions in Energy Systems. *Global Energy Assessment – Toward a Sustainable Future*. Cambridge University Press, Cambridge, UK and New York, NY, USA and the International Institute for Applied Systems Analysis, Laxenburg, Austria, pp. 1173–1202.

87.  Popova, O. H., Small, M. J., McCoy, S. T., Thomas, A. C., Karimi, B., Goodman, A., & Carter, K. M. (2012). Comparative analysis of carbon dioxide storage resource assessment methodologies. *Environmental Geosciences*, *19*(3), 105–124.

88.  Rehr, A. P., Small, M. J., Bradley, P., Fisher, W. S., Vega, A., Black, K., & Stockton, T. (2012). A decision support framework for science-based, multistakeholder deliberation: A coral reef example. *Environmental Management*, *50*(6), 1204–1218.

89.  Ricke, K. L., Rowlands, D. J., Ingram, W. J., Keith, D. W., & Morgan, M. G. (2012). Effectiveness of stratospheric solar-radiation management as a function of climate sensitivity. *Nature Climate Change*, *2*(2), 92–96.

90.  Rose, S., & Apt, J. (2012). Generating wind time series as a hybrid of measured and simulated data. *Wind Energy*, *15*(5), 699–715.

91.  Rose, S., Jaramillo, P., Small, M. J., Grossmann, I., & Apt, J. (2012). Quantifying the hurricane risk to offshore wind turbines. *Proceedings of the National Academy of Sciences*, *109*(9), 3247–3252.

92. Rose, S., Jaramillo, P., Small, M. J., Grossmann, I., & Apt, J. (2012). Quantifying the hurricane risk to offshore wind turbines. *Proceedings of the National Academy of Sciences, 109*(9), 3247–3252.

93. Rose, S., Jaramillo, P., Small, M. J., Grossmann, I., & Apt, J. (2012). Reply to Powell and Cocke: On the probability of catastrophic damage to offshore wind farms from hurricanes in the US Gulf Coast. *Proceedings of the National Academy of Sciences, 109*(33), E2193-E2194.

94. Rubin, E. S. (2012). Understanding the pitfalls of CCS cost estimates. *International Journal of Greenhouse Gas Control, 10*, 181–190.

95. Rubin, E. S., & Zhai, H. (2012). The cost of carbon capture and storage for natural gas combined cycle power plants. *Environmental Science & Technology, 46*(6), 3076–3084.

96. Rubin, E. S., Mantripragada, H., Marks, A., Versteeg, P., & Kitchin, J. (2012). The outlook for improved carbon capture technology. *Progress in Energy and Combustion Science, 38*(5), 630–671.

97. Schläpfer, F., & Fischhoff, B. (2012). Task familiarity and contextual cues predict hypothetical bias in a meta-analysis of stated preference studies. *Ecological Economics, 81*, 44–47.

98. Siler-Evans, K., Azevedo, I. L., & Morgan, M. G. (2012). Marginal emissions factors for the US electricity system. *Environmental Science & Technology, 46*(9), 4742–4748.

99. Siler-Evans, K., Morgan, M. G., & Azevedo, I. L. (2012). Distributed cogeneration for commercial buildings: Can we make the economics work?. *Energy Policy, 42*, 580–590.

100. Thomas, B. A., Azevedo, I. L., & Morgan, G. (2012). Edison Revisited: Should we use DC circuits for lighting in commercial buildings?. *Energy Policy, 45*, 399–411.

101. Traut, E., Hendrickson, C., Klampfl, E., Liu, Y., & Michalek, J. J. (2012). Optimal design and allocation of electrified vehicles and dedicated charging infrastructure for minimum life cycle greenhouse gas emissions and cost. *Energy Policy, 51*, 524–534.

102. Vignola, R., McDaniels, T. L., & Scholz, R. W. (2012). Negotiation analysis for mechanisms to deliver ecosystem services: The case of soil conservation in Costa Rica. *Ecological Economics, 75*, 22–31.

103. Wagner, S. J., & Rubin, E. S. (2012). Economic implications of thermal energy storage for concentrated solar thermal power. *Renewable Energy, 61*, 81–95.

104. Yang, M., Patiño-Echeverri, D., & Yang, F. (2012). Wind power generation in China: Understanding the mismatch between capacity and generation. *Renewable Energy, 41*, 145–151.

105. Yeh, S., & Rubin, E. S. (2012). A review of uncertainties in technology experience curves. *Energy Economics, 34*(3), 762–771.

## *2013*

106. Abdulla, A., Azevedo, I. L., & Morgan, M. G. (2013). Expert assessments of the cost of light water small modular reactors. *Proceedings of the National Academy of Sciences, 110*(24), 9686–9691.

107. Apt, J., & Secomandi, N. (2013). Preface to the special issue on energy modeling. *Socio-Economic Planning Sciences,* Vol. 47, No. 2, 2013, 75.

108. Bandyopadhyay, R., & Patiño-Echeverri, D. (2013). Optimal size and joint operation of wind power and CCS coal plant with amine storage. In *IIE Annual Conference. Proceedings* (p. 367). Institute of Industrial and Systems Engineers (IISE).

109. Bruine de Bruin, W, & Bostrom, A. (2013). Assessing what to address in science communication. *Proceedings of the National Academy of Sciences, 110*(Supplement 3), 14062–14068.

110. Davis, A. L., & Krishnamurti, T. (2013). The problems and solutions of predicting participation in energy efficiency programs. *Applied Energy, 111,* 277–287.

111. Davis, A. L., Krishnamurti, T., Fischhoff, B., & de Bruin, W. B. (2013). Setting a standard for electricity pilot studies. *Energy Policy, 62,* 401–409.

112. Detels, R., Gulliford, M., Karim, Q. A., & Tan, C. C. (2013). Risk perception and communication. *Oxford Textbook of Public Health, sixth edition.*

113. Fischhoff, B. (2013). The sciences of science communication. *Proceedings of the National Academy of Sciences, 110*(Supplement 3), 14033–14039.

114. Fleishman-Mayer, L. A., & de Bruin, W. B. (2013). The 'Mental Models' methodology for developing communications. *Effective Risk Communication,* 165.

115. Gattuso, J. P., Mach, K. J., & Morgan, G. (2013). Ocean acidification and its impacts: An expert survey. *Climatic Change, 117*(4), 725–738.

116. Grossmann, W. D., Grossmann, I., & Steininger, K. W. (2013). Distributed solar electricity generation across large geographic areas, Part I: A method to optimize site selection, generation and storage. *Renewable and Sustainable Energy Reviews, 25,* 831–843.

117. Halpern, B. S., Longo, C., McLeod, K. L., Cooke, R., Fischhoff, B., Samhouri, J. F., & Scarborough, C. (2013). Elicited preferences for components of ocean health in the California Current. *Marine Policy, 42,* 68–73.

118. Hoşgör, E., Apt, J., & Fischhoff, B. (2013). Incorporating seismic concerns in site selection for enhanced geothermal power generation. *Journal of Risk Research, 16*(8), 1021–1036.

119. Isley, S., Lempert, R., Popper, S., & Vardavas, R. (2013). An evolutionary model of industry transformation and the political sustainability of emission control policies. *RAND TR-1308.*

120. Jenn, A., Azevedo, I. L., & Ferreira, P. (2013). The impact of federal incentives on the adoption of hybrid electric vehicles in the United States. *Energy Economics, 40,* 936–942.

121. Jones, K. B., Bartell, S. J., Nugent, D., Hart, J., & Shrestha, A. (2013). The urban microgrid: Smart legal and regulatory policies to support electric grid resiliency and climate mitigation. *Fordham Urb. LJ, 41*, 1695.

122. Kadane, J. B., & Fischhoff, B. (2013). A cautionary note on global recalibration. *Judgment and Decision Making, 8*(1), 25.

123. Karabasoglu, O., & Michalek, J. (2013). Influence of driving patterns on life cycle cost and emissions of hybrid and plug-in electric vehicle powertrains. *Energy Policy, 60*, 445–461.

124. Krishnamurti, T., Davis, A. L., Wong-Parodi, G., Wang, J., & Canfield, C. (2013). Creating an in-home display: Experimental evidence and guidelines for design. *Applied Energy, 108*, 448–458.

125. Lima Azevedo, I., Morgan, M. G., Palmer, K., & Lave, L. B. (2013). Reducing US residential energy use and CO2 emissions: How much, how soon, and at what cost?. *Environmental Science & Technology, 47*(6), 2502–2511.

126. Mantripragada, H. C., & Rubin, E. S. (2013). Performance, cost and emissions of coal-to-liquids (CTLs) plants using low-quality coals under carbon constraints. *Fuel, 103*, 805–813.

127. Mauch, B., Apt, J., Carvalho, P. M., & Jaramillo, P. (2013). What day-ahead reserves are needed in electric grids with high levels of wind power?. *Environmental Research Letters, 8*(3), 034013.

128. Mauch, B., Apt, J., Carvalho, P. M., & Jaramillo, P. (2013). What day-ahead reserves are needed in electric grids with high levels of wind power?. *Environmental Research Letters, 8*(3), 034013.

129. Mauch, B., Apt, J., Carvalho, P. M., & Small, M. J. (2013). An effective method for modeling wind power forecast uncertainty. *Energy Systems, 4*(4), 393–417.

130. Mitchell, A. L., Small, M., & Casman, E. A. (2013). Surface water withdrawals for Marcellus Shale gas development: Performance of alternative regulatory approaches in the Upper Ohio River Basin. *Environmental Science & Technology, 47*(22), 12669–12678.

131. Moore, J., & Apt, J. (2013). Can hybrid solar-fossil power plants mitigate CO2 at lower cost than PV or CSP?. *Environmental Science & Technology, 47*(6), 2487–2493.

132. Morgan, M. G., Nordhaus, R. R., & Gottlieb, P. (2013). Needed: Research guidelines for solar radiation management. *Issues in Science and Technology, 29*(3), 37–44.

133. Oates, D. L., & Jaramillo, P. (2013). Production cost and air emissions impacts of coal cycling in power systems with large-scale wind penetration. *Environmental Research Letters, 8*(2), 024022.

134. Olson, R., Sriver, R., Chang, W., Haran, M., Urban, N. M., & Keller, K. (2013). What is the effect of unresolved internal climate variability on climate sensitivity estimates?. *Journal of Geophysical Research: Atmospheres, 118*(10), 4348–4358.

135. Patino-Echeverri, D. (2013). Feasibility of flexible technology standards for existing coal-fired power plants and their implications for new technology development. *UCLA Law Review*, *61*, 1896.

136. Patino-Echeverri, D., Burtraw, D., & Palmer, K. (2013). Flexible mandates for investment in new technology. *Journal of Regulatory Economics*, *44*(2), 121–155.

137. Peterson, S. B., & Michalek, J. J. (2013). Cost-effectiveness of plug-in hybrid electric vehicle battery capacity and charging infrastructure investment for reducing US gasoline consumption. *Energy Policy*, *52*, 429–438.

138. Rose, S., Jaramillo, P., Small, M. J., & Apt, J. (2013). Quantifying the hurricane catastrophe risk to offshore wind power. *Risk Analysis*, *33*(12), 2126–2141.

139. Ruckelshaus, M., Doney, S. C., Galindo, H. M., Barry, J. P., Chan, F., Duffy, J. E., … & Talley, L. D. (2013). Securing ocean benefits for society in the face of climate change. *Marine Policy*, *40*, 154–159.

140. Sakti, A., Michalek, J. J., Chun, S. E., & Whitacre, J. F. (2013). A validation study of lithium-ion cell constant c-rate discharge simulation with Battery Design Studio®. *International Journal of Energy Research*, *37*(12), 1562–1568.

141. Schwartz, D., Fischhoff, B., Krishnamurti, T., & Sowell, F. (2013). The Hawthorne effect and energy awareness. *Proceedings of the National Academy of Sciences*, *110*(38), 15242–15246.

142. Siler-Evans, K., Azevedo, I. L., Morgan, M. G., & Apt, J. (2013). Regional variations in the health, environmental, and climate benefits of wind and solar generation. *Proceedings of the National Academy of Sciences*, *110*(29), 11768–11773.

143. Tam, J., & McDaniels, T. L. (2013). Understanding individual risk perceptions and preferences for climate change adaptations in biological conservation. *Environmental Science & Policy*, *27*, 114–123.

144. Teehan, P., & Kandlikar, M. (2013). Comparing embodied greenhouse gas emissions of modern computing and electronics products. *Environmental Science & Technology*, *47*(9), 3997–4003.

145. Thomas, B. A., & Azevedo, I. L. (2013). Estimating direct and indirect rebound effects for US households with input–output analysis Part 1: Theoretical framework. *Ecological Economics*, *86*, 199–210.

146. Thomas, B. A., & Azevedo, I. L. (2013). Estimating direct and indirect rebound effects for US households with input–output analysis. Part 2: Simulation. *Ecological Economics*, *86*, 188–198.

147. Thomas, B. A., & Azevedo, I. L. (2013). Estimating direct and indirect rebound effects for US households with input–output analysis Part 1: Theoretical framework. *Ecological Economics*, *86*, 199–210.

148. Victor, D. G., Morgan, M. G., Apt, J., Steinbruner, J., & Ricke, K. L. (2013). The truth about geoengineering. *New York: Foreign Affairs*.

149. Vignola, R., Klinsky, S., Tam, J., & McDaniels, T. (2013). Public perception, knowledge and policy support for mitigation and adaption to climate change in Costa Rica: Comparisons with North American and European studies. *Mitigation and Adaptation Strategies for Global Change*, *18*(3), 303–323.

150. Vignola, R., McDaniels, T., & Scholz, Roland. (2013). Governance structures for ecosystem-based adaptation: Using policy-network analysis to identify key organizations for bridging information across scales and policy areas. *Environmental Science & Policy, 31*, 71–84. (DOI: 10.1016/j.envsci.2013.03.004).

151. Wong-Parodi, G., de Bruin, W. B., & Canfield, C. (2013). Effects of simplifying outreach materials for energy conservation programs that target low-income consumers. *Energy Policy, 62*, 1157–1164.

152. Zhai, H., & Rubin, E. S. (2013). Comparative performance and cost assessments of coal-and natural-gas-fired power plants under a CO2 emission performance standard regulation. *Energy & Fuels, 27*(8), 4290–4301.

153. Zhai, H., & Rubin, E. S. (2013). Techno-economic assessment of polymer membrane systems for postcombustion carbon capture at coal-fired power plants. *Environmental Science & Technology, 47*(6), 3006–3014.

154. Zhu, D., & Hug-Glanzmann, G. (2013). Coordination of storage and generation in power system frequency control using an H∞ approach. *IET Generation, Transmission & Distribution, 7*(11), 1263–1271.

## *2014*

155. Azevedo, I. M. (2014). Consumer end-use energy efficiency and rebound effects. *Annual Review of Environment and Resources, 39*, 393–418.

156. Azevedo, I. L. (2014). Energy efficiency and rebound effects: A review. *Annual Reviews of Energy and the Resources*, Vol. 39, 2014.

157. Bradbury, K., Pratson, L., & Patiño-Echeverri, D. (2014). Economic viability of energy storage systems based on price arbitrage potential in real-time US electricity markets. *Applied Energy, 114*, 512–519.

158. Bruine de Bruin, W., & Wong-Parodi, G. (2014). The role of initial affective impressions in responses to educational communications: The case of carbon capture and sequestration (CCS). *Journal of Experimental Psychology: Applied, 20*(2), 126.

159. Chang, S. E., McDaniels, T., Fox, J., Dhariwal, R., & Longstaff, H. (2014). Toward disaster-resilient cities: Characterizing resilience of infrastructure systems with expert judgments. *Risk Analysis, 34*(3), 416–434.

160. Daraeepour, A., & Echeverri, D. P. (2014). Day-ahead wind speed prediction by a Neural Network-based model. In *ISGT 2014* (pp. 1–5). IEEE.

161. Davis, A. L., & Fischhoff, B. (2014). Communicating uncertain experimental evidence. *Journal of Experimental Psychology: Learning, Memory, and Cognition, 40*(1), 261.

162. De Bruin, W. B., Wong-Parodi, G., & Morgan, M. G. (2014). Public perceptions of local flood risk and the role of climate change. *Environment Systems and Decisions, 34*(4), 591–599.

163. Fertig, E., Heggedal, A. M., Doorman, G., & Apt, J. (2014). Optimal investment timing and capacity choice for pumped hydropower storage. *Energy Systems, 5*(2), 285–306.

164. Fischhoff, B. (2014). Better decisions: From the lab to the real world. Policy Options, 43(3), 53–55.

165. Fischhoff, B. (2014). Four answers to four questions (about risk communication). *Journal of Risk Research, 17*(10), 1265–1267.

166. Fischhoff, B., & Davis, A. L. (2014). Communicating scientific uncertainty. *Proceedings of the National Academy of Sciences, 111*(Supplement 4), 13664–13671.

167. Gilbraith, N., Azevedo, I. L., & Jaramillo, P. (2014). Evaluating the benefits of commercial building energy codes and improving federal incentives for code adoption. *Environmental Science & Technology, 48*(24), 14121–14130.

168. Gilbraith, N., Azevedo, I. L., & Jaramillo, P. (2014). Regional energy and GHG savings from building codes across the United States. *Environmental Science & Technology, 48*, 14121–14130.

169. Halpern, B. S., Longo, C., Scarborough, C., Hardy, D., Best, B. D., Doney, S. C., … & Samhouri, J. F. (2014). Assessing the health of the US West coast with a regional-scale application of the ocean health index. *PLoS One, 9*(6), e98995.

170. Harish, S. M., Morgan, G. M., & Subrahmanian, E. (2014). When does unreliable grid supply become unacceptable policy? Costs of power supply and outages in rural India. *Energy Policy, 68*, 158–169.

171. Hittinger, E., Apt, J., & Whitacre, J. F. (2014). The effect of variability-mitigating market rules on the operation of wind power plants. *Energy Systems, 5*(4), 737–766.

172. Hoss, F., & Fischbeck, P. S. (2014). Performance and robustness of probabilistic river forecasts computed with quantile regression based on multiple independent variables in the North Central USA. *Hydrology and Earth System Sciences Discussions, 11*, 11281–11333.

173. Hoss, F., Klima, K., & Fischbeck, P. (2014). Ten strategies to systematically exploit all options to cope with anthropogenic climate change. *Environment Systems and Decisions, 34*(4), 578–590.

174. Kern, J. D., Patino-Echeverri, D., & Characklis, G. W. (2014). An integrated reservoir-power system model for evaluating the impacts of wind integration on hydropower resources. *Renewable Energy, 71*, 553–562.

175. Kern, J. D., Patino-Echeverri, D., & Characklis, G. W. (2014). The impacts of wind power integration on sub-daily variation in river flows downstream of hydroelectric dams. *Environmental Science & Technology, 48*(16), 9844–9851.

176. Lamy, J., Azevedo, I. L., & Jaramillo, P. (2014). The role of energy storage in accessing remote wind resources in the Midwest. *Energy Policy, 68*, 123–131.

177. Lueken, R., & Apt, J. (2014). The effects of bulk electricity storage on the PJM market. *Energy Systems, 5*(4), 677–704.

178. Mayer, L. A., Bruine de Bruin, W., & Morgan, M. G. (2014). Informed public choices for low-carbon electricity portfolios using a computer decision tool. *Environmental Science & Technology, 48*(7), 3640–3648.

179. Min, J., Azevedo, I. L., Michalek, J., & de Bruin, W. B. (2014). Labeling energy cost on light bulbs lowers implicit discount rates. *Ecological Economics*, *97*, 42–50.

180. Moore, J., & Apt, J. (2014). Consumer cost effectiveness of CO2 mitigation policies in restructured electricity markets. *Environmental Research Letters*, *9*(10), 104019.

181. Moore, J., Borgert, K., & Apt, J. (2014). Could low carbon capacity standards be more cost effective at reducing CO2 than renewable portfolio standards?. *Energy Procedia*, *63*, 7459–7470.

182. Morgan, M. G. (2014). Use (and abuse) of expert elicitation in support of decision making for public policy. *Proceedings of The National Academy of Sciences*, *111*(20), 7176–7184.

183. Oates, D. L., Versteeg, P., Hittinger, E., & Jaramillo, P. (2014). Profitability of CCS with flue gas bypass and solvent storage. *International Journal of Greenhouse Gas Control*, *27*, 279–288.

184. Peña, I., Azevedo, I. L., & Ferreira, L. A. F. M. (2014). Economic analysis of the profitability of existing wind parks in Portugal. *Energy Economics*, *45*, 353–363.

185. Peña, I., Gonzalez, E., Azevedo, I. L. (2014). Difusión de energía eólica: comparación de políticas de incentivos en Estados unidos y Europa. *Revista Nano Ciencia Y Tecnologia*, *2*(1), 18–25.

186. Ritchie, J., & Dowlatabadi, H. (2014). Understanding the shadow impacts of investment and divestment decisions: Adapting economic input–output models to calculate biophysical factors of financial returns. *Ecological Economics*, *106*, 132–140.

187. Rose, S., & Apt, J. (2014). The cost of curtailing wind turbines for secondary frequency regulation capacity. *Energy Systems*, *5*(3), 407–422.

188. Talati, S., Zhai, H., & Morgan, M. G. (2014). Water impacts of CO2 emission performance standards for fossil fuel-fired power plants. *Environmental Science & Technology*, *48*(20), 11769–11776.

189. Taylor, A. L., Dessai, S., & de Bruin, W. B. (2014). Public perception of climate risk and adaptation in the UK: A review of the literature. *Climate Risk Management*, *4*, 1–16.

190. Taylor, A., de Bruin, W. B., & Dessai, S. (2014). Climate change beliefs and perceptions of weather-related changes in the United Kingdom. *Risk Analysis*, *34*(11), 1995–2004.

191. Thomas, B. A., & Azevedo, I. L. (2014). Should policy-makers allocate funding to vehicle electrification or end-use energy efficiency as a strategy for climate change mitigation and energy reductions? Rethinking electric utilities efficiency programs. *Energy Policy*, *67*, 28–36.

192. Thomas, B. A., Hausfather, Z., & Azevedo, I. L. (2014). Comparing the magnitude of simulated residential rebound effects from electric end-use efficiency across the US. *Environmental Research Letters*, *9*(7), 074010.

193. Thomas, B. A., & Azevedo, I. L. (2014). Should policy-makers allocate funding to vehicle electrification or end-use energy efficiency as a strategy for climate change mitigation and energy reductions? Rethinking electric utilities efficiency programs. *Energy Policy, 67*, 28–36.

194. Vaishnav, P. (2014). Costs and benefits of reducing fuel burn and emissions from taxiing aircraft: Low-hanging fruit?. *Transportation Research Record, 2400*(1), 65–77.

195. Vaishnav, P. (2014). Greenhouse gas emissions from international transport. *Issues Sci. Technol, 30*(2), 25–28.

196. Weis, A., Jaramillo, P., & Michalek, J. (2014). Estimating the potential of controlled plug-in hybrid electric vehicle charging to reduce operational and capacity expansion costs for electric power systems with high wind penetration. *Applied Energy, 115*, 190–204.

197. Wong-Parodi, G., Fischhoff, B., & Strauss, B. (2014). A method to evaluate the usability of interactive climate change impact decision aids. *Climatic Change, 126*(3), 485–493.

## *2015*

198. Abrahams, L. S., Samaras, C., Griffin, W. M., & Matthews, H. S. (2015). Life cycle greenhouse gas emissions from US liquefied natural gas exports: Implications for end uses. *Environmental Science & Technology, 49*(5), 3237–3245.

199. Bradford, K., Abrahams, L., Hegglin, M., & Klima, K. (2015). A heat vulnerability index and adaptation solutions for Pittsburgh, Pennsylvania. *Environmental Science & Technology, 49*(19), 11303–11311.

200. Bronfman, N. C., Cisternas, P. C., López-Vázquez, E., Maza, C. D. L., & Oyanedel, J. C. (2015). Understanding attitudes and pro-environmental behaviors in a Chilean community. *Sustainability, 7*(10), 14133–14152.

201. Bruine de Bruin, W., Krishnamurti, T. (2015). Developing interventions about residential electricity use: Three lessons learned. *Delivering Energy Policy in the UK and the US: A Multi-Disciplinary Reader.*

202. Bruine de Bruin, W., Mayer, L. A., & Morgan, M. G. (2015). Developing communications about CCS: Three lessons learned. *Journal of Risk Research, 18*(6), 699–705.

203. Cooley, S. R., Rheuban, J. E., Hart, D. R., Luu, V., Glover, D. M., Hare, J. A., & Doney, S. C. (2015). An integrated assessment model for helping the United States sea scallop (Placopecten magellanicus) fishery plan ahead for ocean acidification and warming. *PLoS One, 10*(5), e0124145.

204. Dale, A. L., Casman, E. A., Lowry, G. V., Lead, J. R., Viparelli, E., & Baalousha, M. (2015). Modeling nanomaterial environmental fate in Aquatic Systems. *Environmental Science & Technology, 49*(5), 2587–2593.

205. Dale, A. L., Lowry, G. V., & Casman, E. A. (2015). Stream dynamics and chemical transformations control the environmental fate of silver and zinc oxide

nanoparticles in a watershed-scale model. *Environmental Science & Technology*, *49*(12), 7285–7293.

206. Daraeepour, A., Kazempour, S. J., Patiño-Echeverri, D., & Conejo, A. J. (2015). Strategic demand-side response to wind power integration. *IEEE Transactions on Power Systems*, *31*(5), 3495–3505.

207. De Faria, F. A., Jaramillo, P., Sawakuchi, H. O., Richey, J. E., & Barros, N. (2015). Estimating greenhouse gas emissions from future Amazonian hydroelectric reservoirs. *Environmental Research Letters*, *10*(12), 124019.

208. Dewitt, B., Fischhoff, B., Davis, A., & Broomell, S. B. (2015). Environmental risk perception from visual cues: The psychophysics of tornado risk perception. *Environmental Research Letters*, *10*(12), 124009.

209. Faria, F., Klima, K., Posen, I. D., & Azevedo, I. M. (2015). A new approach of science, technology, engineering, and mathematics outreach in climate change, energy, and environmental decision making. *Sustainability: The Journal of Record*, *8*(5), 261–271.

210. Fischhoff, B. (2015). Environmental Cognition, Perception, and Attitudes. *International Encyclopedia of the Social and Behavioral Sciences*, 7. (DOI: 4596–4602. 10.1016/B978–0–08–097086–8.91012–2).

211. Fischhoff, B. (2015). The realities of risk-cost-benefit analysis. *Science*, *350*(6260).

212. Helveston, J. P., Liu, Y., Feit, E. M., Fuchs, E., Klampfl, E., & Michalek, J. J. (2015). Will subsidies drive electric vehicle adoption? Measuring consumer preferences in the US and China. *Transportation Research Part A: Policy and Practice*, *73*, 96–112.

213. Heo, J., McCoy, S. T., & Adams, P. J. (2015). Implications of ammonia emissions from post-combustion carbon capture for airborne particulate matter. *Environmental Science & Technology*, *49*(8), 5142–5150.

214. Hittinger, E. S., & Azevedo, I. M. (2015). Bulk energy storage increases United States electricity system emissions. *Environmental Science & Technology*, *49*(5), 3203–3210.

215. Jenn, A., Azevedo, I. L., & Fischbeck, P. (2015). How will we fund our roads? A case of decreasing revenue from electric vehicles. *Transportation Research Part A: Policy and Practice*, *74*, 136–147.

216. Khalilpour, R., Mumford, K., Zhai, H., Abbas, A., Stevens, G., & Rubin, E. S. (2015). Membrane-based carbon capture from flue gas: A review. *Journal of Cleaner Production*, *103*, 286–300.

217. Klima, K., & Morgan, M. G. (2015). Ice storm frequencies in a warmer climate. *Climatic Change*, *133*(2), 209–222.

218. Markolf, S. A., Klima, K., & Wong, T. L. (2015). Adaptation frameworks used by US decision-makers: A literature review. *Environment Systems and Decisions*, *35*(4), 427–436.

219. Mathis, J. T., Cooley, S. R., Lucey, N., Colt, S., Ekstrom, J., Hurst, T., ... & Feely, R. A. (2015). Ocean acidification risk assessment for Alaska's fishery sector. *Progress in Oceanography*, *136*, 71–91.

220. Min, J., Azevedo, I. L., & Hakkarainen, P. (2015). Assessing regional differences in lighting heat replacement effects in residential buildings across the United States. *Applied Energy, 141*, 12–18.

221. Min, J., Azevedo, I., Morgan, A., Hakkarainen, P. (2015). Net carbon emissions savings and energy reductions from lighting energy efficiency measures when accounting for changes in heating and cooling demands: A regional comparison. *Applied Energy, 414*(1), 12–15.

222. Morgan, M. G. (2015). Our knowledge of the world is often not simple: Policymakers should not duck that fact, but should deal with it. *Risk Analysis, 35*(1), 19–20.

223. Namazu, M., & Dowlatabadi, H. (2015). Characterizing the GHG emission impacts of carsharing: A case of Vancouver. *Environmental Research Letters, 10*(12), 124017.

224. Posen, I. D., Griffin, W. M., Matthews, H. S., & Azevedo, I. L. (2015). Changing the renewable fuel standard to a renewable material standard: Bioethylene case study. *Environmental Science & Technology, 49*(1), 93–102.

225. Prasad, S., Abdulla, A., Morgan, M. G., & Azevedo, I. L. (2015). Nonproliferation improvements and challenges presented by small modular reactors. *Progress in Nuclear Energy, 80*, 102–109.

226. Rose, S., & Apt, J. (2015). What can reanalysis data tell us about wind power?. *Renewable Energy, 83*, 963–969.

227. Rubin, E. S., Azevedo, I. M., Jaramillo, P., & Yeh, S. (2015). A review of learning rates for electricity supply technologies. *Energy Policy, 86*, 198–218.

228. Rubin, E. S., Davison, J. E., & Herzog, H. J. (2015). The cost of CO2 capture and storage. *International Journal of Greenhouse Gas Control, 40*, 378–400.

229. Sakti, A., Michalek, J. J., Fuchs, E. R., & Whitacre, J. F. (2015). A techno-economic analysis and optimization of Li-ion batteries for light-duty passenger vehicle electrification. *Journal of Power Sources, 273*, 966–980.

230. Schwartz, D., Bruine de Bruin, W., Fischhoff, B., & Lave, L. (2015). Advertising energy saving programs: The potential environmental cost of emphasizing monetary savings. *Journal of Experimental Psychology: Applied, 21*(2), 158.

231. Seshadri, A. K. (2015). Nonconvex equilibrium prices in prediction markets. *Decision Analysis, 12*(1), 1–14.

232. Tamayao, M. A. M., Michalek, J. J., Hendrickson, C., & Azevedo, I. M. (2015). Regional variability and uncertainty of electric vehicle life cycle CO2 emissions across the United States. *Environmental Science & Technology, 49*(14), 8844–8855.

233. Tong, F., Jaramillo, P., & Azevedo, I. M. (2015). Comparison of life cycle greenhouse gases from natural gas pathways for medium and heavy-duty vehicles. *Environmental Science & Technology, 49*(12), 7123–7133.

234. Van Den Broek, M., Berghout, N., & Rubin, E. S. (2015). The potential of renewables versus natural gas with CO2 capture and storage for power generation under CO2 constraints. *Renewable and Sustainable Energy Reviews, 49*, 1296–1322.

235. Weis, A., Michalek, J. J., Jaramillo, P., & Lueken, R. (2015). Emissions and cost implications of controlled electric vehicle charging in the US PJM interconnection. *Environmental Science & Technology*, *49*(9), 5813–5819.

236. Williams, N. J., Jaramillo, P., Taneja, J., & Ustun, T. S. (2015). Enabling private sector investment in microgrid-based rural electrification in developing countries: A review. *Renewable and Sustainable Energy Reviews*, *52*, 1268–1281.

237. Yu, X., Moreno-Cruz, J., & Crittenden, J. C. (2015). Regional energy rebound effect: The impact of economy-wide and sector level energy efficiency improvement in Georgia, USA. *Energy Policy*, *87*, 250–259.

238. Yuksel, T., & Michalek, J. J. (2015). Effects of regional temperature on electric vehicle efficiency, range, and emissions in the United States. *Environmental Science & Technology*, *49*(6), 3974–3980.

239. Zhai, H., & Rubin, E. S. (2015). Water impacts of a low-carbon electric power future: Assessment methodology and status. *Current Sustainable/Renewable Energy Reports*, *2*(1), 1–9.

240. Zhai, H., Ou, Y., & Rubin, E. S. (2015). Opportunities for decarbonizing existing US coal-fired power plants via CO2 capture, utilization and storage. *Environmental Science & Technology*, *49*(13), 7571–7579.

## *2016*

241. Alqahtani, B. J., & Patiño-Echeverri, D. (2016). Integrated solar combined cycle power plants: Paving the way for thermal solar. *Applied Energy*, *169*, 927–936.

242. Alqahtani, B. J., Holt, K. M., Patiño-Echeverri, D., & Pratson, L. (2016). Residential solar PV systems in the Carolinas: Opportunities and outcomes. *Environmental Science & Technology*, *50*(4), 2082–2091.

243. Bandyopadhyay, R., & Patiño-Echeverri, D. (2016). An alternate wind power integration mechanism: Coal plants with flexible amine-based CCS. *Renewable Energy*, *85*, 704–713.

244. Haaf, C. G., Morrow, W. R., Azevedo, I. M., Feit, E. M., & Michalek, J. J. (2016). Forecasting light-duty vehicle demand using alternative-specific constants for endogeneity correction versus calibration. *Transportation Research Part B: Methodological*, *84*, 182–210.

245. Hagerman, S., Jaramillo, P., & Morgan, M. G. (2016). Is rooftop solar PV at socket parity without subsidies?. *Energy Policy*, *89*, 84–94.

246. Heo, J., Adams, P. J., & Gao, H. O. (2016). Public health costs of primary PM2.5 and inorganic PM2.5 precursor emissions in the United States. *Environmental Science & Technology*, *50*(11), 6061–6070.

247. Heo, J., Adams, P. J., & Gao, H. O. (2016). Reduced-form modeling of public health impacts of inorganic PM2.5 and precursor emissions. *Atmospheric Environment*, *137*, 80–89.

248. Horner, N., & Azevedo, I. (2016). Power usage effectiveness in data centers: Overloaded and underachieving. *The Electricity Journal*, *29*(4), 61–69.

249. Jenn, A., Azevedo, I. M., & Michalek, J. J. (2016). Alternative fuel vehicle adoption increases fleet gasoline consumption and greenhouse gas emissions under United States corporate average fuel economy policy and greenhouse gas emissions standards. *Environmental Science & Technology*, *50*(5), 2165–2174.

250. Klima, K., & Apt, J. (2016). Geographic smoothing of solar PV: Results from Gujarat. *Environmental Research Letters*, *10*(10), 104001.

251. Lamy, J. V., Jaramillo, P., Azevedo, I. L., & Wiser, R. (2016). Should we build wind farms close to load or invest in transmission to access better wind resources in remote areas? A case study in the MISO region. *Energy Policy*, *96*, 341–350.

252. Lovering, J. R., Yip, A., & Nordhaus, T. (2016). Historical construction costs of global nuclear power reactors. *Energy Policy*, *91*, 371–382.

253. Lueken, R., Apt, J., & Sowell, F. (2016). Robust resource adequacy planning in the face of coal retirements. *Energy Policy*, *88*, 371–388.

254. Lueken, R., Klima, K., Griffin, W. M., & Apt, J. (2016). The climate and health effects of a USA switch from coal to gas electricity generation. *Energy*, *109*, 1160–1166.

255. Mersky, A. C., & Samaras, C. (2016). Fuel economy testing of autonomous vehicles. *Transportation Research Part C: Emerging Technologies*, *65*, 31–48.

256. Ou, Y., Zhai, H., & Rubin, E. S. (2016). Life cycle water use of coal-and natural-gas-fired power plants with and without carbon capture and storage. *International Journal of Greenhouse Gas Control*, *44*, 249–261.

257. Posen, I. D., Jaramillo, P., & Griffin, W. M. (2016). Uncertainty in the life cycle greenhouse gas emissions from US production of three biobased polymer families. *Environmental Science & Technology*, *50*(6), 2846–2858.

258. Rose, S., & Apt, J. (2016). Quantifying sources of uncertainty in reanalysis derived wind speed. *Renewable Energy*, *94*, 157–165.

259. Schweizer, V. J., & Morgan, M. G. (2016). Bounding US electricity demand in 2050. *Technological Forecasting and Social Change*, *105*, 215–223.

260. Teller-Elsberg, J., Sovacool, B., Smith, T., & Laine, E. (2016). Fuel poverty, excess winter deaths, and energy costs in Vermont: Burdensome for whom?. *Energy Policy*, *90*, 81–91.

261. Vaishnav, P., Fischbeck, P. S., Morgan, M. G., & Corbett, J. J. (2016). Shore power for vessels calling at US ports: Benefits and costs. *Environmental Science & Technology*, *50*(3), 1102–1110.

262. Vaishnav, P., Petsonk, A., Avila, R. A. G., Morgan, M. G., & Fischbeck, P. S. (2016). Analysis of a proposed mechanism for carbon-neutral growth in international aviation. *Transportation Research Part D: Transport and Environment*, *45*, 126–138.

263. Weis, A., Jaramillo, P., & Michalek, J. (2016). Consequential life cycle air emissions externalities for plug-in electric vehicles in the PJM interconnection. *Environmental Research Letters*, *11*(2), 024009.

264. Wong-Parodi, G., Krishnamurti, T., Davis, A., Schwartz, D., & Fischhoff, B. (2016). A decision science approach for integrating social science in climate and energy solutions. *Nature Climate Change, 6*(6), 563–569.

265. Yuksel, T., Tamayao, M. A. M., Hendrickson, C., Azevedo, I. M., & Michalek, J. J. (2016). Effect of regional grid mix, driving patterns and climate on the comparative carbon footprint of gasoline and plug-in electric vehicles in the United States. *Environmental Research Letters, 11*(4), 044007.

266. Zhai, H., & Rubin, E. S. (2016). A techno-economic assessment of hybrid cooling systems for coal-and natural-gas-fired power plants with and without carbon capture and storage. *Environmental Science & Technology, 50*(7), 4127–4134.

## *2017*

267. Abrahams, L. S., Samaras, C., Griffin, W. M., & Matthews, H. S. (2017). Effect of crude oil carbon accounting decisions on meeting global climate budgets. *Environment Systems and Decisions, 37*(3), 261–275.

268. Canfield, C., Bruine de Bruin, W., & Wong-Parodi, G. (2017). Perceptions of electricity-use communications: effects of information, format, and individual differences. *Journal of Risk Research, 20*, 1132–1153.

269. Craig, M. T., Jaramillo, P., Zhai, H., & Klima, K. (2017). The economic merits of flexible carbon capture and sequestration as a compliance strategy with the clean power plan. *Environmental Science & Technology, 51*(3), 1102–1109.

270. Drummond, C., & Fischhoff, B. (2017). Development and validation of the scientific reasoning scale. *Journal of Behavioral Decision Making, 30*(1), 26–38.

271. Drummond, C., & Fischhoff, B. (2017). Individuals with greater science literacy and education have more polarized beliefs on controversial science topics. *Proceedings of the National Academy of Sciences, 114*(36), 9587–9592.

272. Fischhoff, B. (2017). Breaking ground for psychological science: The US Food and Drug Administration. *American Psychologist, 72*(2), 118.

273. Fisher, M. J., & Apt, J. (2017). Emissions and economics of behind-the-meter electricity storage. *Environmental Science & Technology, 51*(3), 1094–1101.

274. Ford, M. J., Abdulla, A., & Morgan, M. G. (2017). Evaluating the cost, safety, and proliferation risks of small floating nuclear reactors. *Risk Analysis, 37*(11), 2191–2211.

275. Ford, M. J., Abdulla, A., Morgan, M. G., & Victor, D. G. (2017). Expert assessments of the state of US advanced fission innovation. *Energy Policy, 108*, 194–200.

276. Gentner, D. R., Jathar, S. H., Gordon, T. D., Bahreini, R., Day, D. A., El Haddad, I., … & Robinson, A. L. (2017). Review of urban secondary organic aerosol formation from gasoline and diesel motor vehicle emissions. *Environmental Science & Technology, 51*(3), 1074–1093.

277. Gingerich, D. B., Sun, X., Behrer, A. P., Azevedo, I. L., & Mauter, M. S. (2017). Spatially resolved air-water emissions tradeoffs improve regulatory impact

analyses for electricity generation. *Proceedings of the National Academy of Sciences*, *114*(8), 1862–1867.

278. Handschy, M. A., Rose, S., & Apt, J. (2017). Is it always windy somewhere? Occurrence of low-wind-power events over large areas. *Renewable Energy*, *101*, 1124–1130.

279. Jathar, S. H., Woody, M., Pye, H. O., Baker, K. R., & Robinson, A. L. (2017). Chemical transport model simulations of organic aerosol in southern California: Model evaluation and gasoline and diesel source contributions. *Atmospheric Chemistry and Physics*, *17*(6), 4305–4318.

280. Mangones, S. C., Fischbeck, P., & Jaramillo, P. (2017). Safety-related risk and benefit-cost analysis of crash avoidance systems applied to transit buses: Comparing New York City vs. Bogota, Colombia. *Safety Science*, *91*, 122–131.

281. Markolf, S. A., Matthews, H. S., Azevedo, I. L., & Hendrickson, C. (2017). An integrated approach for estimating greenhouse gas emissions from 100 US metropolitan areas. *Environmental Research Letters*, *12*(2), 024003.

282. Mayfield, E. N., Robinson, A. L., & Cohon, J. L. (2017). System-wide and superemitter policy options for the abatement of methane emissions from the US natural gas system. *Environmental Science & Technology*, *51*(9), 4772–4780.

283. Morgan, M. G., Vaishnav, P., Dowlatabadi, H., & Azevedo, I. L. (2017). Rethinking the social cost of carbon dioxide. *Issues in Science and Technology*, *33*(4), 43–50.

284. Namhata, A., Small, M. J., Dilmore, R. M., Nakles, D. V., & King, S. (2017). Bayesian inference for heterogeneous caprock permeability based on above zone pressure monitoring. *International Journal of Greenhouse Gas Control*, *57*, 89–101.

285. Orak, N. H., & Small, M. J. (2017). Implications of a statistical occurrence model for mixture toxicity estimation. *Human and Ecological Risk Assessment: An International Journal*, *23*(3), 534–549.

286. Peña, I., Azevedo, I. L., & Ferreira, L. A. F. M. (2017). Lessons from wind policy in Portugal. *Energy Policy*, *103*, 193–202.

287. Posen, I. D., Jaramillo, P., Landis, A. E., & Griffin, W. M. (2017). Greenhouse gas mitigation for US plastics production: Energy first, feedstocks later. *Environmental Research Letters*, *12*(3), 034024.

288. Ritchie, J., & Dowlatabadi, H. (2017). The 1000 GtC coal question: Are cases of vastly expanded future coal combustion still plausible?. *Energy Economics*, *65*, 16–31.

289. Roohani, Y. H., Roy, A. A., Heo, J., Robinson, A. L., & Adams, P. J. (2017). Impact of natural gas development in the Marcellus and Utica shales on regional ozone and fine particulate matter levels. *Atmospheric Environment*, *155*, 11–20.

290. Sakti, A., Azevedo, I. M., Fuchs, E. R., Michalek, J. J., Gallagher, K. G., & Whitacre, J. F. (2017). Consistency and robustness of forecasting for emerging

technologies: The case of Li-ion batteries for electric vehicles. *Energy Policy*, *106*, 415–426.

291. Vaishnav, P., Horner, N., & Azevedo, I. L. (2017). Was it worthwhile? Where have the benefits of rooftop solar photovoltaic generation exceeded the cost?. *Environmental Research Letters*, *12*(9), 094015.

292. Wang, Y., Small, M. J., & VanBriesen, J. M. (2017). Assessing the risk associated with increasing bromide in drinking water sources in the Monongahela River, Pennsylvania. *Journal of Environmental Engineering*, *143*(3), 04016089.

293. Welle, P. D., Small, M. J., Doney, S. C., & Azevedo, I. L. (2017). Estimating the effect of multiple environmental stressors on coral bleaching and mortality. *PLoS One*, *12*(5), e0175018.

294. Wong-Parodi, G., Fischhoff, B., & Strauss, B. (2017). Plans and prospects for coastal flooding in four communities affected by Sandy. *Weather, Climate, and Society*, *9*(2), 183–200.

## *2018*

295. Cornelius, A., Bandyopadhyay, R., & Patiño-Echeverri, D. (2018). Assessing environmental, economic, and reliability impacts of flexible ramp products in MISO's electricity market. *Renewable and Sustainable Energy Reviews*, *81*, 2291–2298.

296. Davis, S. J., Lewis, N. S., Shaner, M., Aggarwal, S., Arent, D., Azevedo, I. L., … & Caldeira, K. (2018). Net-zero emissions energy systems. *Science*, *360*(6396).

297. De la Maza, C., Davis, A., Gonzalez, C., Azevedo, I. L. (2018). Understanding cumulative risk perception form judgements and choices: An application to flood risks. *Accepted, Risk Analysis*.

298. Fisher, M., Apt, J., & Sowell, F. (2018). The economics of commercial demand response for spinning reserve. *Energy Systems*, *9*(1), 3–23.

299. Fisher, M., Whitacre, J., & Apt, J. (2018). A simple metric for predicting revenue from electric peak-shaving and optimal battery sizing. *Energy Technology*, *6*(4), 649–657.

300. Glasgo, B., Azevedo, I. L., & Hendrickson, C. (2018). Expert assessments on the future of direct current in buildings. *Environmental Research Letters*, *13*(7), 074004.

301. Hanus, N., Wong-Parodi, G., Hoyos, L., & Rauch, M. (2018). Framing clean energy campaigns to promote civic engagement among parents. *Environmental Research Letters*, *13*(3), 034021.

302. Kaack, L. H., Vaishnav, P., Morgan, M. G., Azevedo, I. L., & Rai, S. (2018). Decarbonizing intraregional freight systems with a focus on modal shift. *Environmental Research Letters*, *13*(8), 083001.

303. Lam, L. T., Branstetter, L., & Azevedo, I. L. (2018). A sunny future: Expert elicitation of China's solar photovoltaic technologies. *Environmental Research Letters*, *13*(3), 034038.

304. Lamy, J. V., & Azevedo, I. L. (2018). Do tidal stream energy projects offer more value than offshore wind farms? A case study in the United Kingdom. *Energy Policy*, *113*, 28–40.

305. Lesic, V., De Bruin, W. B., Davis, M. C., Krishnamurti, T., & Azevedo, I. M. (2018). Consumers' perceptions of energy use and energy savings: A literature review. *Environmental Research Letters*, *13*(3), 033004.

306. Masnadi, M. S., El-Houjeiri, H. M., Schunack, D., Li, Y., Englander, J. G., Badahdah, A., ... & Brandt, A. R. (2018). Global carbon intensity of crude oil production. *Science*, *361*(6405), 851–853.

307. Murphy, S., Apt, J., Moura, J., & Sowell, F. (2018). Resource adequacy risks to the bulk power system in North America. *Applied Energy*, *212*, 1360–1376.

308. Namazu, M., & Dowlatabadi, H. (2018). Vehicle ownership reduction: A comparison of one-way and two-way carsharing systems. *Transport Policy*, *64*, 38–50.

309. Namazu, M., MacKenzie, D., Zerriffi, H., & Dowlatabadi, H. (2018). Is carsharing for everyone? Understanding the diffusion of carsharing services. *Transport Policy*, *63*, 189–199.

310. Namazu, M., Zhao, J., & Dowlatabadi, H. (2018). Nudging for responsible carsharing: Using behavioral economics to change transportation behavior. *Transportation*, *45*(1), 105–119.

311. Prata, R., Carvalho, P. M., & Azevedo, I. L. (2018). Distributional costs of wind energy production in Portugal under the liberalized Iberian market regime. *Energy Policy*, *113*, 500–512.

312. Ritchie, J., & Dowlatabadi, H. (2018). Defining climate change scenario characteristics with a phase space of cumulative primary energy and carbon intensity. *Environmental Research Letters*, *13*(2), 024012.

313. Schivley, G., Samaras, C., & Azevedo, I. L. (2018). Emissions intensity of the U.S. power system. *Environmental Research Letters*, *3*.

314. Seki, S. M., Griffin, W. M., Hendrickson, C., & Matthews, H. S. (2018). Refueling and infrastructure costs of expanding access to E85 in Pennsylvania. *Journal of Infrastructure Systems*, *24*(1), 04017045.

315. Sergi, B., Davis, A., & Azevedo, I. (2018). The effect of providing climate and health information on support for alternative electricity portfolios. *Environmental Research Letters*, *13*(2), 024026.

316. Sherwin, E. D., Henrion, M., & Azevedo, I. M. (2018). Estimation of the year-on-year volatility and the unpredictability of the United States energy system. *Nature Energy*, *3*(4), 341–346.

317. Stolaroff, J. K., Samaras, C., O'Neill, E. R., Lubers, A., Mitchell, A. S., & Ceperley, D. (2018). Energy use and life cycle greenhouse gas emissions of drones for commercial package delivery. *Nature Communications*, *9*(1), 1–13.

318. Tong, F., Jaramillo, P., & Azevedo, I. L., (2018). Should we build a national infrastructure to refuel natural gas-powered trucks? *Journal of Industrial Systems, 11.*

319. Trutnevyte, E., & Azevedo, I. L. (2018). Induced seismicity hazard and risk by enhanced geothermal systems: An expert elicitation approach. *Environmental Research Letters*, *13*(3), 034004.

320. Wong-Parodi, G., Fischhoff, B., & Strauss, B. (2018). Effect of risk and protective decision aids on flood preparation in vulnerable communities. *Weather, Climate, and Society*, *10*(3), 401–417.

## *2019*

321. Abdulla, A., Vaishnav, P., Sergi, B., & Victor, D. G. (2019). Limits to deployment of nuclear power for decarbonization: Insights from public opinion. *Energy Policy*, *129*, 1339–1346.

322. Allen, T., Wells, E., & Klima, K. (2019). Culture and cognition: Understanding public perceptions of risk and (in) action. *IBM Journal of Research and Development*, *64*(1/2), 11:1–11:17.

323. Alqahtani, B. J., & Patiño-Echeverri, D. (2019). Combined effects of policies to increase energy efficiency and distributed solar generation: A case study of the Carolinas. *Energy Policy*, *134*, 110936.

324. Baik, S., Davis, A. L., & Morgan, M. G. (2019). Illustration of a method to incorporate preference uncertainty in benefit–cost analysis. *Risk Analysis*, *39*(11), 2359–2368.

325. Bicalho, T., Sauer, I., & Patiño-Echeverri, D. (2019). Quality of data for estimating GHG emissions in biofuel regulations is unknown: A review of default values related to sugarcane and corn ethanol. *Journal of Cleaner Production*, *239*, 117903.

326. Bostrom, A., Hayes, A. L., & Crosman, K. M. (2019). Efficacy, action, and support for reducing climate change risks. *Risk Analysis*, *39*(4), 805–828.

327. Camilleri, A. R., Larrick, R. P., Hossain, S., & Patino-Echeverri, D. (2019). Consumers underestimate the emissions associated with food but are aided by labels. *Nature Climate Change*, *9*(1), 53–58.

328. Daraeepour, A., Patino-Echeverri, D., & Conejo, A. J. (2019). Economic and environmental implications of different approaches to hedge against wind production uncertainty in two-settlement electricity markets: A PJM case study. *Energy Economics*, *80*, 336–354.

329. Davis, A., Wong-Parodi, G., & Krishnamurti, T. (2019). Neither a borrower nor a lender be: Beyond cost in energy efficiency decision-making among office buildings in the United States. *Energy Research & Social Science*, *47*, 37–45.

330. de Bruin, W. B., & Morgan, M. G. (2019). Reflections on an interdisciplinary collaboration to inform public understanding of climate change, mitigation, and impacts. *Proceedings of the National Academy of Sciences*, *116*(16), 7676–7683.

331. De La Maza, C., Davis, A., Gonzalez, C., & Azevedo, I. (2019). Understanding cumulative risk perception from judgments and choices: An application to flood risks. *Risk Analysis*, *39*(2), 488–504.

332.  Deetjen, T. A., & Azevedo, I. L. (2019). Reduced-order dispatch model for simulating marginal emissions factors for the United States power sector. *Environmental Science & Technology, 53*(17), 10506–10513.

333.  Donti, P. L., Kolter, J. Z., & Azevedo, I. L. (2019). How much are we saving after all? Characterizing the effects of commonly varying assumptions on emissions and damage estimates in PJM. *Environmental Science & Technology, 53*(16), 9905–9914.

334.  Fisher, M., Apt, J., & Whitacre, J. F. (2019). Can flow batteries scale in the behind-the-meter commercial and industrial market? A techno-economic comparison of storage technologies in California. *Journal of Power Sources, 420*, 1–8.

335.  Hanus, N. L., Wong-Parodi, G., Vaishnav, P. T., Darghouth, N. R., & Azevedo, I. L. (2019). Solar PV as a mitigation strategy for the US education sector. *Environmental Research Letters, 14*(4), 044004.

336.  Harcourt, R., de Bruin, W. B., Dessai, S., & Taylor, A. (2019). Investing in a good pair of wellies: How do non-experts interpret the expert terminology of climate change impacts and adaptation?. *Climatic Change, 155*(2), 257–272.

337.  Helveston, J. P., Seki, S. M., Min, J., Fairman, E., Boni, A. A., Michalek, J. J., & Azevedo, I. M. (2019). Choice at the pump: Measuring preferences for lower-carbon combustion fuels. *Environmental Research Letters, 14*(8), 084035.

338.  Jenn, A., Azevedo, I. L., & Michalek, J. J. (2019). Alternative-fuel-vehicle policy interactions increase US greenhouse gas emissions. *Transportation Research Part A: Policy and Practice, 124*, 396–407.

339.  Kaack, L. H. (2019). *Challenges and Prospects for Data-Driven Climate Change Mitigation* (Doctoral dissertation, Carnegie Mellon University).

340.  Kaack, L. H., Chen, G. H., & Morgan, M. G. (2019, July). Truck traffic monitoring with satellite images. In *Proceedings of the 2nd ACM SIGCAS Conference on Computing and Sustainable Societies* (pp. 155–164).

341.  Keen, J. F., & Apt, J. (2019). Can solar PV reliably reduce loading on distribution networks?. In review at *Applied Energy*.

342.  Keen, J. F., & Apt, J. (2019). How much capacity deferral value can targeted solar deployment create in Pennsylvania?. *Energy Policy, 134*, 110902.

343.  Khan, A., Harper, C. D., Hendrickson, C. T., & Samaras, C. (2019). Net-societal and net-private benefits of some existing vehicle crash avoidance technologies. *Accident Analysis & Prevention, 125*, 207–216.

344.  Lempert, R., Zhao, J., & Dowlatabadi, H. (2019). Convenience, savings, or lifestyle? Distinct motivations and travel patterns of one-way and two-way carsharing members in Vancouver, Canada. *Transportation Research Part D: Transport and Environment, 71*, 141–152.

345.  Lesic, V., Glasgo, B., Krishnamurti, T., de Bruin, W. B., Davis, M., & Azevedo, I. L. (2019). Comparing consumer perceptions of appliances' electricity use to appliances' actual direct-metered consumption. *Environmental Research Communications, 1*(11), 111002.

346. Li, M., Patiño-Echeverri, D., & Zhang, J. J. (2019). Policies to promote energy efficiency and air emissions reductions in China's electric power generation sector during the 11th and 12th five-year plan periods: Achievements, remaining challenges, and opportunities. *Energy Policy*, *125*, 429–444.

347. Li, M., Shan, R., Hernandez, M., Mallampalli, V., & Patiño-Echeverri, D. (2019). Effects of population, urbanization, household size, and income on electric appliance adoption in the Chinese residential sector towards 2050. *Applied Energy*, *236*, 293–306.

348. Mayfield, E. N., Cohon, J. L., Muller, N. Z., Azevedo, I. M., & Robinson, A. L. (2019). Cumulative environmental and employment impacts of the shale gas boom. *Nature Sustainability*, *2*(12), 1122–1131.

349. Mayfield, E. N., Cohon, J. L., Muller, N. Z., Azevedo, I. M., & Robinson, A. L. (2019). Quantifying the social equity state of an energy system: Environmental and labor market equity of the shale gas boom in Appalachia. *Environmental Research Letters*, *14*(12), 124072.

350. Mohan, A., Sripad, S., Vaishnav, P., & Viswanathan, V. (2019). Automation is no barrier to light vehicle electrification. *arXiv preprint arXiv:1908.08920*.

351. Murphy, S., Apt, J., Moura, J., & Sowell, F. (2019). Resource adequacy risks to the bulk power system in North America. *Applied Energy*, *212*, 1360–1376.

352. Murphy, S., Sowell, F., & Apt, J. (2019). A time-dependent model of generator failures and recoveries captures correlated events and quantifies temperature dependence. *Applied Energy*, *253*, 113513.

353. Murphy, S., Sowell, F., Apt, J. (2019). A model of correlated power plant failures and recoveries. Revised and resubmitted after first review at *Applied Energy*.

354. Reed, L., Morgan, M. G., Vaishnav, P., & Armanios, D. E. (2019). Converting existing transmission corridors to HVDC is an overlooked option for increasing transmission capacity. *Proceedings of the National Academy of Sciences*, *116*(28), 13879–13884.

355. Roca, J. B., Vaishnav, P., Laureijs, R. E., Mendonça, J., & Fuchs, E. R. (2019). Technology cost drivers for a potential transition to decentralized manufacturing. *Additive Manufacturing*, *28*, 136–151.

356. Schwetschenau, S. E., VanBriesen, J. M., & Cohon, J. L. (2019). Integrated multiobjective optimization and simulation model applied to drinking water treatment placement in the context of existing infrastructure. *Journal of Water Resources Planning and Management*, *145*(11), 04019048.

357. Sergi, B., Azevedo, I., Xia, T., Davis, A., & Xu, J. (2019). Support for emissions reductions based on immediate and long-term pollution exposure in China. *Ecological Economics*, *158*, 26–33.

358. Sherwin, E. D., Henrion, M., & Azevedo, I. M. (2019). Publisher correction: Estimation of the year-on-year volatility and the unpredictability of the United States energy system. *Nature Energy*, *3*(4), 341–346.

359. Sun, X., Gingerich, D. B., Azevedo, I. L., & Mauter, M. S. (2019). Trace element mass flow rates from US coal fired power plants. *Environmental Science & Technology*, *53*(10), 5585–5595.

360. Taylor, A., Dessai, S., & Bruine de Bruin, W. (2019). Public priorities and expectations of climate change impacts in the United Kingdom. *Journal of Risk Research*, *22*(2), 150–160.

361. Thind, M. P., Tessum, C. W., Azevedo, I. L., & Marshall, J. D. (2019). Fine particulate air pollution from electricity generation in the US: Health impacts by race, income, and geography. *Environmental Science & Technology*, *53*(23), 14010–14019.

362. Tong, F., Azevedo, I., & Jaramillo, P. (2019). Economic viability of a natural gas refueling infrastructure for long-haul trucks. *Journal of Infrastructure Systems*, *25*(1), 04018039.

363. Tschofen, P., Azevedo, I. L., & Muller, N. Z. (2019). Fine particulate matter damages and value added in the US economy. *Proceedings of the National Academy of Sciences*, *116*(40), 19857–19862.

364. Wang, X., Virguez, E., Chen, L., Duan, K., Dong, Q., Ma, H., … & Wang, H. (2019). New index for runoff variability analysis in rainfall driven rivers in southeastern United States. *Journal of Hydrologic Engineering*, *24*(12), 05019031.

365. Wang, X., Virguez, E., Kern, J., Chen, L., Mei, Y., Patiño-Echeverri, D., & Wang, H. (2019). Integrating wind, photovoltaic, and large hydropower during the reservoir refilling period. *Energy Conversion and Management*, *198*, 111778.

366. Wang, X., Virguez, E., Xiao, W., Mei, Y., Patiño-Echeverri, D., & Wang, H. (2019). Clustering and dispatching hydro, wind, and photovoltaic power resources with multiobjective optimization of power generation fluctuations: A case study in southwestern China. *Energy*, *189*, 116250.

367. Ward, J. W., Michalek, J. J., Azevedo, I. L., Samaras, C., & Ferreira, P. (2019). Effects of on-demand ridesourcing on vehicle ownership, fuel consumption, vehicle miles traveled, and emissions per capita in US States. *Transportation Research Part C: Emerging Technologies*, *108*, 289–301.

368. Whiston, M. M., Azevedo, I. L., Litster, S., Whitefoot, K. S., Samaras, C., & Whitacre, J. F. (2019). Expert assessments of the cost and expected future performance of proton exchange membrane fuel cells for vehicles. *Proceedings of the National Academy of Sciences*, *116*(11), 4899–4904.

369. Whiston, M. M., Azevedo, I. M., Litster, S., Samaras, C., Whitefoot, K. S., & Whitacre, J. F. (2019). Meeting US solid oxide fuel cell targets. *Joule*, *3*(9), 2060–2065.

## *2020*

370. Adekanye, O. G., Davis, A., & Azevedo, I. L. (2020). Federal policy, local policy, and green building certifications in the US. *Energy and Buildings*, *209*, 109700.

371. Alarfaj, A. F., Griffin, W. M., & Samaras, C. (2020). Decarbonizing US passenger vehicle transport under electrification and automation uncertainty has a travel budget. *Environmental Research Letters*, *15*(9), 0940c2.

372. Baik, S., Davis, A. L., Park, J. W., Sirinterlikci, S., & Morgan, M. G. (2020). Estimating what US residential customers are willing to pay for resilience to large electricity outages of long duration. *Nature Energy*, *5*(3), 250–258.

373. Burgess, M. G., Ritchie, J., Shapland, J., & Pielke, R. (2020). IPCC baseline scenarios have over-projected CO2 emissions and economic growth. *Environmental Research Letters*, *16*(1), 014016.

374. Deetjen, T. A., & Azevedo, I. L. (2020). Climate and health benefits of rapid coal-to-gas fuel switching in the US power sector offset methane leakage and production cost increases. *Environmental Science & Technology*, *54*(18), 11494–11505.

375. Freeman, G. M., Apt, J., & Moura, J. (2020). What causes natural gas fuel shortages at US power plants?. *Energy Policy*, *147*, 111805.

376. Glasgo, B., Khan, N., & Azevedo, I. L. (2020). Simulating a residential building stock to support regional efficiency policy. *Applied Energy*, *261*, 114223.

377. Kause, A., Bruine de Bruin, W., Fung, F., Taylor, A., & Lowe, J. (2020). Visualizations of projected rainfall change in the United Kingdom: An interview study about user perceptions. *Sustainability*, *12*(7), 2955.

378. Kirchen, K., Harbert, W., Apt, J., & Morgan, M. G. (2020). A solar-centric approach to improving estimates of exposure processes for coronal mass ejections. *Risk Analysis*, *40*(5), 1020–1039.

379. Lamy, J., de Bruin, W. B., Azevedo, I. M., & Morgan, M. G. (2020). Keep wind projects close? A case study of distance, culture, and cost in offshore and onshore wind energy siting. *Energy Research & Social Science*, *63*, 101377.

380. Lavin, L., Murphy, S., Sergi, B., & Apt, J. (2020). Dynamic operating reserve procurement improves scarcity pricing in PJM. *Energy Policy*, *147*, 111857.

381. Lovering, J. R., Abdulla, A., & Morgan, G. (2020). Expert assessments of strategies to enhance global nuclear security. *Energy Policy*, *139*, 111306.

382. Mohan, A., Sripad, S., Vaishnav, P., & Viswanathan, V. (2020). Trade-offs between automation and light vehicle electrification. *Nature Energy*, *5*(7), 543–549.

383. Murphy, S., Lavin, L., & Apt, J. (2020). Resource adequacy implications of temperature-dependent electric generator availability. *Applied Energy*, *262*, 114424.

384. Rath, M., & Morgan, M. G. (2020). Assessment of a hybrid system that uses small modular reactors (SMRs) to back up intermittent renewables and desalinate water. *Progress in Nuclear Energy*, *122*, 103269.

385. Reed, L., Dworkin, M., Vaishnav, P., & Morgan, M. G. (2020). Expanding transmission capacity: Examples of regulatory paths for five alternative strategies. *The Electricity Journal*, *33*(6), 106770.

386. Sergi, B. J., Adams, P. J., Muller, N. Z., Robinson, A. L., Davis, S. J., Marshall, J. D., & Azevedo, I. L. (2020). Optimizing emissions reductions from the US power sector for climate and health benefits. *Environmental Science & Technology*, *54*(12), 7513–7523.

387. Sergi, B., Azevedo, I., Davis, S. J., & Muller, N. Z. (2020). Regional and county flows of particulate matter damage in the US. *Environmental Research Letters*, *15*(10), 104073.

388. Sherwin, E. D., & Azevedo, I. M. (2020). Characterizing the association between low-income electric subsidies and the intra-day timing of electricity consumption. *Environmental Research Letters*, *15*(9), 094089.

389. Tong, F., & Azevedo, I. M. (2020). What are the best combinations of fuel-vehicle technologies to mitigate climate change and air pollution effects across the United States?. *Environmental Research Letters*, *15*(7), 074046.

## *2021*

390. Dryden. R. M., & Morgan M.G. (2021). A simple strategy to communicate about climate attribution. *Bulletin of the American Meteorological Society*, *101*(6), E949-E953, June 2020. And in print version: Using Spinner boards to explain climate attribution. *BAMS* 102 (1), 27–29, January 2021.

391. Dryden, R., Morgan, M. G., & Broomell, S. (2020). Lay detection of unusual patterns in the frequency of hurricanes. *Weather, Climate, and Society*, *12*(3), 597–609.

392. Freeman, G. M., Apt, J., Blumsack, S., & Coleman, T. (2021). Could on-site fuel storage economically reduce power plant-gas grid dependence in pipeline constrained areas like New England?. *The Electricity Journal*, *34*(5), 106956.

393. Lathwal, P., Vaishnav, P., & Morgan, M. G. (2021). Environmental and health consequences of shore power for vessels calling at major ports in India. *Environmental Research Letters*, *16*(6), 064042.

394. Lavin, L., & Apt, J. (2021). The importance of peak pricing in realizing system benefits from distributed storage. *Energy Policy*, *157*, 112484.

395. Savage, T., Davis, A., Fischhoff, B., & Morgan, M. G. (2021). A strategy to improve expert technology forecasts. *Proceedings of the National Academy of Sciences*, *118*(21).

## *2022*

396. Bohman, A. D., Abdulla, A., & Morgan, M. G. (2022). Individual and collective strategies to limit the impacts of large power outages of long duration. *Risk Analysis*, *42*(3), 544–560.

397. Lavi, Y., & Apt, J. (2022). Using PV inverters for voltage support at night can lower grid costs. *Energy Reports*, *8*, 6347–6354.

398. Smillie, S., Muller, N., Griffin, W. M., & Apt, J. (2022). Greenhouse Gas Estimates of LNG Exports Must Include Global Market Effects. *Environmental Science & Technology*, *56*(2), 1194–1201.

At the time this book went to press a number of additional CEDM-supported papers were in preparation for, or in review at, a variety of journals.

## Books and Book Chapters

*2010*

1. Morgan, M. G. (2010). Technology and Policy. *Holistic Engineering Education: Beyond Technology.* Grasso, D., & Burkins, M. (eds.), Springer, 271–282.

*2011*

2. Boyle, M., & Dowlatabadi, H., (2011). Anticipatory Adaptation in Marginalized Communities Within Developed Countries. *Climate Change Adaptation in Developed Nations: From Theory to Practice*, Ford, D. J., & Berrang-Ford, L. (eds.), Springer Netherlands: Dordrecht, 461–473.

3. Cook, C. L., & Dowlatabadi, H., (2011). Learning Adaptation: Climate-Related Risk Management in the Insurance Industry. *Climate Change Adaptation in Developed Nations: From Theory to Practice*, Ford, D. J., & Berrang-Ford, L. (eds.), Springer Netherlands: Dordrecht, 255–265.

4. Morgan, M. G. (2011). Technically Focused Policy Analysis. *The Science of Science Policy: A Handbook.* Husbands-Fealing, K., Lane, J., Marburger III, J., & Ship, S. (eds.), Stanford University Press, 120–130.

*2012*

5. Morgan. M. G., McCoy, S., & 15 others. (2012). Carbon Capture and Sequestration: Removing the legal and regulatory barriers. RFF Press/Routledge, New York, 274.

*2013*

6. Seybolt, T., Aronson, J., & Fischhoff, B. (2013). Counting civilian casualties: An introduction to recording and estimating nonmilitary deaths in conflict. Oxford: Oxford University Press.

*2014*

7. Apt, J., Jaramillo, P., & 19 others. (2014). Variable Renewable Energy and the Electricity Grid. *RFF Press/Routledge*, New York.

8. Doney, S., Rosenberg, A. A., Alexander, M., Chavez, F., Harvell, C. D., Hofmann, G., Orbach, M., & Ruckelshaus, M. (2014). Climate Change Impacts in the United States: The Third National Climate Assessment. *U.S. Global Change Research Program*, 557–578.

9. Walsh, J., Wuebbles, D., Hayhoe, K., Kossin, J., Kunkel, K., Stephens, G., Thorne, P., Vose, R., Wehner, M., Willis, J., Anderson, D., Doney, S., Feely, R., Hennon., P., Kharin, V., Knutson, T., Landerer, F., Lenton, T., Kennedy, J., & Somerville, R. (2014). Climate Change Impacts in the United States: The Third National Climate Assessment. *U.S. Global Change Research Program*, 19–67.

10. Wilhelms, E. A., & Reyna, V. F. (2014). Individual differences in decision-making competence across the lifespan. *Neuroeconomics, Judgment and Decision Making*, New York: Psychology Press, 219–236.

### 2016

11. Donahue, N. M., Posner, L. N., Westervelt, D. M., Li, Z., Shrivastava, M., Presto, A. A., Sullivan, R. C., Adams, P. J., Pandis, S. N., & Robinson, A. L. (2016). Where did this particle come from? Sources of particle number and mass for human exposure estimates. *Issues in Environmental Science & Technology*, Royal Society of Chemistry, 35–71.
12. Griffin, W. M., Saville, B. A., & MacLean, H. L. (2016). Ethanol Use in the United States: Status, Threats and the Potential Future. *Global Bioethanol*, Academic Press, 34–62.
13. Saville, B. A., Griffin, W. M., & MacLean, H. L. (2016). Ethanol Production Technologies in the US: Status and Future Developments. *Global Bioethanol*, Academic Press, 163–180.

### 2017

14. Morgan, M. G. (2017). *Theory and practice in policy analysis*. Cambridge University Press.

### Theses

### 2010

1. Barradale, M. J. (2010). *Practitioner Perspectives Matter: Public Policy and Private Investment in the U.S. Electric Power Sector.* University of California at Berkeley.
2. Hassan, M. N. A. (2010). *Life Cycle GHG Emissions from Malaysian Oil Palm Bioenergy Development: The Impact on Transportation Sector's Energy Security*. Carnegie Mellon University.
3. Mazzi, E. (2010). *An Integrated Assessment of Climate Change Policy, Air Quality and Traffic Safety for Passenger Cars in the UK*. University of British Columbia.
4. Thomas, B. A. (2010). *Edison Revisited: An Assessment of Direct Current Circuits for Lighting in Commercial Buildings*. Carnegie Mellon University.
5. Schweizer, V. (2010). *Developing Useful Long-Term Energy Projections in the Face of Climate Change*. Carnegie Mellon University.

### 2011

6. Fleishman, L. (2011). *Informed Public Decision-Making About Low-Carbon Electricity Generation*. Carnegie Mellon University.
7. Klima, K. (2011). *Does Tropical Cyclone Modification Make Sense? A decision analytic perspective*. Carnegie Mellon University.

8.     Ricke, K. (2011). *Characterizing Impacts and Implications of Proposals for Solar Radiation Management, aka Geoengineering.* Carnegie Mellon University.

9.     Wagner, S. (2011). *Environmental and Economic Implications of Thermal Energy Storage for Concentrated Solar Power Plants.* Carnegie Mellon University.

10.    Wong-Parodi, G. (2011). *Perspectives on Carbon Capture and Sequestration in the United States.* Supported by the University of California at Berkeley and Supervised by the University of British Columbia.

11.    Vignola, R. (2011). *Decision-processes across scales regarding the management of ecosystems' goods and services for ecosystem-based adaptation to climate change.* ETH Zurich, Partly Supervised by the University of British Columbia.

## *2012*

12.    Evans, K. S. (2012). *Evaluating Interventions in the U.S. Electricity System: Assessments of energy efficiency, renewable energy, and small-scale cogeneration.* Carnegie Mellon University.

13.    Gouge, B. (2012). *Modeling and Mitigating the Climate and Health Impacts of Emissions from Public Transportation Bus Fleets: An Integrated Approach to Sustainable Public Transportation.* University of British Columbia.

14.    Hassan, M. N. A. (2012). *GHG emissions and costs of developing biomass energy in Malaysia: Implications of energy security in the transportation and electricity sector.* Carnegie Mellon University.

15.    Hittinger, E. (2012). *Energy storage on the grid and the short-term variability of wind.* Carnegie Mellon University.

16.    Horowitz, S. (2012). *Topics in Residential Electric Demand Response.* Carnegie Mellon University.

17.    Kumar, A. (2012). *Studies in Climate Prediction and Decision.* Carnegie Mellon University.

18.    Lueken, C. (2012). *Integrating Variable and Intermittent Renewables into the Electric Grid: An evaluation of challenges and potential solutions.* Carnegie Mellon University.

19.    Mauch, B. (2012). *Managing Wind Power Forecast Uncertainty in Electric Grids.* Carnegie Mellon University.

20.    Mullins, K. (2012). *Evaluating Biomass Energy Policy in the Face of Emissions Reductions Uncertainty and Feedstock Supply Risk.* Carnegie Mellon University.

21.    Thomas, B. (2012). *Energy efficiency and rebound effects in the US: Implications for renewables investment and emissions abatement.* Carnegie Mellon University.

## *2013*

22.    Fertig, E. (2013). *Facilitating the Development and Integration of Low-Carbon Energy Technologies.* Carnegie Mellon University.

23.   Hosgor, E. (2013). *Residential energy profiling: A statistical study using publicly available data on Gainesville, FL, building stock.* Carnegie Mellon University.

24.   Huimin, T. (2013). *A Transition to Energy-Efficient Lighting Systems in the US Residential Sector: An Assessment of Consumer Preferences and Perceptions.* Carnegie Mellon University.

25.   Mercer, A. (2013). *An Examination of Emerging Public and Expert Judgements of Solar Radiation Management.* University of Calgary.

26.   Mitchell, A. (2013). *Analysis of Health and Environmental Risks Associated with Marcellus Shale Development.* Carnegie Mellon University.

27.   Olson, R. (2013). *How well can Historical Temperature Observations Constrain Climate Sensitivity?* The Pennsylvania State University.

28.   Rose, S. (2013). *Assessing the Costs and Risks of Novel Wind Turbine Applications.* Carnegie Mellon University.

29.   Schwietzke, S. (2013). *Atmospheric Impacts of Biofuel and Natural Gas Life Cycle Greenhouse Gas Emissions and Policy Implications.* Carnegie Mellon University.

## *2014*

30.   Abdulla, A. (2014). *Estimating the Benefits, Costs and Risks of Small Nuclear Reactors.* Carnegie Mellon University.

31.   Borgert, K. (2014). *Oxyfuel Carbon Capture for Pulverized Coal: Techno-Economic Model Creation and Evaluation Amongst Alternatives.* Carnegie Mellon University.

32.   Hoss, F. (2014). *Uncertainty in river forecasts: Quantification and implications for decision-making in emergency management.* Carnegie Mellon University.

33.   Isley, S. (2014). *The Political Sustainability of Carbon Control Policies in an Evolutionary Economics Setting.* Pardee RAND Graduate School.

34.   Jenn, A. (2014). *Advanced and alternative fuel vehicle policies: Regulations and incentives in the United States.* Carnegie Mellon University.

35.   Lueken, R. (2014). *Reducing Carbon Intensity in Restructured Markets: Challenges and Potential Solutions.* Carnegie Mellon University.

36.   Meyer, R. (2014). *Analysis of selected regulatory interventions to improve energy efficiency.* Carnegie Mellon University.

37.   Min, J. (2014). *Energy efficient lighting: Consumer preferences, choices, and system wide effects.* Carnegie Mellon University.

38.   Moore, J. (2014). *Cost Effectiveness of CO2 Mitigation Technologies and Policies in the Electricity Sector.* Carnegie Mellon University.

39.   Pena-Cabra, I. (2014). *Retrospective and prospective analysis of policy incentives for wind power in Portugal.* Carnegie Mellon University.

40.   Schnitzer, D. (2014). *Microgrids and High-Quality Central Grid Alternatives: Challenges and Imperatives Elucidated by Case Studies and Simulation.* Carnegie Mellon University.

41.   Tamayo, M. A. (2014). *Urbanization and vehicle electrification in the U.S.: CO2 estimation and climate policy implications.* Carnegie Mellon University.

## 2015

42. Heo, J. (2015). *Evaluation of air quality impacts on society: Methods and application.* Carnegie Mellon University.
43. Oates, D. L. (2015). *Low carbon policy and technology in the power sector. Evaluating economic and environmental effects.* Carnegie Mellon University.
44. Vaishnav, P. (2015). *Reducing Pollution From Aviation and Ocean Shipping.* Carnegie Mellon University.
45. Weis, A. (2015). *Electric vehicles and the grid: Interactions and environmental and health impacts.* Carnegie Mellon University.

## 2016

46. Bandyopadhyay, R. (2016). *Coal-fired Power Plants with Flexible Amine-based CCS and Co-located Wind Power: Environmental, Economic and Reliability Outcomes.* Duke University.
47. Canfield, C. (2016). *Using vigilance to quantify human behavior for phishing risk.* Carnegie Mellon University.
48. Faria, F. (2016). *Hydropower development in the Brazilian Amazon.* Carnegie Mellon University.
49. Gilbraith, N. (2016). *Powering the information age: Metrics, social cost optimization strategies, and indirect effects related to data center energy use.* Carnegie Mellon University.
50. Hagerman, S. (2016). *Economics of Behind-The Meter Solar PV and Energy Storage.* Carnegie Mellon University.
51. Helveston, J. (2016). *Driving vehicles innovation and electrification in China's automotive industry: Markets, policy, and technology trajectories.* Carnegie Mellon University.
52. Horner, N. (2016). *Powering the information age: Metrics, social cost optimization strategies, and indirect effects related to data center energy use.* Carnegie Mellon University.
53. Lamy, J. (2016). *Optimal locations for siting wind projects: Technical challenges, economics, and public preferences.* Carnegie Mellon University.
54. Necefer, L. (2016). *Development of a decision aid for energy resource management for the Navajo incorporating environmental cultural values.* Carnegie Mellon University.
55. Posen, D. (2016). *Fuel, feedstock, or neither? – Evaluating tradeoffs in the use of biomass for greenhouse gas mitigation.* Carnegie Mellon University.
56. Ryan, T. (2016). *Case-studies in the economics of ancillary services of power systems in support of high wind penetrations.* Carnegie Mellon University.
57. Seki, S. (2016). *Evaluating the economic, environmental and policy impacts of ethanol as a transportation fuel in Pennsylvania.* Carnegie Mellon University.
58. Talati, S. (2016). *The Future of Low Carbon Electric Power Generation: An assessment of economic viability and water impacts under climate change and mitigation policies.* Carnegie Mellon University.

59.   Tong, F. (2016). *The good, the bad, and the ugly: Economic and environmental implications of using natural gas to power on-road vehicles in the United States.* Carnegie Mellon University.

## 2017

60.   Fisher, M. J. (2017). *Integrating Demand-Side Resources into the Electric Grid: Economic and Environmental Considerations.* Carnegie Mellon University.
61.   Glasgo, B. (2017). *Assessing the feasibility of residential DC buildings.* Carnegie Mellon University.
62.   Lam, L. (2017). *Wind innovation in China and in the United States.* Carnegie Mellon University.
63.   Welle, P. (2017). *Remotely Sensed Data for High Resolution Agro-Environmental Policy Analysis.* Carnegie Mellon University.

## 2018

64.   Baik, S. (2018). *An improved method to assess the value of assuring limited local electric service in the event of major grid outages.* Carnegie Mellon University.
65.   Carless, T. (2018). *Framing a New Nuclear Renaissance Through Environmental Competitiveness, Community Characteristics, and Cost Mitigation Through Passive Safety.* Carnegie Mellon University.
66.   Ciez, R. (2018). *Battery energy storage for maturing markets: performance, cost, perceptions, and environmental impacts.* Carnegie Mellon University.
67.   Hanus, N. L. (2018). *An Engineering and Behavioral Sciences Approach to Understand and Inform Energy Efficiency and Renewable Energy Decision-Making.* Carnegie Mellon University.
68.   Zhang, X. (2018). *Evaluating Indirect GHG Emissions in Biofuel and Implications for Renewable Energy Policies.* Carnegie Mellon University.

## 2019

69.   Dryden Steratore, R. L. (2019). *Public Understanding of Climate Science, Extreme Weather and Climate Attribution.* Carnegie Mellon University.
70.   Freeman, G. M. (2019). *Power Plant-Gas Grid Dependence.* Carnegie Mellon University.
71.   Kaack, L. H. (2019). *Challenges and Prospects for Data-Driven Climate Change Mitigation.* Carnegie Mellon University.
72.   Keen, J. F. (2019). *Stakeholder Costs and Benefits of Distributed Energy Resources on Distribution Networks.* Carnegie Mellon University.
73.   Murphy, S. J. (2019). *Correlated Generator Failures and Power System Reliability.* Carnegie Mellon University.
74.   Sergi, B. J. (2019). *Integrating climate and health damages in decision-making for the electric power sector.* Carnegie Mellon University.

75. Sherwin, E. D. (2019). *Decisions and Uncertainties in the US Energy System: Electrofuels and Other Applications.* Carnegie Mellon University.

### 2020

76. Lovering, J. (2020). *Evaluating changing paradigms across the nuclear industry.* Carnegie Mellon University.
77. Rath, M. J. (2020). *Future Pathways to U.S. Decarbonization Through Nuclear and Other Energy Systems.* Carnegie Mellon University.
78. Reed, L.B. (2020). *HVDC: New Opportunities to Expand Transmission Capacity.* Carnegie Mellon University.

### 2021

79. Anderson, J. (2021). *Marching to the beat of an absent drummer: Carbon Dioxide Emissions Reduction in the U.S. Power Sector.* Carnegie Mellon University.
80. Lathwal, P. (2021). *Essays in Environmental, Climate, and Public Health Impacts of Freight Transportation.* Carnegie Mellon University.
81. Lavin, L. (2021). *Data and technology-driven improvements to electricity market design.* Carnegie Mellon University.

## CDMC

The following is an almost complete chronological list of publications, book chapters, and theses that were produced with support from the CDMC Center. Not included are roughly 40 reports and papers in conference proceedings which can be found online in a more complete list at www.cmu.edu/epp/climate/.

### *Journals*

### *1999*

1. McDaniels, T. L., Gregory, R. S., & Fields, D. (1999). Democratizing risk management: Successful public involvement in local water management decisions. *Risk Analysis, 19*(3), 497–510.

### *2001*

2. Gregory, R., McDaniels, T., & Fields, D. (2001). Decision aiding, not dispute resolution: A new perspective for environmental negotiation. *Journal of Policy Analysis and Management, 20*(3), 415–432.
3. Keeney, R. L., & McDaniels, T. L. (2001). A framework to guide thinking and analysis regarding climate change policies. *Risk Analysis, 21*(6), 989–1000.

### 2003

4.  McDaniels, T. L., Gregory, R., Arvai, J., & Chuenpagdee, R. (2003). Decision structuring to alleviate embedding in environmental valuation. *Ecological Economics*, 46(1), 33–46.

### 2004

5.  Boyle, M., Gibson, R. B., & Curran, D. (2004). If not here, then perhaps not anywhere: Urban growth management as a tool for sustainability planning in British Columbia's capital regional district. *Local Environment*, 9(1), 21–43.
6.  Dowlatabadi, H., Boyle, M., Kandlikar, M., & Rowley, S. (2004). Learning from History: Lessons for cumulative effects assessment and planning. *Meridian*, Fall/Winter, 6–12.
7.  Johnson, T. L., & Keith, D. W. (2004). Fossil electricity and CO2 sequestration: How natural gas prices, initial conditions and retrofits determine the cost of controlling CO2 emissions. *Energy Policy*, 32(3), 367–382.
8.  Keith, D. W., DeCarolis, J. F., Denkenberger, D. C., Lenschow, D. H., Malyshev, S. L., Pacala, S., & Rasch, P. J. (2004). The influence of large-scale wind power on global climate. *Proceedings of the National Academy of Sciences*, 101(46), 16115–16120.
9.  Mastrandrea, M. D., & Schneider, S. H. (2004). Probabilistic integrated assessment of "dangerous" climate change. *Science*, 304(5670), 571–575.
10. McDaniels, T., & Gregory, R. (2004). Learning as an objective within structured decision processes for managing environmental risks. *Environmental Science & Technology*, 38, 7, 1921–1926.
11. Palmgren, C. R., Morgan, M. G., Bruine de Bruin, W., & Keith, D. W. (2004). Initial public perceptions of deep geological and oceanic disposal of carbon dioxide.
12. Zickfeld, K., Slawig, T., & Rahmstorf, S. (2004). A low-order model for the response of the Atlantic thermohaline circulation to climate change. *Ocean Dynamics*, 54(1), 8–26.

### 2005

13. Apt, J. (2005). Competition has not lowered US industrial electricity prices. *The Electricity Journal*, 18(2), 52–61.
14. DeCarolis, J. F., & Keith, D. W. (2005). The costs of wind's variability: Is there a threshold?. *The Electricity Journal*, 18(1), 69–77.
15. Gregory, R., Fischhoff, B., & McDaniels, T. (2005). Acceptable input: Using decision analysis to guide public policy deliberations. *Decision Analysis*, 2(1), 4–16.
16. Hamouda, L., Hipel, K. W., Kilgour, D. M., Noakes, D. J., Fang, L., & McDaniels, T. (2005). The salmon aquaculture conflict in British Columbia: A graph model analysis. *Ocean & Coastal Management*, 48(7–8), 571–587.

17. Keith, D.W., Giardina, J.A., Morgan, M. G., & Wilson, E. J. (2005). Regulating the underground injection of CO2. *Environmental Science & Technology*, 499A–504A.

18. Keith, D. W., Hassanzadeh, H., & Pooladi-Darvish, M. (2005). Reservoir engineering to accelerate dissolution of stored CO2 in brines. In *Greenhouse Gas Control Technologies 7* (pp. 2163–2167). Elsevier Science Ltd.

19. McDaniels, T. L., Dowlatabadi, H., & Stevens, S. (2005). Multiple scales and regulatory gaps in environmental change: The case of salmon aquaculture. *Global Environmental Change, 15*(1), 9–21.

20. McDaniels, T., & Trousdale, W. (2005). Evaluating losses of traditional native values with multi-attribute value assessment. *Ecological Economics*, 55(2), pp. 173–186.

21. Rahmstorf, S., & Zickfeld, K. (2005). Thermohaline circulation changes: A question of risk assessment. *Climatic Change, 68*(1–2), 241–247.

22. Stolaroff, J. K., Lowry, G. V., & Keith, D. W. (2005). Using CaO-and MgO-rich industrial waste streams for carbon sequestration. *Energy Conversion and Management, 46*(5), 687–699.

## *2006*

23. Apt, J., & Fischhoff, B. (2006). Power and people. *The Electricity Journal, 19*(9), 17–25.

24. Apt, J., Lave, L. B., & Morgan, M. G. (2006). Power play: A more reliable US electric system. *Issues in Science and Technology, 22*(4), 51–58.

25. Chang, S., McDaniels, T., Longstaff, H., & Wilmot, S. (2006). Fostering disaster resilience through addressing infrastructure interdependencies. *Plan Canada, 46*, 33–36.

26. DeCarolis, J. F., & Keith, D. W. (2006). The economics of large-scale wind power in a carbon constrained world. *Energy Policy, 34*(4), 395–410.

27. Dowlatabadi, H., & Oravetz, M. (2006). Understanding trends in energy intensity: A simple model of technical change. *Energy Policy, 34*(17).

28. Dowlatabadi, H. (2006). A Peek Past Peak Oil. *UBC Reports, 52*(4).

29. Farrell, A. E., & Brandt, A. R. (2006). Risks of the oil transition. *Environmental Research Letters, 1*(1), 014004.

30. Farrell, A. E., Plevin, R. J., Turner, B. T., Jones, A. D., O'hare, M., & Kammen, D. M. (2006). Ethanol can contribute to energy and environmental goals. *Science, 311*(5760), 506–508.

31. Gerwing, K., & McDaniels, T. (2006). Listening to the salmon people: Coastal First Nations' objectives regarding salmon aquaculture in British Columbia. *Society and Natural Resources, 19*(3), 259–273.

32. Lave, L. B., & Apt, J. (2006). Planning for natural disasters in a stochastic world. *Journal of Risk and Uncertainty, 33*(1), 117–130.

33. McDaniels, T. L., Keen, P. L., & Dowlatabadi, H. (2006). Expert judgments regarding risks associated with salmon aquaculture practices in British Columbia. *Journal of Risk Research, 9*(7), 775–800.

34.   McDaniels, T., Longstaff, H., & Dowlatabadi, H. (2006). A value-based framework for risk management decisions involving multiple scales: A salmon aquaculture example. *Environmental Science & Policy*, *9*(5), 423–438.

35.   Morgan, M. G., Adams, P. J., & Keith, D. W. (2006). Elicitation of expert judgments of aerosol forcing. *Climatic Change*, *75*(1), 195–214.

36.   Shepherd, P., Tansey, J., & Dowlatabadi, H. (2006). Context matters: What shapes adaptation to water stress in the Okanagan?. *Climatic Change*, *78*(1), 31–62.

37.   Stolaroff, J. K., Keith, D. W., & Lowry, G. V. (2006). A pilot-scale prototype contactor for $CO_2$ capture from ambient air: Cost and energy requirements. *Environmental Science & Pollution Research*, 13(6).

## *2007*

38.   Apt, J. (2007). The spectrum of power from wind turbines. *Journal of Power Sources*, *169*(2), 369–374.

39.   Apt, J., Keith, D. W., & Morgan, M. G. (2007). Promoting low-carbon electricity production. *Issues in Science and Technology*, *23*(3), 37–43.

40.   Azevedo, I. L., Attari, S., Flath, B., & Samaras, C. (2007). An open letter to the 2008 presidential candidates on energy and sustainability issues. *USA Today*, November 20.

41.   Bergerson, J. A., & Lave, L. B. (2007). Baseload coal investment decisions under uncertain carbon legislation. *Environmental Science & Technology*, *41*(10), 3431–3436.

42.   Dowlatabadi, H. (2007). On integration of policies for climate and global change. *Mitigation and Adaptation Strategies for Global Change*, *12*(5), 651–663.

43.   Hanova, J., & Dowlatabadi, H. (2007). Strategic GHG reduction through the use of ground source heat pump technology. *Environmental Research Letters*, *2*(4), 044001.

44.   Hanova, J., Dowlatabadi, H., & Mueller, L. (2007). Ground source heat pump systems in Canada. *Resources for the Future*, 1–37.

45.   Klinsky, S. (2007). Mapping emergence: Network analysis of climate change media coverage. *Integrated Assessment*, *7*(1).

46.   Kunreuther, H. C., & Michel-Kerjan, E. O. (2007). Climate change, insurability of large-scale disasters and the emerging liability challenge. *National Bureau of Economic Research*, 12821.

47.   Mazzi, E. A., & Dowlatabadi, H. (2007). Air quality impacts of climate mitigation: UK policy and passenger vehicle choice. *Environmental Science & Technology*, 41, pp. 387–392.

48.   Mazzi, E., & Dowlatabadi, H. (2007). Mortality and morbidity from climate policy-accelerated diesel car sales in the UK. *Environmental Science & Technology*, (in review).

49.   McDaniels, T., Chang, S., Peterson, K., Mikawoz, J., & Reed, D. (2007). Empirical framework for characterizing infrastructure failure interdependencies. *Journal of Infrastructure Systems*, *13*(3), 175–184.

50. Morgan, M. G. (2007). Moving to a low-carbon future: Perspectives on nuclear and alternative power sources. *Health Physics*, *93*(5), 568–570.

51. Patiño-Echeverri, D., Morel, B., Apt, J., & Chen, C. (2007). Should a coal-fired power plant be replaced or retrofitted?. *Environmental Science & Technology*, 41(23), pp. 7980–7986.

52. Reinelt, P. S., & Keith, D. W. (2007). Carbon capture retrofits and the cost of regulatory uncertainty. *The Energy Journal*, *28*(4).

53. Wilson, C., & McDaniels, T. (2007). Linking climate change adaptation, mitigation and sustainable development through structured decision-making tools. *Climate Policy*. Special issue linking adaptation and mitigation, 7.4.

54. Wilson, C., & Dowlatabadi, H. (2007). Models of decision making and residential energy use. *Annual Review of Environment and Resources*, *32*, 169–203.

55. Wilson, C., & McDaniels, T. (2007). Structured decision-making to link climate change and sustainable development. *Climate Policy*, 7(4), 353–370.

56. Zerriffi, H., Dowlatabadi, H., & Farrell, A. (2007). Incorporating stress in electric power systems reliability models. *Energy Policy*, *35*(1), 61–75.

57. Zickfeld, K., Levermann, A., Morgan, M. G., Kuhlbrodt, T., Rahmstorf, S., & Keith, D. W. (2007). Expert judgements on the response of the Atlantic meridional overturning circulation to climate change. *Climatic Change*, *82*(3), 235–265.

## 2008

58. Barradale, M. J. (2008). Impact of policy uncertainty on renewable energy investment: Wind power and PTC. *United States Association for Energy Economists WP*, 08–003.

59. Bornik, Z. B., & Dowlatabadi, H. (2008). Genomics in Cyprus: Challenging the social norms. *Technology in Society*, *30*(1), 84–93.

60. Curtright, A. E., & Apt, J. (2008). The character of power output from utility-scale photovoltaic systems. *Progress in Photovoltaics: Research and Applications*, *16*(3), 241–247.

61. Curtright, A. E., Morgan, M. G., & Keith, D. W. (2008). Expert assessments of future photovoltaic technologies. *Environmental Science & Technology*, *42*(24), 9031–9038.

62. Dowlatabadi, H., & Cook, C. (2008). Climate risk management & institutional learning. *Integrated Assessment*, *8*(1), 151–163.

63. Galland, D., & McDaniels, T. (2008). Are new industry policies precautionary? The case of salmon aquaculture siting policy in British Columbia. *Environmental Science & Policy*, *11*(6), 517–532.

64. Lemoine, D. M., Kammen, D. M., & Farrell, A. E. (2008). An innovation and policy agenda for commercially competitive plug-in hybrid electric vehicles. *Environmental Research Letters*, *3*(1), 014003.

65.   Lenton, T. M., Held, H., Kriegler, E., Hall, J. W., Lucht, W., Rahmstorf, S., & Schellnhuber, H. J. (2008). Tipping elements in the Earth's climate system. *Proceedings of the national Academy of Sciences, 105*(6), 1786–1793.

66.   Mastrandrea, M. D., & Schneider, S. H. (2008). The rising tide: Time to adapt to climate change. *The Boston Review Special Report, 33*(6), November/December, 7–10.

67.   McDaniels, T., Chang, S., Cole, D., Mikawoz, J., & Longstaff, H. (2008). Fostering resilience to extreme events within infrastructure systems: Characterizing decision contexts for mitigation and adaptation. *Global Environmental Change, 18*(2), 310–318.

68.   Michel-Kerjan, E., & Morlaye, F. (2008). Extreme events, global warming, and insurance-linked securities: How to trigger the "tipping point." *The Geneva Papers on Risk and Insurance-Issues and Practice, 33*(1), 153–176.

69.   Mills, T., Blackwell, B., McDaniels, T., Gregory, R., & Ohlson, D. (2008). Mountain Pine Beetles and Climate Change: Using expert perspectives for structured decision-making in forest policy. *Forest Professional Magazine*, May-June.

70.   Morgan, M. G., & Keith, D. W. (2008). Improving the way we think about projecting future energy use and emissions of carbon dioxide. *Climatic Change, 90*(3), 189–215.

71.   Newcomer, A., & Apt, J. (2008). Implications of generator siting for CO2 pipeline infrastructure. *Energy Policy, 36*(5), 1776–1787.

72.   Newcomer, A., Blumsack, S. A., Apt, J., Lave, L. B., & Morgan, M. G. (2008). Short run effects of a price on carbon dioxide emissions from U.S. electric generators. *Environmental Science & Technology, 42*(9), 3139–3144.

73.   Samaras, C., & Meisterling, K. (2008). Life cycle assessment of greenhouse gas emissions from plug-in hybrid vehicles: Implications for policy. *Environmental Science & Technology, 42*(9), pp. 3170–3176.

74.   Stephens, J. C., & Keith, D. W. (2008). Assessing geochemical carbon management. *Climatic Change, 90*(3), 217–242.

75.   Stolaroff, J. K., Keith, D. W., & Lowry, G. V. (2008). Carbon dioxide capture from atmospheric air using sodium hydroxide spray. *Environmental Science & Technology, 42*(8), 2728–2735.

76.   Walawalkar, R., Blumsack, S., Apt, J., & Fernands, S. (2008). An economic welfare analysis of demand response in the PJM electricity market. *Energy Policy, 36*(10), 3692–3702.

77.   Wilson, E. J., Morgan, M. G., Apt, J., Bonner, M., Bunting, C., Gode, J., … & Wright, I. W. (2008). Regulating the geological sequestration of CO2. *Environmental Science & Technology*, 42(8), 2718–2722.

## *2009*

78.   Grossmann, I. (2009). Atlantic hurricane risks: Preparing for the plausible. *Environmental Science & Technology, 43*(20), 7604–7608.

79. Grossmann, I., & Klotzbach, P. J. (2009). A review of North Atlantic modes of natural variability and their driving mechanisms. *Journal of Geophysical Research: Atmospheres, 114*(D24).

80. Grossmann, W. D., Steininger, K., Grossmann, I., & Magaard, L. (2009). Indicators on economic risk from global climate change. *Environmental Science & Technology, 43*(16), 6421–6426.

81. Held, H., Kriegler, E., Lessmann, K., & Edenhofer, O. (2009). Efficient climate policies under technology and climate uncertainty. *Energy Economics, 31,* S50-S61.

82. Jaramillo, P., Samaras, C., Wakeley, H., & Meisterling, K. (2009). Greenhouse gas implications of using coal for transportation: Life cycle assessment of coal-to-liquids, plug-in hybrids, and hydrogen pathways. *Energy Policy, 37*(7), 2689–2695.

83. Klinsky, S., & Dowlatabadi, H. (2009). Conceptualizations of justice in climate policy. *Climate Policy, 9*(1), 88–108.

84. Kriegler, E. (2009). Updating under unknown unknowns: An extension of Bayes' rule. *International Journal of Approximate Reasoning, 50*(4), 583–596.

85. Kunreuther, H. C., & Michel-Kerjan, E. (2009). *Long Term Insurance and Climate Change.* Working Paper # 2009–03–13, Wharton Risk Management and Decision Processes Center, University of Pennsylvania.

86. Meisterling, K., Samaras, C., & Schweizer, V. (2009). Decisions to reduce greenhouse gases from agriculture and product transport: LCA case study of organic and conventional wheat. *Journal of Cleaner Production, 17*(2), 222–230.

87. Morgan, M. G. (2009). Best Practice Approaches for Characterizing, Communicating and Incorporating Scientific Uncertainty in Climate Decision Making: Synthesis and Assessment Product 5.2 Report (Vol. 5). US Climate Change Science Program.

88. Newcomer, A., & Apt, J. (2009). Near-term implications of a ban on new coal-fired power plants in the United States. *Environmental Science & Technology, 43*(11), 3995–4001.

89. Patiño-Echeverri, D., Fischbeck, P., & Kriegler, E. (2009). Economic and environmental costs of regulatory uncertainty for coal-fired power plants. *Environmental Science & Technology, 43*(3), 578–584.

90. Smith, J. B., Schneider, S. H., Oppenheimer, M., Yohe, G. W., Hare, W., Mastrandrea, M. D., … & van Ypersele, J. P. (2009). Assessing dangerous climate change through an update of the Intergovernmental Panel on Climate Change (IPCC) "reasons for concern." *Proceedings of the National Academy of Sciences, 106*(11), 4133–4137.

91. Victor, D. G., Morgan, M. G., Apt, J., & Steinbruner, J. (2009). The geoengineering option: A last resort against global warming. *Foreign Affairs, 88,* 64.

92. Weber, C. L., Jaramillo, P., Marriott, J., & Samaras, C. (2009). Uncertainty and variability in accounting for grid electricity in life cycle assessment. In *2009 IEEE International Symposium on Sustainable Systems and Technology* (pp. 1–8). IEEE.

## 2010

93. Fleishman, L. A., De Bruin, W. B., & Morgan, M. G. (2010). Informed public preferences for electricity portfolios with CCS and other low-carbon technologies. *Risk Analysis: An International Journal, 30*(9), 1399–1410.

94. Gresham, R. L., McCoy, S. T., Apt, J., & Morgan, M. G. (2010). Implications of compensating property owners for geologic sequestration of CO2. *Environmental Science & Technology, 44*(8), 2897–2903.

95. Hagerman, S. M., Dowlatabadi, H., & Satterfield, T. (2010). Observations on drivers and dynamics of environmental policy change: Insights from 150 years of forest management in British Columbia. *Ecology and Society, 15*(1).

96. Hagerman, S., Dowlatabadi, H., Chan, K. M., & Satterfield, T. (2010). Integrative propositions for adapting conservation policy to the impacts of climate change. *Global Environmental Change, 20*(2), 351–362.

97. Hagerman, S., Dowlatabadi, H., Satterfield, T., & McDaniels, T. (2010). Expert views on biodiversity conservation in an era of climate change. *Global Environmental Change, 20*(1), 192–207.

98. Hagerman, S. M., Satterfield, T. S., & Dowlatabadi, H. (2010). Adapting conservation policy to the impacts of climate change: Promotion, ambivalence and resistance at the WCC. *Conservation and Society, 8*, 298–311.

99. Hoppock, D. C., & Patiño-Echeverri, D. (2010). Cost of wind energy: Comparing distant wind resources to local resources in the Midwestern United States. *Environmental Science & Technology, 44*(22), 8758–8765.

100. Jacob, C., McDaniels, T., & Hinch, S. (2010). Indigenous culture and adaptation to climate change: Sockeye salmon and the St'át'imc people. *Mitigation and Adaptation Strategies for Global Change, 15*(8), 859–876.

101. Keith, D. W. (2010). Photophoretic levitation of engineered aerosols for geoengineering. *Proceedings of the National Academy of Sciences, 107*(38), 16428–16431.

102. Keith, D. W., Parson, E., & Morgan, M. G. (2010). Research on global sun block needed now. *Nature, 463*(7280), 426–427.

103. McDaniels, T., Wilmot, S., Healey, M., & Hinch, S. (2010). Vulnerability of Fraser River sockeye salmon to climate change: A life cycle perspective using expert judgments. *Journal of Environmental Management, 91*(12), 2771–2780.

104. Michel-Kerjan, E. O., & Kousky, C. (2010). Come rain or shine: Evidence on flood insurance purchases in Florida. *Journal of Risk and Insurance, 77*(2), 369–397.

105. Reynolds, T. W., Bostrom, A., Read, D., & Morgan, M. G. (2010). Now what do people know about global climate change? Survey studies of educated laypeople. *Risk Analysis: An International Journal, 30*(10), 1520–1538.

106. Ricke, K. L., Morgan, M. G., & Allen, M. R. (2010). Regional climate response to solar-radiation management. *Nature Geoscience, 3*(8), 537–541.

## 2011

107. Fleishman, L. A., Bruine de Bruin, W., & Morgan, M. G. (2011). The value of CCS public opinion research: A letter in response to Malone, Dooley and Bradbury (2010). Moving from misinformation derived from public attitude surveys on carbon dioxide capture and storage towards realistic stakeholder involvement. *International Journal of Greenhouse Gas Control.*

108. Grossmann, I., & Morgan, M. G. (2011). Tropical cyclones, climate change, and scientific uncertainty: What do we know, what does it mean, and what should be done?. *Climatic Change, 108*(3), 543–579.

109. Klima, K., Morgan, M. G., Grossmann, I., & Emanuel, K. (2011). Does it make sense to modify tropical cyclones? A decision-analytic assessment. *Environmental Science & Technology, 45*(10), 4242–4248.

110. Mitchell, A. L., & Casman, E. A. (2011). Economic incentives and regulatory framework for shale gas well site reclamation in Pennsylvania. *Environmental Science & Technology, 45*(22), 9506–9514.

111. Moreno-Cruz, J. B., Ricke, K. L., & Keith, D. W. (2011). A simple model to account for regional inequalities in the effectiveness of solar radiation management. *Climatic Change, 110*(3), 649–668.

112. Mullins, K. A., Griffin, W. M., & Matthews, H. S. (2010). Policy implications of uncertainty in modeled life-cycle greenhouse gas emissions of Biofuels. *Environmental Science & Technology, 45*(1), 132–138.

113. Schwietzke, S., Griffin, W. M., & Matthews, H. S. (2011). Relevance of emissions timing in biofuel greenhouse gases and climate impacts. *Environmental Science & Technology, 45*(19), 8197–8203.

114. Wagner, S. J. &. Rubin, E. S. (2011). Economic implications of thermal energy storage for concentrated solar thermal power. *Proceedings of the World Renewable Energy Congress*, Linköping, Sweden, May 8–13.

## 2012

115. Klima, K., Bruine de Bruin, W., Morgan, M. G., & Grossmann, I. (2012). Public perceptions of hurricane modification. *Risk Analysis: An International Journal, 32*(7), 1194–1206.

116. Klima, K., Lin, N., Emanuel, K., Morgan, M. G., & Grossmann, I. (2012). Hurricane modification and adaptation in Miami-Dade County, Florida. *Environmental Science & Technology, 46*(2), 636–642.

117. McDaniels, T., Mills, T., Ohlson, D., & Gregory, R. (2012). Exploring robust alternatives for climate adaptation in forest-land management through expert judgments. *Risk Analysis*, December, 32 (12), 2098–2112.

118. Patiño-Echeverri, D., & Hoppock, D. C. (2012). Reducing the energy penalty costs of postcombustion CCS systems with amine-storage. *Environmental Science & Technology, 46*(2), 1243–1252.

## 2014

119. Chang, S. E., McDaniels, T., Fox, J., Dhariwal, R., & Longstaff, H. (2014). Toward disaster-resilient cities: Characterizing resilience of infrastructure systems with expert judgments. *Risk analysis, 34*(3), 416-434.

## Books and Book Chapters

### 2004

1. Dowlatabadi, H., Boyle, M., Rowley, S., & Kandalikar, M. (2004). Bridging the gap between project-level assessments and regional development dynamics: A methodology for estimating cumulative effects. *Research supported by Canadian Environmental Agency's Research and Development Program. Ottawa, ON: Canadian Environmental Assessment Agency.*

### 2005

2. Gregory, R., & McDaniels, T. (2005). Improving Environmental Decision Processes. *Decision Making For The Environment: Social And Behavioral Science Research Priorities (23–40).* National Academies Press.

### 2006

3. Farrell, A. E., & Hanemann, W. M. (2006). *Field Notes on the Political Economy of California Climate Change Policy.* H. Selin and S. D. VanDeveer (eds.), Woodrow Wilson International Center for Scholars.
4. Kunreuther, H. (2006). *Reflections on US Disaster Insurance Policy for the 21st Century* (No. w12449). National Bureau of Economic Research.
5. Ogushi, Y., Kandlikar, M., & Dowlatabadi, H. (2006). *Assessing Product Life Cycle Strategies in the Japanese Market.* Springer, 448.

### 2010

6. Boyle, M., & Dowlatabadi, H. (2010). *Anticipatory Adaptation in Marginalised Communities within Developed Countries.* McGill-Queens.
7. Cook, C., & Dowlatabadi, H. (2010). *Learning Adaptation: Climate Risk Management in the Insurance Industry.* McGill-Queens.

## Theses

### 2006

1. Plevin, R. (2006). *California Policy Should Distinguish Biofuels by Differential Global Warming Effects.* M.S. thesis, Energy and Resources Group, UC Berkeley, Berkeley, CA.

2. Stolaroff, J.K. (2006). *Capturing CO2 from Ambient Air: A feasibility assessment using experimental, engineering-economic, and model-based analysis.* Department of Engineering and Public Policy, Carnegie Mellon University, Pittsburgh, PA, Ph.D. thesis.

## 2007

3. Boyle, M. (2007). *Community of Mittimatalik: Input into the preparation of the socio-economic impact assessment for the potential Mary River Mine.* University of British Columbia. Also Boyle, M (2007). *Pond Inlet Socio-Economic Study Information Comments for QIA Negotiating Team.* University of British Columbia, Vancouver, 9pp. (confidential report).

## 2009

4. Azevedo, I. L. (2009). *Energy Efficiency in the U.S. Residential Sector: An engineering and economic assessment of opportunities for large energy savings and greenhouse gas emissions reductions.* Department of Engineering and Public Policy, Carnegie Mellon University, Pittsburgh, PA, Ph.D. thesis.
5. Elmieh, N. (2009). *An Integrated Assessment of Public Health Responses to the Spread of the West Nile Virus.* Institute for Resources, Environment & Sustainability, University of British Columbia, British Columbia, Canada, Ph.D. thesis.
6. Hagerman, S.M. (2009). *Adapting Conservation Policy to the Impacts of Climate Change: An integrated examination of ecological and social dimensions of change.* Institute for Resources, Environment & Sustainability, University of British Columbia, British Columbia, Canada, Ph.D. thesis.

## 2010

7. Barradale, M. J. (2010). *Practitioner perspectives matter: Public policy and private investment in the US electric power sector.* Energy and Resources Department, UC Berkeley.
8. Mazzi, E. (2010). *An integrated assessment of climate mitigation policy, air quality and traffic safety for passenger cars in the UK.* University of British Columbia, Ph.D. thesis.
9. Schweizer, V. J. (2010). *Developing useful long-term energy projections in the face of climate change.* Department of Engineering and Public Policy, Carnegie Mellon University, Ph.D. thesis.

## 2011

10. Klima, K. (2011). *Does Tropical Cyclone Modification Make Sense? A Decision Analytic Perspective.* Engineering and Public Policy, Carnegie Mellon University, Ph.D. thesis.

11.  Ricke, K. L. (2011). *Characterizing Impacts and Implications of Proposals for Solar Radiation Management, a Form of Climate Engineering.* Engineering and Public Policy, Carnegie Mellon University, Ph.D. thesis.

## 2012

12.  Thomas, B. A. (2012). *Energy efficiency and rebound effects in the United States: Implications for renewables investment and emissions abatement.* Carnegie Mellon University, Ph.D. thesis.

## HDGC

The following is an almost complete chronological list of publications, book chapters, and theses that were produced with support from the HDGC Center. Not included are roughly 60 reports and papers in conference proceedings which can be found online in a more complete list at www.cmu.edu/epp/climate/.

## *Journals*

### 1995

1.  Longwell, J. P., Rubin, E. S., & Wilson, J. (1995). Coal: Energy for the future. *Progress in Energy and Combustion Science, 21*(4), 269–360.

### 1996

2.  Axelrod, L. J., & McDaniels, T. (1996). Individual Judgement, Social Choices and Sustainability. In *Sustainability and Human Choices in the Lower Fraser Basin: Resolving the Dissonance.*
3.  Dowlatabadi, H. (1996). *Adaptive Strategies for Climate Change Mitigation: Implications for Policy Design and Timing.* Department of Engineering and Public Policy.
4.  Dowlatabadi, H., Lave, L. B., & Russell, A. G. (1996). A free lunch at higher CAFE? A review of economic, environmental and social benefits. *Energy Policy, 24*(3), 253–264.
5.  Jasanoff, S. (1996). Beyond epistemology: Relativism and engagement in the politics of science. *Social Studies of Science, 26*(2), 393–418.
6.  Jasanoff, S. (1996). Is science socially constructed—And can it still inform public policy?. *Science and Engineering Ethics, 2*(3), 263–276.
7.  Jasanoff, S. (1996). The dilemma of environmental democracy. *Issues in Science and Technology, 13*(1), 63–70.
8.  Kandlikar, M. (1996). Indices for comparing greenhouse gas emissions: Integrating science and economics. *Energy Economics, 18*(4), 265–281.

9. Keith, D. W. (1996). When is it appropriate to combine expert judgments?. *Climatic Change, 33*(2), 139–143.

10. Morgan, M. G., & Dowlatabadi, H. (1996). Learning from integrated assessment of climate change. *Climatic Change, 34*(3), 337–368.

11. Ramachandran, G., & Kandlikar, M. (1996). Bayesian analysis for inversion of aerosol size distribution data. *Journal Of Aerosol Science, 27*(7), 1099–1112.

12. Risbey, J. S., & Entekhabi, D. (1996). Observed Sacramento Basin streamflow response to precipitation and temperature changes and its relevance to climate impact studies. *Journal of Hydrology, 184*(3–4), 209–223.

13. Risbey, J. S., & Stone, P. H. (1996). A case study of the adequacy of GCM simulations for input to regional climate change assessments. *Journal of Climate, 9*(7), 1441–1467.

14. Risbey, J., Kandlikar, M., & Patwardhan, A. (1996). Assessing integrated assessments. *Climatic Change, 34*(3), 369–395.

## 1997

15. Agrawala, S. (1997). Explaining the Evolution of the IPCC Structure and Process. Cambridge: Harvard University, *31*. ENRP Discussion Paper E-97–05.

16. Cavanagh, N., & McDaniels, T. L. (1997). Perceptions of ecological risks associated with eutrophication sources in the lower Fraser River Basin, British Columbia. *Canadian Water Resources Journal, 22*(4), 433–444.

17. Corbett, J. J., & Fischbeck, P. (1997). Emissions from ships. *Science, 278*(5339), 823–824.

18. Dowlatabadi, H. (1997). *Adaptive management of climate change mitigation: A strategy for coping with uncertainty*. Pittsburgh: Center for Integrated Study of the Human Dimensions of Global Change, Carnegie Mellon University. Discussion Paper.

19. Dowlatabadi, H. (1997). Assessing the health impacts of climate change: An editorial essay. *Climatic Change, 35*.

20. Fischhoff, B., & Downs, J. (1997). Accentuate the relevant. *Psychological Science, 8*(3), 154–158.

21. Fischhoff, B., & Downs, J. (1997). Overt and covert communication about emerging foodborne pathogens. *Emerging Infectious Diseases, 3*, 489–495.

22. Florig, H. K. (1997). Peer reviewed: China's air pollution risks. *Environmental Science & Technology, 31*(6), 274A–279A.

23. Franz, W. E. (1997). The development of an international agenda for climate change: Connecting science to policy. Cambridge: Harvard University, *33*. ENRP Working Paper E-97–07.

24. Gitelman, A. I., Risbey, J. S., Kass, R. E., & Rosen, R. D. (1997). Trends in the surface meridional temperature gradient. *Geophysical Research Letters, 24*(10), 1243–1246.

25. Jasanoff, S. (1997). NGOs and the environment: From knowledge to action. *Third World Quarterly, 18*(3), 579–594.

26. Jasanoff, S., Colwell, R., Dresselhaus, M. S., Goldman, R. D., Greenwood, M. R. C., Huang, A. S., … & Wexler, N. (1997). Conversations with the community: AAAS at the millennium. *Science, 278*(5346), 2066–2067.

27. Jenni, K., & Loewenstein, G. (1997). Explaining the identifiable victim effect. *Journal of Risk and Uncertainty, 14*(3), 235–257.

28. Kandlikar, M. (1997). Bayesian inversion for reconciling uncertainties in global mass balances. *Tellus B: Chemical and Physical Meteorology, 49*(2), 123–135.

29. Kovacs, D. C., Small, M. J., Davidson, C. I., & Fischhoff, B. (1997). Behavioral factors affecting exposure potential for household cleaning products. *Journal of Exposure Analysis and Environmental Epidemiology, 7*(4), 505–520.

30. Long, M., & Iles, A. (1997). *Assessing Climate Change Impacts: Co-Evolution of Knowledge, Communities, and Methodologies*. Belfer Center for Science and International Affairs, John F. Kennedy School of Government, Harvard University.

31. Mahasenan, N., Watts, R. G., & Dowlatabadi, H. (1997). Low-frequency oscillations in temperature-proxy records and implications for recent climate change. *Geophysical Research Letters, 24*(5), 563–566.

32. McDaniels, T. L., Axelrod, L. J., Cavanagh, N. S., & Slovic, P. (1997). Characterizing perception of ecological risk to water environments. *Risk Analysis, 17*(3), 341–352.

33. Morgan, M. G., & Dowlatabadi, H. (1997). Viewpoint, Peer Reviewed: Energy Technology R&D Essential to Curb Global Warming. *Environmental Science & Technology, 31*(12), 574A–575A.

34. Patt, A. (1997). *Assessing Extreme Outcomes: The Strategic Treatment of Low Probability Impacts of Climate Change*. Cambridge: Harvard University, *35*. ENRP Discussion Paper E-97–10, Kennedy School of Government.

35. Sagar, A., & Kandlikar, M. (1997). Knowledge, rhetoric and power: International politics of climate change. *Economic and Political Weekly*, 3139–3148.

36. West, J. J., Hope, C., & Lane, S. N. (1997). Climate change and energy policy: The impacts and implications of aerosols. *Energy Policy, 25*(11), 923–939.

37. Yates, D. N. (1997). Approaches to continental scale runoff for integrated assessment models. *Journal of Hydrology, 201*(1–4), 289–310.

38. Yohe, G. W. (1997). Uncertainty, short-term hedging and the tolerable window approach. *Global Environmental Change, 7*(4), 303–315.

## 1998

39. Cowan, R., & Miller, J. H. (1998). Technological standards with local externalities and decentralized behaviour. *Journal of Evolutionary Economics, 8*(3), 285–296.

40. Dowlatabadi, H, West, J. J., & Patwardhan, A. (1998). Lessons from assessing impacts of sea level rise. In *The assessment of climate change damages*. Paris: International Energy Agency, 105–110.

41. Dowlatabadi, H. (1998). Bumping against a gas ceiling. *Climatic Change. 46*(3), 391–407.

42. Dowlatabadi, H. (1998). Sensitivity of climate change mitigation estimates to assumptions about technical change. *Energy Economics, 20*(5–6), 473–493.

43. Farrell, A., & Hart, M. (1998). What does sustainability really mean?: The search for useful indicators. *Environment: Science and Policy for Sustainable Development, 40*(9), 4–31.

44. Fischhoff, B. (1998). Communicate unto others. *Reliability Engineering and System Safety, 59*, 63–72.

45. Fischhoff, B. (1998). Diagnosing stigma. *Reliability Engineering and System Safety, 1*(59), 47–48.

46. Fischhoff, B. (1998). Scientific management of science? Submitted for review.

47. Fischhoff, B., Downs, J. S., & de Bruin, W. B. (1998). Adolescent vulnerability: A framework for behavioral interventions. *Applied and Preventive Psychology, 7*(2), 77–94.

48. Fischhoff, B., Riley, D., Kovacs, D. C., & Small, M. (1998). What information belongs in a warning?. *Psychology & Marketing, 15*(7), 663–686.

49. Frederick, S., & B. Fischhoff. (1998). Scope insensitivity in elicited values. *Risk Decision and Policy, 3*, 109–124.

50. Jaffrezo, J. L., Davidson, C. I., Kuhns, H. D., Bergin, M. H., Hillamo, R., Maenhaut, W., … & Harris, J. M. (1998). Biomass burning signatures in the atmosphere of central Greenland. *Journal of Geophysical Research: Atmospheres, 103*(D23), 31067–31078.

51. Jaffrezo, J. L., Davidson, C. I., & Strader, R. (1998). Direct observation of the seasonal variation of aerosols over the Greenland Ice Sheet. Submitted to *Journal of Geophysical Research* for review.

52. Jasanoff, S. (1998). Science and judgment in environmental standard setting. *Applied Measurement in Education, 11*(1), 107–120.

53. Jasanoff, S. (1998). The political science of risk perception. *Reliability Engineering & System Safety, 59*(1), 91–99.

54. Jones, S. A., Fischhoff, B., & Lach, D. (1998). An integrated impact assessment of the effects of climate change on the Pacific Northwest salmon fishery. *Impact Assessment and Project Appraisal, 16*(3), 227–237.

55. Kalagnanam, J., Kandlikar, M., & Linville, C. (1998). *Importance Ranking of Model Uncertainties: A Robust Approach*. IBM Thomas J. Watson Research Division.

56. Kandlikar, M. (1998). Economic value of weather and climate forecasts (Book Review). *Environment and Development Economics. 3*, 541–543.

57. Kuhns, H., Davidson, C., Bolzan, J., McConnell, J., Dibb, J., Hart, V., Jaffrezo, J. L., Bergin, M., & Robertson, I. (1998). Spatial and temporal variability

of chemical concentrations in ice core records from central Greenland. Submitted to *Journal of Geophysical Research* for review.

58. Lave, L. B., & Shevliakova, E. (1998). Potential damages from climate changes in the US. *Energy & Environment, 9*(4), 349–363.

59. Lookman, A. A., & Rubin, E. S. (1998). Barriers to adopting least-cost particulate control strategies for Indian power plants. *Energy Policy, 26*(14), 1053–1063.

60. Lookman, A. A., & Rubin, E. S. (1998). Barriers to adopting least-cost particulate control strategies for Indian power plants. *Energy Policy, 26*(14), 1053–1063.

61. McDaniels, T. L. (1998). Systemic blind spots: Implications for communicating ecological risk. *Human and Ecological Risk Assessment: An International Journal, 4*(3), 633–638.

62. McDaniels, T. L. (1998). Ten propositions for untangling descriptive and prescriptive lessons in risk perception findings. *Reliability Engineering & System Safety, 59*(1), 129–134.

63. McDaniels, T. L., & Roessler, C. (1998). Multiattribute elicitation of wilderness preservation benefits: A constructive approach. *Ecological Economics, 27*(3), 299–312.

64. McDaniels, T. L., Axelrod, L. J., & Cavanagh, N. (1998). Public perceptions regarding water quality and attitudes toward water conservation in the lower Fraser Basin. *Water Resources Research, 34*(5), 1299–1306.

65. Miller, J. H. (1998). Active nonlinear tests (ANTs) of complex simulation models. *Management Science, 44*(6), 820–830.

66. Morgan, M. G., & Tierney, S. (1998). Research support for the power industry. *Issues in Science and Technology, 15*(1), 81–87.

67. Parson, E. A., & Keith, D. W. (1998). Fossil fuels without $CO_2$ emissions. *Science, 282*, 1053–1054.

68. Risbey, J. S. (1998). Sensitivities of water supply planning decisions to streamflow and climate scenario uncertainties. *Water Policy, 1*(3), 321–340.

69. Schneider, S. H., Turner, B. L., & Garriga, H. M. (1998). Imaginable surprise in global change science. *Journal of Risk Research, 1*(2), 165–185.

70. Shackley, S., Risbey, J., & Kandlikar, M. (1998). Science and the contested problem of climate change: A tale of two models. *Energy and Environment, 9*.

71. Strahilevitz, M. A., & Loewenstein, G. (1998). The effect of ownership history on the valuation of objects. *Journal of Consumer Research, 25*(3), 276–289.

72. West, J. J., Pilinis, C., Nenes, A., & Pandis, S. N. (1998). Marginal direct climate forcing by atmospheric aerosols. *Atmospheric Environment, 32*(14–15), 2531–2542.

73. Yates, D. N., & Strzepek, K. M. (1998). An assessment of integrated climate change impacts on the agricultural economy of Egypt. *Climatic Change, 38*(3), 261–287.

74. Yohe, G., Malinowski, T., & Yohe, M. (1998). Fixing global carbon emissions: Choosing the best target year. *Energy Policy, 26*(3), 219–231.

## 1999

75. Axelrod, L. J., McDaniels, T., & Slovic, P. (1999). Perceptions of ecological risk from natural hazards. *Journal of Risk Research, 2*(1), 31–53.

76. Casman, E. A., Morgan, M. G., & Dowlatabadi, H. (1999). Mixed levels of uncertainty in complex policy models. *Risk Analysis, 19*(1), 33–42.

77. Cavanagh, N., McDaniels, T., Axelrod, L., & Slovic, P. (2000). Perceived ecological risks to water environments from selected forest industry activities. *Forest Science, 46*(3), 344–355.

78. Corbett, J. J., Fischbeck, P. S., & Pandis, S. N. (1999). Global nitrogen and sulfur inventories for oceangoing ships. *Journal of Geophysical Research: Atmospheres, 104*(D3), 3457–3470.

79. Corbett, J. J., Fischbeck, P. S., & Pandis, S. N. (1999). Global nitrogen and sulfur inventories for oceangoing ships. *Journal of Geophysical Research: Atmospheres, 104*(D3), 3457–3470.

80. Dahinden, U., Querol, C., & Nilsson, M. (1999). *Using computer models in participatory integrated assessment – Experiences gathered in the ULYSSES project and recommendations for further steps.* Duebendorf: Swiss Federal Institute for Environmental Science and Technology (EAWAG). Uylsses Working Paper.

81. Dowlatabadi, H, & Oravetz, M. (1999). Understanding trends in energy intensity: A simple model of technical change. *Energy Policy.* forthcoming.

82. Dowlatabadi, H. (1999). Climate change thresholds and guardrails for emissions. *Climatic Change, 41*(3–4), 297.

83. Dowlatabadi, H. (1999). Cultural content of integrated assessments & models. *Environmental Modelling and Assessment.* forthcoming.

84. Farrell, A. (1999). Sustainability and decision-making: The EPA's Sustainable Development Challenge Grant Program. *Review of Policy Research, 16*(3-4), 36–74.

85. Farrell, A., Carter, R., & Raufer, R. (1999). The NOx budget: Market-based control of tropospheric ozone in the northeastern United States. *Resource and Energy Economics, 21*(2), 103–124.

86. Fischhoff, B. (1999). If trust is so good, why isn't there more of it. *Social Trust and the Management of Risk. Earthscan, London.*

87. Fischhoff, B. (1999). Why (cancer) risk communication can be hard. *JNCI Monographs, 1999*(25), 7–13.

88. Fischhoff, B., Welch, N., & Frederick, S. (1999). Construal processes in preference elicitation. *Journal of Risk and Uncertainty.* forthcoming.

89. Gitelman, A. I., Risbey, J. S., Kass, R. E., & Rosen, R. D. (1999). Sensitivity of a meridional temperature gradient index to latitudinal domain. *Journal of Geophysical Research: Atmospheres, 104*(D14), 16709–16717.

90. Hsee, C. K., Loewenstein, G. F., Blount, S., & Bazerman, M. H. (1999). Preference reversals between joint and separate evaluations of options: A review and theoretical analysis. *Psychological Bulletin, 125*(5), 576.

91. Jones, S. A., Fischhoff, B., & Lach, D. (1999). Evaluating the science-policy interface for climate change research. *Climatic Change, 43*(3), 581–599.

92. Kandlikar, M., & Ramachandran, G. (1999). Inverse methods for analysing aerosol spectrometer measurements: A critical review. *Journal of Aerosol Science, 30*(4), 413–437.

93. Kandlikar, M., & Sagar, A. (1999). Climate change research and analysis in India: An integrated assessment of a South–North divide. *Global Environmental Change, 9*(2), 119–138.

94. Keating, T. J., & Farrell, A. (1999). Transboundary environmental assessment: Lessons from the ozone transport assessment group. *National Center for Environmental Decision-Making Research.*

95. Keeney, R. L., & McDaniels, T. L. (1999). Identifying and structuring values to guide integrated resource planning at BC Gas. *Operations Research, 47*(5), 651–662.

96. Keith, D. W. (1999). Systematic errors in climate prediction: Managing applied research. *Nature.* Forthcoming.

97. McDaniels, T. (1999). An analysis of the Tatshenshini-Alsek Wilderness preservation decision. *Journal of Environmental Management, 57*, 123–141.

98. McDaniels, T., & Trousdale, W. (1999). Value-focused thinking in a difficult context: Planning tourism for Guimaras, Philippines. *Interfaces, 29*(4), 58–70.

99. McDaniels, T., Gregory, R., & Fields, D. (1999). Democratizing risk management: Successful public involvement in an electric utility water management decision. *Risk Analysis, 19*(3), 497–510.

100. McDaniels, T., & Thomas, K. (1999). Eliciting public preferences for local land use alternatives: A structured value referendum with approval voting. *Journal of Policy Analysis and Management, 18*(2), 264–280.

101. Morel, B., Yeh, S., & Cifuentes, L. (1999). Statistical distributions for air pollution applied to the study of the particulate problem in Santiago. *Atmospheric Environment, 33*(16), 2575–2585.

102. Morgan, M. G., Kandlikar, M., Risbey, J., & Dowlatabadi, H. (1999). Why conventional tools for policy analysis are often inadequate for problems of global change. *Climatic Change, 41*(3–4), 271.

103. Read, D., Loewenstein, G., Kalyanaraman, S., & Bivolaru, A. (1999). Mixing virtue and vice: The combined effects of hyperbolic discounting and diversification. *Journal of Behavioral Decision Making, 12*(4), 257–273.

104. Risbey, J, Milind K., Graetz, D., & Dowlatabadi, H. (1999). Scale and contextual issues in agricultural adaptation to climate variability and change. *Mitigation and Adaptation Strategies for Global Change, 4*(2), 137–165.

105. Shackley, S., Risbey, J., Stone, P., & Wynne, B. (1999). An interdisciplinary study of flux adjustments in coupled atmosphere-ocean general circulation models. *Climatic Change, 43*, 413–454.

106. Strachan, N, & Dowlatabadi, H. (1999). An engineering economic analysis of a decentralized energy technology: UK engine cogeneration. *The Energy Journal.* submitted.

107. Strachan, N., & Dowlatabadi, H. (1999). Looking beyond the usual barriers to adoption of a decentralized energy technology: Supplier strategies. *The Energy Journal*. submitted.

108. Strachan, N., & Dowlatabadi, H. (1999). The adoption of a decentralized energy technology: The case of UK engine cogeneration. *ACEEE Summer Study on Energy Efficiency in Industry, Saratoga Springs, NY.*

109. Strachan, Neil, & Dowlatabadi, H. (1999). An engineering economic analysis of a decentralized energy technology: UK engine cogeneration. *The Energy Journal*. submitted.

110. Tol, R. S. (1999). Kyoto, efficiency, and cost-effectiveness: Applications of FUND. *The Energy Journal, 20*(Special Issue-The Cost of the Kyoto Protocol: A Multi-Model Evaluation).

111. Tol, R. S. (1999). Spatial and temporal efficiency in climate policy: Applications of FUND. *Environmental and Resource Economics, 14*(1), 33–49.

112. Tol, R. S. (1999). The marginal costs of greenhouse gas emissions. *The Energy Journal, 20*(1).

113. Tol, R. S. J. (1999). Safe policies in an uncertain climate: Applications of FUND. *Global Environmental Change*. forthcoming.

114. Tol, R. S. J. (1999). Time discounting and optimal control of climate change: An application of FUND. *Climatic Change, 41*(3–4), 351–362.

115. Yohe, G. W. (1999). The tolerable windows approach: Lessons and limitations. *Climatic Change, 41*(3–4), 283.

116. Yohe, G., & Dowlatabadi, H. (1999). Risk and uncertainties, analysis and evaluation: Lessons for adaptation and integration. *Mitigation and Adaptation Strategies for Global Change, 4*(3), 319–329.

117. Yohe, G., & Jacobsen, M. (1999). Meeting concentration targets in the post-Kyoto world: Does Kyoto further a least cost strategy?. *Mitigation and Adaptation Strategies for Global Change, 4*(1), 1–23.

## *2000*

118. Cavanagh, N., McDaniels, T., Axelrod, L., & Slovic, P. (2000). Perceived ecological risks to water environments from selected forest industry activities. *Forest Science, 46*(3), 344–355.

119. Cebon, P., & Risbey, J. (2000). Four views of the "regional" in regional environmental change. *Global Environmental Change*. submitted.

120. Farrell, A. (2000). The NOx budget: A look at the first year. *The Electricity Journal,* March, 83–93.

121. Fernández, R. J., & Reynolds, J. F. (2000). Potential growth and drought tolerance of eight desert grasses: Lack of a trade-off?. *Oecologia, 123*(1), 90–98.

122. Fischhoff, B. (2000). Informed consent for eliciting environmental values. *Environmental Science & Technology, 34*(8), 1439–1444.

123. Fischhoff, B., Parker, A. M., de Bruin, W. B., Downs, J., Palmgren, C., Dawes, R., & Manski, C. F. (2000). Teen expectations for significant life events. *The Public Opinion Quarterly*, *64*(2), 189–205.

124. Loewenstein, G. (2000). Willpower: A decision-theorist's perspective. *Law and Philosophy*, 51–76.

125. Read, D., Loewenstein, G., Rabin, M., Keren, G., & Laibson, D. (2000). Choice bracketing in *Elicitation of Preferences*, 171–202, Springer.

126. Reynolds, J. F., Kemp, P. R., & Tenhunen, J. D. (2000). Effects of rainfall variability on patterns of evapotranspiration and soil water distribution in the Chihuahuan Desert: A modeling analysis. *Plant Ecology*, *150*(1–2).

127. Risbey, J. S., Kandlikar, M., & Karoly, D. J. (2000). A protocol to articulate and quantify uncertainties in climate change detection and attribution. *Climate Research*, *16*(1), 61–78.

128. Teitelbaum, D., & Dowlatabadi, H. (2000). A computational model of technological innovation at the firm level. *Computational & Mathematical Organization Theory*, *6*(3), 227–247.

129. Yohe, G. (2000). Assessing the role of adaptation in evaluating vulnerability to climate change. *Climatic Change*, *46*(3), 371–390.

130. Yohe, G., & Toth, F. L. (2000). Adaptation and the guardrail approach to tolerable climate change. *Climatic Change*, *45*(1), 103–128.

## 2001

131. Loewenstein, G. F., Weber, E. U., Hsee, C. K., & Welch, N. (2001). Risk as feelings. *Psychological Bulletin*, *127*(2), 267.

132. West, J. J., Small, M. J., & Dowlatabadi, H. (2001). Storms, investor decisions, and the economic impacts of sea level rise. *Climatic Change*, *48*(2), 317–342.

133. West, J. J., Small, M. J., & Dowlatabadi, H. (2001). Storms, investor decisions, and the economic impacts of sea level rise. *Climatic Change*, *48*(2), 317–342.

## 2002

134. Risbey, J. S., Lamb, P. J., Miller, R. L., Morgan, M. C., & Roe, G. H. (2002). Exploring the structure of regional climate scenarios by combining synoptic and dynamic guidance and GCM output. *Journal of Climate*, *15*(9), 1036–1050.

## 2006

135. Dowlatabadi, H., & Oravetz, M. A. (2006). US long-term energy intensity: Backcast and projection. *Energy Policy*, *34*(17), 3245–3256.

## Books and Book Chapters

### 1996

1. Jasanoff, S. (1996). Science and Norms in Global Environmental Regimes. In *Earthly Goods: Environmental Change and Social Justice*, edited by F. O. Hampson and J. Reppy. Ithaca and London: Cornell University Press.

### 1997

2. Fischhoff, B., Lanir, Z., & Johnson, S. (1997). Risky Lessons: Conditions for Organizational Learning. In *Technological Learning, Oversights, and Foresights*, edited by Shapira, Z., Garud, R., and Nayyar, P. New York: Cambridge University Press, 306–324.
3. Fischhoff, B. (1997). Ranking Risks. In *Environment and Ethics: Psychological Contributions*, edited by Bazerman, M., Messick, D., Tenbrunsel, A., Wade-Benzoni, K. San Francisco: Jossey-Bass, 342–371.
4. Fischhoff, B. (1997). What Do Psychologists Want? Contingent Valuation as a Special Case of Asking Questions. In *Determining the Value of Nonmarketed Goods*, edited by Kopp, R. J., Pommerehe, W. W., and Schwarz, N. New York: Plenum, 189–217.
5. Fischhoff, B., Bostrom, A., & Quadrel, M. J. (1997). Risk perception and communication. In *Oxford Textbook of Public Health*, edited by R. Detels, J. McEwen, and G. Omenn. London: Oxford University Press. 987–1002.
6. Jasanoff, S. (1997). Compelling knowledge in public decisions. *Saving the Seas: Values, Scientists, and International Governance*, edited by L. A. Brooks and S. VanDeveer, College Park: Maryland Sea Grant. 229–252.
7. Jasanoff, S. (1997). Harmonization: The Politics of Reasoning Together. In *The Politics of Chemical Risk: Scenarios for a Regulatory Future*, edited by R. Bal and W. Halffman. Dordrecht: Kluwer, 173–194.
8. Kandlikar, M. (1997). The Potential Role of GIS in Integrated Assessments of Global Change. In *Transatlantic Perspectives*, edited by Craglia and Onsrud. London: Taylor and Francis.
9. Loewenstein, G., & Frederick, S. (1997). Predicting Reactions to Environmental Change. In *Environment, Ethics, and Behavior,* edited by M. Bazerman, D. Messick, A. Tenbrunsel, and K. Wade-Benzoni. San Francisco: New Lexington Press.
10. Shevliakova, E. (1997). Modeling of Global Biosphere-Atmosphere Interactions. In *Elements of Change*, edited by J. Hassol and J. Katzenberger, 118–127.

### 1998

11. Frederick, S., Loewenstein, G., Diener, E., Schwartz, N., & Kahneman, D. (1998). Hedonic Adaptation. In *Foundations of Hedonic Psychology: Scientific Perspectives on Enjoyment and Suffering*, edited by E. Diener, N. Schwartz, and D. Kahneman. New York: Russell Sage Foundation.

12.  Geoghegan, J., Pritchard, L., Ogneva-Himmelberger, Y., Chowdhury, R. R., Sanderson, S., & Turner, B. L. I. (1998). "Socializing the pixel" and "Pixelizing the social" in land-use and land-cover change. In *People and Pixels*. National Academy Press, 51–69.

13.  Jasanoff, S, & Wynne, B. (1998). Science and Decision-making. In *Human Choice & Climate Change: The Societal Framework*, edited by Rayner, Steve, Malone, Elizabeth. Columbus: Battelle Press, 1:1–87.

14.  Jasanoff, S. (1998). Contingent Knowledge: Implications for Implementation and Compliance. In *Engaging Countries: Strengthening Compliance with International Environmental Accords,* edited by H. Jacobson and E. Brown Weiss. Cambridge: MIT Press.

15.  Loewenstein, G. & Schkade, D. (1998). Wouldn't it be Nice? Predicting Future Feelings. In *Foundations of Hedonic Psychology: Scientific Perspectives on Enjoyment and Suffering*, edited by E. Diener, N. Schwartz, and D. Kahneman. New York: Russell Sage Foundation.

16.  Loewenstein, G., Prelec, D., & Weber, R. (1998). What Me Worry? A psychological perspective on economic aspects of retirement. In *Psychological Perspective on Retirement*, edited by Henry Aaron. New York: Russell Sage Foundation and The Brookings Institute. forthcoming.

17.  Meyer, W. B., Butzer, K. W., Downing, T. E., Wenzel, G. W., Wescoat, J. L., & Turner II, B. L. (1998). Reasoning By Analogy. In *Human Choice and Climate Change Vol. 3 – Tools for Policy Analysis,* edited by S. Raynor and E. L. Malone. Columbus: Battelle Memorial Press, 218–289.

18.  Reynolds, J. F., Fernndez, R. J., & Kemp, P. R. (1998). Drylands and Global Change: Rainfall Variability and Sustainable Rangeland Production. In *Proceedings of the 12th Toyota Conference: Challenge of Plant and Agricultural Sciences to the Crisis of Biosphere on the Earth in the 21st Century*, edited by K. N. Watanabe and A Komanine. Austin: Landes Biosciences.

19.  West, J. J., & Dowlatabadi, H. (1998). On assessing the economic impacts of sea level rise on developed coasts. In *Climate, Change and Risk*, edited by T. E. Downing, A. A. Olsthoorn, and R. S. J. Tol. London: Routledge.

### *1999*

20.  Dowlatabadi, H. (1999). Integrated Assessment: Implications of Uncertainty. In *Encyclopedia of Life Support Systems*. Oxford: Oxford University Press.

21.  Dowlatabadi, H. (1999). Integrated Assessment. In *Encyclopedia of Global Change*.

22.  Fischhoff, B. (1999). Decision Making. In *Encyclopedia of Psychology*. Washington, DC: American Psychological Association and Oxford University Press. forthcoming.

23.  Fischhoff, B. (1999). Defining Stigma. In *Risk, Media, and Stigma*, edited by H. Kunreuther and P. Slovic. forthcoming.

24. Fischhoff, B. (1999). Judgment Heuristics. In *The MIT Encyclopedia of the Cognitive Sciences*, edited by R. Wilson and F. Keil, 421–423. Cambridge: MIT Press.

25. Fischhoff, B. (1999). Learning From Experience. In *Handbook of Forecasting Principles*, edited by J. S. Armstrong. Norwell: Kluwer. forthcoming.

26. Keith, D. W. (1999). Geoengineering. In *Encyclopedia of Global Change*. New York: Oxford University Press. forthcoming.

27. McDaniels, T. (1999). Valuation. In *Encyclopedia of Global Change*. Oxford: Oxford University Press. forthcoming.

28. Patwardhan, A. (1999). Climate Change: Shifting the Burden. In *The Hindu Survey of the Environment '99*, 127–132.

29. Reynolds, J. F. (1999). Desertification. In *Encyclopedia of Biodiversity,* edited by Levin, S., Gaily, G., Grifo, F., Lubechenco, J., Mooney, H. A., Schulze, E., & Tilman, D. San Diego: Academic Press.

## *2000*

30. Clark, W. C., Jaeger, J., Eijndhoven, L, & Dickson, N. M. (2000). *Learning to Manage Global Environmental Risks: A Comparative History of Social Responses to Climate Change, Ozone Depletion, and Acid Rain.* Cambridge: MIT Press.

## *Theses*

### *1996*

1. Shevliakova, E. (1996). Application of Statistical Methods for Modeling Impacts of Climate Change on Terrestrial Distribution of Vegetation. Pittsburgh: Carnegie Mellon University, Department of Engineering & Public Policy. Thesis.

### *1998*

2. Teitelbaum, D. (1998). Technological Change and Pollution Control: An Adaptive Agent-Based Analysis. Pittsburgh: Department of Engineering & Public Policy, Carnegie Mellon University. Thesis.

3. West, J. J. (1998). Studies in natural and human system response relevant to global environmental change. Ph.D. Dissertation submitted in the Civil & Environmental Engineering, and Engineering and Public Policy Departments at Carnegie Mellon University, Pittsburgh, PA.

### *1999*

4. Frederick, S. (1999). Time Preferences. Pittsburgh: Carnegie Mellon University. Department of Social and Decision Sciences. Ph.D. Thesis.

# APPENDIX 3

# GROUPS WITH WHICH ONE OR MORE OF OUR CENTERS HAVE COLLABORATED

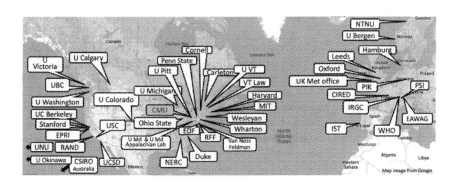

| Institution | Involvement with our Centers |
|---|---|
| Carleton U. | After time at UCSD and then back again at CMU, Ahmed Abdulla moved to a faculty position at Carleton University where he has continued to collaborate with us. |
| CIRED | *Centre international de recherche sur l'environnement et le développement.* We collaborated with Minh Ha Duong using his various models of energy transition. |
| Cornell U. | Hadi Dowlatabadi and Elena Shevliakova worked with David Weinstein on forest issues. |
| CSIRO | The Commonwealth Scientific and Industrial Research Organization. We collaborated with Dean Greatz and others in Australia on integrated assessment of global change impacts on agriculture. |
| Duke U. | Dalia Patiño-Echeverri has long been a co-investigator with our Centers. Hadi collaborated with Jim Reynolds at Duke on the issue of desertification. |

| Institution | Involvement with our Centers |
|---|---|
| EAWAG | *The Swiss Federal Institute of Aquatic Science and Technology.* We collaborated on a variety of topics with both Carlo Jaeger and Peter Reichert. |
| EDF | *Environmental Defense Fund.* Parth Vaishnav collaborated with investigators at EDF on strategies to control GHG emissions from aircraft. |
| EPRI | *The Electric Power Research Institute.* Lou Pitelka collaborated with us on expert elicitation of ecologists. Over the years we also had substantial interactions with several others at EPRI including Rich Richels. |
| Hamburg U. | We collaborated with Richard Toll and involved other folks from Hamburg in our work on malaria. |
| IRGC | *The International Risk Governance Council.* We collaborated with IRGC on policy issues related to solar radiation management (SRM). |
| IST | *Instituto Superior Técnico.* Under the CMU-Portugal program we collaborated with Joanna Mendonça on the use of floating dynamically positioned offshore wind. We also worked with P.M.S. Carvalho on modeling wind power forecast uncertainty, on distribution grid reconfiguration, and on wind farm storage. |
| Leeds U. | After she left CMU to move to Leeds, Wändi Bruine de Bruin continued as a Center co-investigator at Leeds. Later she moved to USC. |
| MIT | After he left the University of Colorado and moved to MIT, Ken Strzepek continued to collaborate with us on a number of issues related to climate change and water including water in Africa. |
| NERC | *North American Electric Reliability Corporation.* We collaborated with Tom Coleman and John Moura on reliability of the electric and natural gas systems. |
| NTNU | *Norwegian University of Science and Technology.* We collaborated with Ane Marte Heggedal and Gerard Doorman on investments for pumped hydropower storage. |
| Ohio State U. | We collaborated with Ratan Lal on soil carbon. |
| Oxford U. | We collaborated with Daniel Rowlands, William Ingram, Miles Allen, and others using the climateprediction.net system to study SRM. |
| Penn State U. | We collaborated with Seth Blumsack on natural gas resilience. |
| PIK | *Potsdam Institute for Climate Research.* Stefan Rahmstorf and Kirsten Zickfeld collaborated with us on expert elicitation of climate science. Carlo Jaeger collaborated with us on issues in climate policy. |
| PSI | *The Paul Scherrer Institute.* Together with IRGC, we collaborated with PSI to run a workshop on the topic of small modular reactors. |
| RAND Corp. | Rob Lempert collaborated with us on methods in robust decision making. |

| Institution | Involvement with our Centers |
|---|---|
| RFF | *Resources for the Future.* We collaborated with Karen Palmer, Dallas Burtraw, and Ray Kopp on a variety of issues related to climate and energy policy. |
| Stanford U. | After she left CMU and moved to Stanford, Inês Azevedo continue as a Center co-investigator. |
| U. Bergen | Ann Bostrom collaborated on a number of topics with Gisela Böhm at the University of Bergen. |
| U. Calgary | After leaving CMU to take up a position at the University of Calgary, David Keith continued to collaborate with us. |
| U. Colorado | While he was at the University of Colorado, Ken Strzepek collaborated with us on a number of issues related to climate change and water, including water in Africa. |
| U. Maryland | We collaborated with John Steinbruner on geoengineering governance. |
| U. Maryland Appalachian Lab | After leaving EPRI to direct the University of Maryland Appalachian Laboratory, Lou Pitelka continued to collaborate with us on expert elicitation of ecologists. |
| U. Michigan | After he left CMU to take up a faculty position at the University of Michigan, Parth Vaishnav continued to collaborate with us as a Center co-investigator. |
| U. Okinawa | We collaborated with Mahesh Bandi on the physics of variability in wind and solar power. |
| U. Pittsburgh | We collaborated with Kathrin Kirchen and William Harbert on the risks of solar coronal mass ejections. |
| U. Victoria | After leaving PIK to move to the University of Victoria, Kirsten Zickfeld continued to collaborate with us on expert elicitation of climate science. |
| U. Vt | *University of Vermont.* We collaborated with Paul Hines and others on a review of wind integration studies. |
| U. Washington | Ann Bostrom has long been a co-investigator with our Centers. |
| UBC | *University of British Columbia.* After leaving CMU, both Hadi Dowlatabadi and Tim McDaniels have long been Center co-investigators. We also collaborated with Milind Kandlikar and others. |
| UC Berkeley | Alex Farrell worked with Hisham Zerriffi on distributed electricity networks, and their resilience. Alex's graduate student, Gabriel Wong-Parodi worked with Isha Ray, Tim McDaniels, and Hadi to complete her PhD on understanding the key determinants of public attitudes about CCS and then joined us in EPP. |
| UCSD | *University of California at San Diego.* We worked with David Victor on issues involving geoengineering. After he left CMU to take up a staff position at UC San Diego, Ahmed Abdulla continued to collaborate with us as a Center co-investigator. |
| UK Met Office | While she was at Leeds, Wändi Bruine de Bruin collaborated with Jason Lowe and Fai Fung at the U.K. Met Office. |
| UNU | *United Nations University.* We collaborated on the AIM model as well as on endogenous energy efficiency. |

| Institution | Involvement with our Centers |
|---|---|
| USC | *University of Southern California.* After she left Leeds, Wändi Bruine de Bruin continued as a Center co-investigator at USC. |
| Van Ness Feldman | We collaborated with Bob Nordhaus and others on the regulation of carbon capture and deep geologic sequestration and on governance of research on solar radiation management. |
| VT Law | Michael Dworkin has been a Center co-investigator. |
| WHO | *World Health Organization.* We collaborated with WHO on our workshop and subsequent book on the environmental determinants of malaria. |

# INDEX

Because the names of the fourteen co-authors, and of Lester Lave, appear throughout this book, no index entries have been included for them.

Printed in the United States
by Baker & Taylor Publisher Services